T0183858

Lecture Notes in Computer Science   11783

Xiaoxin Tang · Quan Chen · Pradip Bose ·
Weiming Zheng · Jean-Luc Gaudiot (Eds.)

Network and Parallel Computing

16th IFIP WG 10.3 International Conference, NPC 2019
Hohhot, China, August 23–24, 2019
Proceedings

 Springer

Editors
Xiaoxin Tang
Shanghai University of Finance
and Economics
Shanghai, China

Pradip Bose
IBM T. J. Watson Research Center
Yorktown Heights, NY, USA

Jean-Luc Gaudiot
University of California
Irvine, CA, USA

Quan Chen
Shanghai Jiao Tong University
Shanghai, China

Weiming Zheng
Tsinghua University
Beijing, China

ISSN 0302-9743 ISSN 1611-3349 (electronic)
Lecture Notes in Computer Science
ISBN 978-3-030-30708-0 ISBN 978-3-030-30709-7 (eBook)
https://doi.org/10.1007/978-3-030-30709-7

LNCS Sublibrary: SL1 – Theoretical Computer Science and General Issues

This Springer imprint is published by the registered company Springer Nature Switzerland AG
The registered company address is: Gewerbestrasse 11, 6330 Cham, Switzerland

Preface

These proceedings contain the papers presented at the 2019 IFIP International Conference on Network and Parallel Computing (NPC 2019), held in Hohhot, Inner Mongolia, China, from August 23–24, 2019. The goal of the conference is to establish an international forum for engineers and scientists to present their ideas and experiences in network and parallel computing.

A total of 107 submissions were received in response to our call for papers. These papers originate from Australia, Asia (China, Japan), Europe, and North America (USA). Each submission was sent to at least three reviewers. Each paper was judged according to its originality, innovation, readability, and relevance to the expected audience. Based on the reviews received, 36 full papers (about 33%), including 8 papers published as Special Issue papers of the *International Journal of Parallel Programming*, 5 papers published as Speicial Issue papers of Tsinghua Science and Technology, and 23 papers published as LNCS proceedings were retained. A number of strong papers that could not be accepted to the full-paper track were considered for the short-paper tracks. Finally, we selected 14 short papers. These papers cover traditional areas of network and parallel computing, including parallel applications, distributed algorithms, parallel architectures, software environments, and distributed tools.

We share the view that, during the past decade, the tools and cultures of high-performance computing and big data analytics are diverging to the detriment of both, and the international community should find a unified path that can best serve the need of a broad spectrum of major application areas. Unlike other tools, which are limited to particular scientific domains, computational modeling and data analytics are applicable to all areas of science and engineering, as they breathe life into the underlying mathematics of scientific models.

We sincerely appreciate the work and effort of the authors in preparing their submissions for review, and addressing the reviewers' comments before submitting the camera-ready copies of their accepted papers, and attending the conference to present and discuss their work. We also want to thank every member of the NPC 2019 Organizing Committee and Steering Committee for their help in putting together such an exciting program. Finally, we thank all the attendees.

August 2019

Xiaoxin Tang
Quan Chen
Pradip Bose
Weiming Zheng
Jean-Luc Gaudiot

Organization

Organizing Committee

General Co-chairs

Jean-Luc Gaudiot	University of California Irvine, USA
Weiming Zheng	Tsinghua University, China

Program Co-chairs

Pradip Bose	IBM T.J. Watson Research Center, USA
Quan Chen	Shanghai Jiao Tong University, China

Publications Chair

Xiaoxin Tang	Shanghai University of Finance and Economics, China

Publicity Chairs

Takatsugu Ono	Kyushu University, Japan
Stéphane Zuckerman	Université de Cergy-Pontoise, France
Karthik V. Swaminathan	IBM T.J. Watson Research Center, USA

Web Chair

Zijun Li	Shanghai Jiao Tong University, China

Advisory Committee

Steering Committee

Kemal Ebcioglu (Chair)	Global Supercomputing, USA
Hai Jin (Vice Chair)	Huazhong University of Science and Technology, China
Chen Ding	University of Rochester, USA
Jack Dongarra	University of Tennessee, USA
Guang R. Gao	University of Delaware, USA
Zhiwei Xu	Institute of Computing Technology, China
Tony Hey	Science and Technology Facilities Council, UK
Guojie Li	Institute of Computing Technology, China
Yoichi Muraoka	Waseda University, Japan
Viktor Prasanna	University of Southern California, USA
Daniel Reed	University of Iowa, USA
Weisong Shi	Wayne State University, USA
Ninghui Sun	Institute of Computing Technology, China

Program Committee

Jidong Zhai	Tsinghua University, China
Feng Zhang	Renmin University of China, China
Guoyang Chen	Alibaba Group US Inc., USA
Dezun Dong	National University of Defense Technology, China
Kejiang Ye	SIAT, Chinese Academy of Sciences, China
Yunlan Wang	Northwestern Polytechnical University, China
Shanjiang Tang	Nanyang Technological University, Singapore
Zheng Wang	Lancaster University, UK
Dingwen Tao	The University of Alabama, USA
Li Shen	National University of Defense Technology, China
Lin Gu	Huazhong University of Science and Technology, China
Zhibin Yu	Shenzhen Institute of Advanced Technology, China
Yungang Bao	Institute of Computing Technology (ICT), CAS, China
Zhijia Zhao	University of California Riverside, USA
Stéphane Zuckerman	ETIS Laboratory, France
Weihua Zhang	Fudan University, China
Dejun Jiang	Institute of Computing Technology, CAS, China
Wenli Zheng	Shanghai Jiao Tong University, China
Deze Zeng	The University of Aizu, Japan
Jingwen Leng	Shanghai Jiao Tong University, China
Chao Li	Shanghai Jiao Tong University, China
Yu Hua	Huazhong University of Science and Technology, China
Xiaochun Ye	Institute of Computing Technology, CAS, China
Hailong Yang	Beihang University, China
Guangzhong Sun	University of Science and Technology of China, China
Lei Wang	National University of Defense Technology, China
Shuang Song	The University of Texas at Austin, USA
Shengli Pan	China University of Geosciencs, China
Huimin Cui	Institute of Computing Technology, CAS, China
Xuanhua Shi	Huazhong University of Science and Technology, China
Muzhou Xiong	China University of Geosciences, China
Ang Li	Pacific Northwest National Lab, USA
Di Wu	Sun Yat-Sen University, China
Amelie Chi Zhou	Shenzhen University, China
Parimala Thulasiram	University of Manitoba, Canada
Bin Ren	Pacific Northwest National Laboratory, USA
Dong Li	University of California, Merced, USA
Chuliang Weng	Huawei Shannon Lab, China
Yingwei Luo	Peking University, China
Songwen Pei	University of Shanghai for Science and Technology, China

Tao Zhang	Shanghai University, China
Bo Wu	Colorado School of Mines, USA
Keiji Kimura	Waseda University, Japan

Contents

Graph Computing

GraphScSh: Efficient I/O Scheduling and Graph Sharing
for Concurrent Graph Processing. 3
  Shang Liu, Zhan Shi, Dan Feng, Shuo Chen, Fang Wang,
  and Yamei Peng

Game-Based Multi-MD with QoS Computation Offloading for Mobile
Edge Computing of Limited Computation Capacity. 16
  Junyan Hu, Chubo Liu, Kenli Li, and Keqin Li

NOC and Networks

KLSAT: An Application Mapping Algorithm Based on Kernighan–Lin
Partition and Simulated Annealing for a Specific WK-Recursive
NoC Architecture 31
  XiaoJun Wang, Feng Shi, and Hong Zhang

Modeling and Analysis of the Latency-Based Congestion Control
Algorithm DX 43
  Wanchun Jiang, Lijuan Peng, Chang Ruan, Jia Wu, and Jianxin Wang

Distributed Quality-Aware Resource Allocation for Video Transmission
in Wireless Networks. 56
  Chao He, Zhidong Xie, and Chang Tian

Neural Networks

PRTSM: Hardware Data Arrangement Mechanisms for Convolutional
Layer Computation on the Systolic Array. 69
  Shuquan Wang, Lei Wang, Shiming Li, Tian Shuo, Shasha Guo,
  Ziyang Kang, Shuzheng Zhang, and Weixia Xu

PParabel: Parallel Partitioned Label Trees for Extreme Classification. 82
  Jiaqi Lu, Jun Zheng, and Wenxin Hu

Statistical Analysis and Prediction of Parking Behavior 93
  Ningxuan Feng, Feng Zhang, Jiazao Lin, Jidong Zhai, and Xiaoyong Du

Big Data+Cloud

ASTracer: An Efficient Tracing Tool for HDFS with Adaptive Sampling. . . .   107
    Yang Song, Yunchun Li, Shuhan Wu, Hailong Yang, and Wei Li

BGElasor: Elastic-Scaling Framework for Distributed Streaming Processing
with Deep Neural Network.   120
    Weimin Mu, Zongze Jin, Junwei Wang, Weilin Zhu, and Weiping Wang

High Performance DDoS Attack Detection System Based
on Distribution Statistics   132
    Xia Xie, Jinpeng Li, Xiaoyang Hu, Hai Jin, Hanhua Chen, Xiaojing Ma,
    and Hong Huang

DDP-B: A Distributed Dynamic Parallel Framework for Meta-genomics
Binary Similarity.   143
    Mengxian Chi, Xu Jin, Feng Li, and Hong An

Optimal Resource Allocation Through Joint VM Selection and Placement
in Private Clouds   156
    Hongkun Chen, Feilong Tang, Linghe Kong, Wenchao Xu,
    Xingjun Zhang, and Yanqin Yang

A Parallel Multi-keyword Top-k Search Scheme over Encrypted
Cloud Data.   169
    Maohu Yang, Hua Dai, Jingjing Bao, Xun Yi, and Geng Yang

N-Docker: A NVM-HDD Hybrid Docker Storage Framework to Improve
Docker Performance   182
    Lin Gu, Qizhi Tang, Song Wu, Hai Jin, Yingxi Zhang, Guoqiang Shi,
    Tingyu Lin, and Jia Rao

HPC

MMSR: A Multi-model Super Resolution Framework   197
    Ninghui Yuan, Zhihao Zhu, Xinzhou Wu, and Li Shen

HiPower: A High-Performance RDMA Acceleration Solution
for Distributed Transaction Processing.   209
    Runhua Zhang, Yang Cheng, Jinkun Geng, Shuai Wang, Kaihui Gao,
    and Guowei Shen

Emerging Topics

LDAPRoam: A Generic Solution for Both Web-Based
and Non-Web-Based Federate Access   225
    Qi Feng and Wei Peng

Characterizing Perception Module Performance and Robustness
in Production-Scale Autonomous Driving System   235
   Alessandro Toschi, Mustafa Sanic, Jingwen Leng, Quan Chen,
   Chunlin Wang, and Minyi Guo

Memory and File System

Spindle: A Write-Optimized NVM Cache for Journaling File System   251
   Ge Yan, Kaixin Huang, and Linpeng Huang

Two-Erasure Codes from 3-Plexes.   264
   Liping Yi, Rebecca J. Stones, and Gang Wang

Deep Fusion: A Software Scheduling Method for Memory Access
Optimization.   277
   Yimin Zhuang, Shaohui Peng, Xiaobing Chen, Shengyuan Zhou,
   Tian Zhi, Wei Li, and Shaoli Liu

Optimizing Data Placement on Hierarchical Storage Architecture
via Machine Learning   289
   Peng Cheng, Yutong Lu, Yunfei Du, Zhiguang Chen, and Yang Liu

Short Papers

I/O Optimizations Based on Workload Characteristics for Parallel
File Systems.   305
   Bing Wei, Limin Xiao, Bingyu Zhou, Guangjun Qin, Baicheng Yan,
   and Zhisheng Huo

Energy Consumption of IT System in Cloud Data Center: Architecture,
Factors and Prediction   311
   Haowei Lin, Xiaolong Xu, and Xinheng Wang

Efficient Processing of Convolutional Neural Networks on SW26010   316
   Yi Zhang, Bing Shu, Yan Yin, Yawei Zhou, Shaodi Li, and Junmin Wu

ADMMLIB: A Library of Communication-Efficient AD-ADMM
for Distributed Machine Learning   322
   Jinyang Xie and Yongmei Lei

Energy-Aware Resource Scheduling with Fault-Tolerance
in Edge Computing.   327
   Yanfen Xue, Guisheng Fan, Huiqun Yu, and Huaiying Sun

DIN: A Bio-Inspired Distributed Intelligence Networking.   333
   Yufeng Li, Yankang Du, Chenhong Cao, and Han Qiu

A DAG Refactor Based Automatic Execution Optimization Mechanism
for Spark   338
    Hang Zhao, Yu Rao, Donghua Li, Jie Tang, and Shaoshan Liu

BTS: Balanced Task Scheduling Strategy Based on Multi-resource
Prediction and Allocation in Cloud Environment.   345
    Yongzhong Sun, Kejiang Ye, Wenbo Wang, and Cheng-Zhong Xu

DAFL: Deep Adaptive Feature Learning for Network Anomaly Detection . . .   350
    *Shujian Ji, Tongzheng Sun, Kejiang Ye, Wenbo Wang,
    and Cheng-Zhong Xu*

SIRM: Shift Insensitive Racetrack Main Memory   355
    Hongbin Zhang, Bo Wei, Youyou Lu, and Jiwu Shu

PDRM: A Probability Distribution Based Resource Management
for Batch Workloads in Heterogeneous Cluster.   361
    Jun Zhou, Dan Feng, and Fang Wang

Collaborating CPUs and MICs for Large-Scale LBM Multiphase
Flow Simulations   366
    Chuanfu Xu, Xi Wang, Dali Li, Yonggang Che, and Zhenghua Wang

Multiple Algorithms Against Multiple Hardware Architectures:
Data-Driven Exploration on Deep Convolution Neural Network   371
    *Chongyang Xu, Zhongzhi Luan, Lan Gao, Rui Wang, Han Zhang,
    Lianyi Zhang, Yi Liu, and Depei Qian*

A Parallel Retinex Image Enhancement Algorithm Based on OpenMP.   376
    *Shixiong Cheng, Bin Liu, Dongjian He, Jinrong He, Yuancheng Li,
    and Yanning Du*

Author Index   383

Graph Computing

GraphScSh: Efficient I/O Scheduling and Graph Sharing for Concurrent Graph Processing

Shang Liu[1], Zhan Shi[1(✉)] [iD], Dan Feng[1], Shuo Chen[1], Fang Wang[1], and Yamei Peng[2]

[1] Wuhan National Laboratory for Optoelectronics,
Huazhong University of Science and Technology, Wuhan, China
{upup,zshi,dfeng,shuochen,wangfang}@hust.edu.cn
[2] Didi, Inc., Beijing, China
1014280613@qq.com

Abstract. With the increasing need for analyzing graph data, graph systems have to efficiently deal with concurrent graph processing (CGP) jobs. However, existing platforms are inherently designed for a single job, they incur the high cost when CGP jobs are executed. In this work, we observed that existing systems do not allow CGP jobs to share graph structure data of each iteration, introducing redundant accesses to same graph. Moreover, all the graphs are real-world graphs with highly skewed power-law degree distributions. The gain from extending multiple external storage devices is diminishing rapidly, which needs reasonable schedulings to balance I/O pressure into each storage. Following this direction, we propose GraphScSh that handles CGP jobs efficiently on a single machine, which focuses on reducing I/O conflict and sharing graph structure data among CGP jobs. We apply a CGP balanced partition method to break graphs into multiple partitions that are stored in multiple external storage devices. Additionally, we present a CGP I/O scheduling method, so that I/O conflict can be reduced and graph data can be shared among multiple jobs. We have implemented GraphScSh in C++ and the experiment shows that GraphScSh outperforms existing out-of-core systems by up to 82%.

Keywords: Graph processing · CGP jobs · Graph sharing · I/O scheduling

1 Introduction

In the past decade, graph analysis has become important in a large variety of domains. Due to the increasing need to analyze graph structure data, it is common that Concurrent Graph Processing (CGP) jobs are executed on same processing platforms, in order to acquire different information from same graphs.

© IFIP International Federation for Information Processing 2019
Published by Springer Nature Switzerland AG 2019
X. Tang et al. (Eds.): NPC 2019, LNCS 11783, pp. 3–15, 2019.
https://doi.org/10.1007/978-3-030-30709-7_1

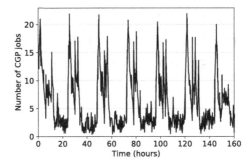

Fig. 1. The number of CGP jobs.

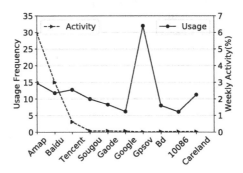

Fig. 2. The utilization of map apps.

For example, Facebook uses Apache Giraph [6] to execute various graph algorithms, such as the variants of PageRank [12], SSSP [10], etc. Figure 1 depicts the number of CGP jobs over a large Chinese social network [17]. The stable distribution shows that more than 83.4% of the time has at least two CGP jobs executed simultaneously. At the peak time, over 20 CGP jobs are submitted to the same platform. Also, Fig. 2 shows the usage of Chinese map Apps in a week of 2017. We can observe that each map App is used by each user more than five times within a week. Particularly, Amap App [2] ranks the first and handles over 10 billion route plannings every week, that is to say, it is used more than 60 thousand times per minute on average.

The existing processing systems can process a single graph job efficiently. They improve the efficiency either by fully utilizing the sequential usage of memory bandwidth, or by achieving a better data locality and less redundant data accesses, like GraphChi [8], X-Stream [13], GridGraph [20] and Graphene [9], PreEdge [11], etc. However, these systems are usually designed for a single graph processing job, which are much more inefficient when executing multiple CGP jobs. The inefficiencies include I/O conflict and repeated access to same graph structure data.

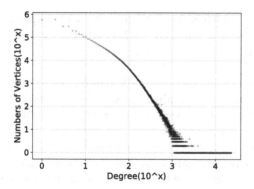

Fig. 3. Power-law degree distribution.

I/O Conflict: When multiple CGP jobs are executed over same graph, it is commonplace that these jobs visit same partition data, resulting in I/O conflict among multiple jobs. Fortunately, extending multiple external storage devices is possible to reduce this conflict, which can distribute multiple I/O of CGP jobs to multiple external storage devices. However, graphs derived from real-world phenomena, like social networks and the web, typically have highly skewed power-law degree distributions [1], which implies that a small subset of vertices connects to a large fraction of the graph. Figure 3 depicts the pow-degree distribution of graph from LiveJournal [14], which is a free online community with almost 10 million members. The highly skewed characteristic of graph challenges the above assumption and make it more difficult. Although using multiple storage devices reduces I/O conflict, this conflict is still the bottleneck of overall performance.

Data Access Problems: Graph processing jobs are usually operated on two types of data [9]: graph structure data and graph state data. The graph structure data mainly consists of vertices, edges, and the information associated with each edge. The graph state data, such as ranking scores for PageRank, is computed within each iteration and consumed in the next iteration. The graph structure data usually occupies a large volume of the memory, whose proportions are varying from 71% to 83% for different datasets [19]. However, existing graph platforms do not allow CGP jobs to share the graph structure data in memory, resulting in redundant access to the graph from external storage. Furthermore, existing out-of-core systems leverage various mechanisms to utilize the sequential usage of memory bandwidth and achieve a better data locality, such as PSW in GraphChi, Edge-Centric in X-Stream and 2-level hierarchical partitioning in GridGraph, etc. Unfortunately, CGP jobs destroy these optimized mechanisms above, increasing overhead of randomized access significantly.

In this paper, we propose GraphScSh, a graph processing system based on multiple external storage devices. Our design concentrates on reducing I/O conflict and sharing the graph structure data among CGP jobs. Specifically, the graph structure data is divided into multiple external storage devices evenly by CGP balanced partition method. The subgraph of each partition can match the size of memory well, which reduces the overhead of frequent swap operations. Furthermore, we present a new CGP I/O scheduling method based on multiple external storage and graph sharing, so that I/O conflict can be reduced and the graph can be shared among multiple CGP jobs.

The system GraphScSh has been implemented in C++. To demonstrate the efficiency of our solutions, we conducted extensive experiments with our system GraphScSh and compared its performance with state-of-the-art systems Grid-Graph over different combinations of CGP jobs. The experiments show that overall performance of GraphScSh outperforms GridGraph by up to 82%.

The rest of this paper is organized as follows. The design details of GraphScSh are presented in Sect. 2, including CGP balanced partition schema, and CGP I/O scheduling method. Section 3 gives the specific implementation of our system GraphScSh, followed by experimental evaluation in Sect. 4. We then describe related work in Sect. 5 and conclude in Sect. 6.

2   Our Proposed Approach

To reduce the I/O conflict and the redundant access to graph efficiently, we propose GraphScSh based on multiple external storage devices, which is designed to reduce I/O conflict and share the graph structure data among CGP jobs.

2.1   CGP Balanced Partition

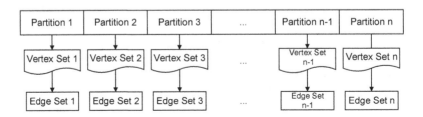

Fig. 4. Partitioning schema of GraphScSh.

The existing partitioning methods are usually designed for a single job. When CGP jobs are executed, we cannot make sure that partitioning size of all jobs match the size of memory, resulting in frequently swap-in and swap-out operations. We propose a new partitioning method to process CGP jobs, as shown in Fig. 4.

The graph is divided into n partitions, and each partition includes a vertex set and an edge set. Within a vertex set, the index id of vertices is continuous. The edge set of a partition consists of all edges whose source vertex is in the partition's vertex set. When GraphScSh executes graph algorithms, each partition size depends on both memory configuration and number of CGP jobs, so that data of each vertex set can be fit into memory. Additionally, GraphScSh leverages multiple external devices to store the graph data. For the load balance, different partitions are stored in multiple storage devices and the number of edges for each partition is same. The position $disk\_id$ of each partition in multiple external storage can be described as,

$$disk\_id = partition\_id\%disk\_num \qquad (1)$$

where $partition\_id$ is the id of graph partition, $disk\_num$ is the number of external storage.

2.2 CGP I/O Scheduling

Based on the above partitioning method, we break graph structure data into multiple partitions evenly which are stored in multiple external storage devices. To reduce the I/O conflict and share the graph among CGP jobs, we propose a CGP I/O scheduling method based on CGP Balanced Partition method. The scheduling method includes two strategies for load balance and graph sharing.

First, we count the total number of jobs in each external storage and select one external storage that has the fewest jobs as the target, for loading balance. During execution of CGP jobs, system records $partition\_id$ that each job visits. The position of graph partition is computed according to the mapping between partitions and the external storage. For example, there are n jobs executed, where m jobs visit the first external storage for graph, and $(n - 1 - m)$ jobs access the second external storage. If $m > (n - 1)/2$, the second one will be selected as the target, otherwise the first will be targeted. Assume that the number of external storage is k, where the number of jobs is $n - 1, n - 2, ..., n - k$, the storage with the fewest jobs will be targeted.

Second, we leverage synchronous field to reduce total number of I/O as much as possible to share the same graph, as Fig. 5 shows. The sync field mainly records information about the mapping from graphs to memory, including mapping address $mmap\_addr$ [18], the number $edge\_num$ of edges, and the descriptor fd of file. In addition, the field must include the total number $unit\_num$ of jobs and determines whether to remove the mapping of partition according to it. Specifically, according to $unit\_num$, the system decides if partition data has been mapped into the memory according to the sync field. If $unit\_num = 0$, the partition is not visited by jobs and should be filled into memory through mapping. Otherwise, the partition has been loaded into memory by other jobs, and the current job visits partition by the address of field.

unit_num	mmap_addr	edge_num	fd

Fig. 5. Sync field of graph.

The specific process of CGP I/O scheduling method includes several steps. Suppose that the number of the external storage is k, the concurrent graph job is A, the I/O scheduling of CGP jobs contains the following steps:

- According to synchronous information of CGP jobs and mapping information between partition and disk, the system counts the number of jobs executed in each external storage as n_1, n_2, ..., n_k, respectively.
- According to synchronous information of CGP jobs and mapping information between partition and disk, the system counts the number of partitions in each external storage, as s_1, s_2, ..., s_k, respectively, and records $partition\_id$.
- The system sorts the external storage according to the values of n_1, n_2, ..., n_k. Then the corresponding id of the external storage is added into set U, where the number of jobs in each external storage is in ascending order.
- The system decides each external storage of U one by one. If the set s_i of one external storage i contains a partition that has not been accessed, the external storage i is selected as the target.
- If the partition data in memory has been processed by job A, A will visit each storage in U to find the data which has not been used. If the data exists, the corresponding external storage will be as the target and the current iteration ends.

Assume that the total execution time of a graph job is T, its computation time is T_c and its I/O wait time is T_w. When N jobs are executed on the same graph, the computation time of jobs is $T_{C1}, T_{C2}...T_{CN}$ respectively, and I/O wait time is T_w. The total execution time of existing systems can be described as,

$$T_o = max(T_{C1}, T_{C2}, ...T_{CN}) + NT_w \qquad (2)$$

where $T_{C-MAX} = max(T_{C1}, T_{C2}, ...T_{CN})$. So the total time can be described as,

$$T_o = T_{C-MAX} + NT_w \qquad (3)$$

Suppose that the number of external storage devices is D. Based on loading balancing, the I/O pressure is balanced into each external storage. Therefore, the number of jobs running on each device is N/D. The new total execution time can be described as,

$$T_{multi-disks} = T_{C-MAX} + T_w * N/D \qquad (4)$$

the total number of I/O is from NT_W to $N/D * T_W$. The new total execution time is described as,

$$T_G = max(T_{C-MAX}, T_w) \qquad (5)$$

We can see that the new I/O Scheduling outperforms the existing methods by up to $(N - N/D)$ theoretically.

3   GraphScSh Implementation

We have implemented our system GraphScSh in C++. Figure 6 illustrates the modules of GraphScSh, including graph management, mapping management, data structure, operation module, and graph algorithms. We mainly focus on two parts in this section: operation module and graph algorithms.

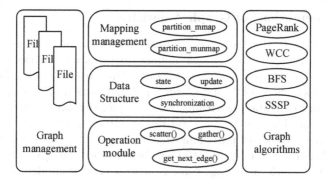

Fig. 6. Modules of GraphScSh.

3.1   Operation Module

The function of this module is achieved by operations of Scatter and Gather. In *Scatter* phase, it accesses to graph in streaming way by function *get_next_edge()* and generates the updated information according to state data. In *Gather* phase, it read updated data and updates the state data. The Traversal operation is the kernel operation and implements by the function *get_next_edge()*. First, the function needs to determine partitions of graph whether to be visited. If false, the next edge data will be accessed. Then, *get_next_edge()* decides all partition of this iteration whether to be visited. If true, the next iteration will be started. If false, the function findNextPartition() will be activated to find the next partition to visit. The implementation details of FindNextPartition are described in Algorithm 1.

Algorithm 1. Details of FindNextPartition

Input:
    The partition set of graph *unaccess_partition*;
    The set of external storage U;
    The visited partition set of graph $s_1, s_2, ..., s_k$;
Output:
    The next partition to be visited *partition_index*;
1: **for** i in U **do**
2:    **if** $\exists p \in unaccess\_partition, p \in s_i$ **then**
3:       $partition\_index = p$;
4:       return;
5:    **else**
6:       continue;
7:    **end if**
8: **end for**
9: **for** p in *unaccess_partition* **do**
10:    $partition\_index = p$;
11:    break;
12: **end for**

3.2 Implementations of Graph Algorithm

We define *Graph* as the base class, which provides a programming interface for graph algorithms. Class *Graph* defines five virtual functions, including *initUnit*() for initialization, *output*() for outputting result, *reset*() for cleaning after one iteration, *Scatter*(), and *Gather*(). The function *initUnit*() initializes the related work of graph algorithms, for example, the out-degree of each vertex in PageRank. The function *reset*() resets partition sets that workers have visited, and the number of partitions that each external storage has accessed. Algorithms 2 and 3 give examples to show how to implement graph algorithms on GraphScSh, which uses edge-centric Scatter-Gather model to run graph algorithms.

Algorithm 2. PageRank Scatter

1: **for** each edge e of graph **do**
2:      update_t upt;
3:      **if** $update\_bitset[e.dst]=$ false **then**
4:          $upt.id = e.dst$;
5:          $upt.value = e.src.value/e.src.degree$;
6:          add upt to $update\_buf$;
7:      **end if**
8: **end for**

Algorithm 3. PageRank Gather

1: **for** each update u of upt_buffer **do**
2:      **if** $update\_bitset[u.id]=$ false **then**
3:          $\_aux[upt.id].tmp+ = u.value$;
4:      **end if**
5: **end for**
6: **for**   each element ele of $\_aux$ **do**
7:      **if** $update\_bitset[ele.index - start]=$false **then**
8:          $tmp = init\_wgt + 0.85 * ele.tmp$;
9:          **if** $fabs(ele.tmp - tm) < 0.00000001$ **then**
10:              $update\_bitset[ele.index - start]=$true;
11:          **else**
12:              $ele.res = tmp$;
13:          **end if**
14:          $ele.tmp = 0$;
15:      **end if**
16: **end for**

4   Experimental Evaluation

4.1   Experiment Environment and Datasets

The hardware platform used in our experiments is a single machine containing 6-core 1.60 GHz Intel(R) Xeon(R) CPU E5-2603. Its memory is 8 GB and has two SSDs with 300 GB. The program is compiled with g++ version 11.0.

In our experiments, four popular graph algorithms are employed as benchmarks: (1) breadth-first search (BFS) [3]; (2) PageRank (PR) [12]; (3) weakly connected component (WCC) [7]; (4) single-source shortest path (SSSP) [10]. The datasets used for these graph algorithms are real-world graphs and generated graphs described in Table 1. Where Twitter [14] is from online social networks and edges represent interactions between people. R-MAT, SW, and ER are generated based on power-law [4], small-world model [15] and ER model [5] respectively.

Table 1. Data sets properties

DataSets	Vertexes	Edges	Average degree	Description
Twitter	61.6 M	1.5 B	23.8	Social networks from Twitter
R-MAT26	67.1 M	1.1 B	16	Power-law degree distributions
ER26	67.1 M	1.1 B	16	Random degree distributions

Table 2. Execution time of algorithms on GridGraph(s)

Data Sets	BFS	WCC	PageRank
Twitter	768.84	883.07	3630.38
R-MAT26	409.27	349.27	2389
ER26	482.22	223	596.33

4.2   Comparison with GridGraph

To compare the performance of GridGraph and GraphScSh, we simultaneously submit multiple jobs to each system. The partition number of GraphScSh is set same as GridGraph, and different datasets have a different number of partitions. The execution time of various graph processing algorithms has been computed, as Table 2 depicted. For better comparing the performance of systems, CGP jobs consist of two graph algorithms with same converge speed based on different datasets. To acquire better integrity, experiments are designed under different degree of parallelism (DOP) [16].

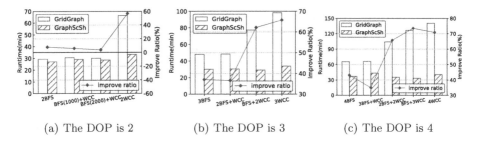

(a) The DOP is 2   (b) The DOP is 3   (c) The DOP is 4

Fig. 7. The comparison of runtime on Twitter for GraphScSh and GridGraph

Twitter: First, for graph dataset Twitter, we evaluate the total execution time and the speed-up ratio of various CGP jobs (e.g. the DOP is 2, 3 and 4, respectively, as Fig. 7(a), (b) and (c). In general, for different combinations of CGP jobs, the execution time of GraphScSh is less than that of GridGraph, and the speed-up ratio grows up as DOP increases. Under the same DOP but a different combination, the longer execution time of CGP jobs is, the greater GraphScSh outperforms GridGraph. Specifically, when two systems are executed on dataset Twitter, the combinations of 2WCC, 3WCC, and 4WCC are accelerated by 56.93%, 65.75%, and 70.8% respectively. Because CGP jobs are executed on the GridGraph, resulting in the I/O conflict greatly.

RMAT26: Next, we execute different combinations of CGP jobs on RMAT26 to compare GridGraph and GraphScSh, as Fig. 8(b) and (c) show. When the DOP is 3 or 4, the performance of GraphScSh is better than that of GridGraph. In particular, with the increase DOP, the speed-up ratio grows up gradually. For example, GraphScSh outperforms GridGraph by 34%, 40.5% and 45.6% under the combinations of 2BFS, 3BFS, and 4BFS, respectively.

ER26: Besides, from Fig. 9(a), (b) and (c), we can observe that the total execution time of GraphScSh is much less than those of GridGraph over dataset ER26. For example, for the combinations of 2PR, 3PR and 4PR, GraphScSh outperforms GridGraph by 64.67%, 76.03% and 82%, respectively. Under the same DOP, the difference that GraphScSh executes different combinations of CGP jobs is smaller than that of GridGraph. It also means that GraphScSh with GSSC and MSGL is suitable to cope with CGP jobs.

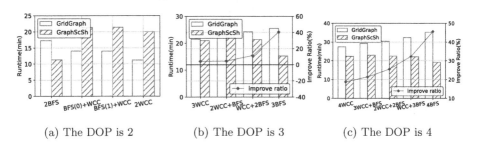

(a) The DOP is 2   (b) The DOP is 3   (c) The DOP is 4

Fig. 8. The comparison of runtime on R-MAT26 for GraphScSh and GridGraph

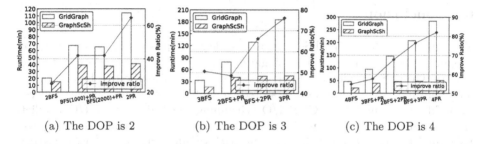

Fig. 9. The comparison of runtime on ER26 for GraphScSh and GridGraph

5   Related Work

With the explosion of graph scale, lots of graph processing systems are created to achieve high efficiency for graph analysis. They improve the efficiency either by a prefetcher for graph algorithms, or by full utilizing the sequential usage of memory bandwidth.

PrefEdge [11] is a prefetcher for graph algorithms that parallelises requests to derive maximum throughput from SSDs. PrefEdge combines a judicious distribution of graph state between main memory and SSDs with an innovative read-ahead algorithm to prefetch needed data in parallel. GraphChi [8] a disk-based system for computing efficiently on graphs with billions of edges. By using a novel parallel sliding windows method, GraphChi is able to execute several advanced data mining, graph mining, and machine learning algorithms on very large graphs, using just a single consumer-level computer. X-Stream [13] is novel in using an edge-centric rather than a vertex-centric implementation of this model, and streaming completely unordered edge lists rather than performing random access. GridGraph [20] is an out-of-core graph engine using a grid representation for large-scale graphs by partitioning vertices and edges to 1D chunks and 2D blocks respectively, which can be produced efficiently through a lightweight range-based shuffling.

Unfortunately, when CGP jobs are executed on these systems above, they incur the extra high cost (e.g., inefficient memory use and high fault tolerance cost). Following this observation, Seraph [17] is designed to handle with CGP jobs based on a decoupled data model, which allows multiple concurrent jobs to share graph structure data in memory [19]. Based on this observation that there are strong spatial and temporal correlations among the data accesses issued by different CGP jobs because these concurrently running jobs usually need to repeatedly traverse the shared graph structure for the iterative processing of each vertex, CGraph [19] proposed a correlations-aware execution model. Together with a core-subgraph based scheduling algorithm, CGraph enables these CGP jobs to efficiently share the graph structure data in memory and their accesses by fully exploiting such correlations.

6   Conclusion

This paper introduces GraphScSh, a large scale graph processing system that can support CGP jobs running on a single machine with multiple external storage devices. GraphScSh adopts a CGP balanced partition method to break graphs into multiple partitions that are stored in multiple external storage devices. In addition, we present a CGP I/O scheduling method, so that I/O conflict can be reduced and the same graph can be shared among multiple CGP jobs. Experimental results depict that our approach significantly outperforms existing out-of-core systems when running CGP jobs. In the future, we will research to further optimize our solution with a snapshot mechanism for efficient graph processing.

Acknowledgments. This work is supported by NSFC No. 61772216, 61821003, U1705261, Wuhan Application Basic Research Project No. 2017010201010103, Fund from Science, Technology and Innovation Commission of Shenzhen Municipality No. JCYJ20170307172248636, Fundamental Research Funds for the Central Universities.

References

1. Dasgupta, A., Hopcroft, J.E., McSherry, F.: Spectral analysis of random graphs with skewed degree distributions. In: FOCS 2004 (2004)
2. BDR: http://www.bigdata-research.cn/content/201801/635.html
3. Beamer, S., Asanovic, K., Patterson, D.A.: Direction-optimizing breadth-first search. In: SC 2012 (2012)
4. Chakrabarti, D., Zhan, Y., Faloutsos, C.: R-MAT: a recursive model for graph mining. In: SDM 2004 (2004)
5. Erdös, P., Rényi, A.: On random graphs. Publicationes Mathematicae Debrecen **6**, 290 (1959)
6. Apache Giraph: http://giraph.apache.org/
7. Khayyat, Z., Awara, K., Alonazi, A., Jamjoom, H., Williams, D., Kalnis, P.: Mizan: a system for dynamic load balancing in large-scale graph processing (2013)
8. Kyrola, A., Blelloch, G., Guestrin, C.: GraphChi: large-scale graph computation on just a PC. In: OSDI 2012 (2012)
9. Liu, H., Huang, H.H.: Graphene: fine-grained IO management for graph computing. In: FAST 2017 (2017)
10. Maleki, S., Nguyen, D., Lenharth, A., Garzarán, M.J., Padua, D.A., Pingali, K.: DSMR: a parallel algorithm for single-source shortest path problem. In: ICS (2016)
11. Nilakant, K., Dalibard, V., Roy, A., Yoneki, E.: PrefEdge: SSD prefetcher for large-scale graph traversal (2014)
12. Page, L., Brin, S., Motwani, R., Winograd, T.: The PageRank citation ranking: bringing order to the web. Technical report (1999)
13. Roy, A., Mihailovic, I., Zwaenepoel, W.: X-stream: edge-centric graph processing using streaming partitions. In: SOSP 2013 (2013)
14. SNAP: http://snap.stanford.edu/data/index.html
15. Watts, D., Strogatz, S.: Collective dynamics of small world networks. Nature **393**, 440–442 (1998)
16. Wikipedia: https://en.wikipedia.org/wiki

17. Xue, J., Yang, Z., Qu, Z., Hou, S., Dai, Y.: Seraph: an efficient, low-cost system for concurrent graph processing. In: HPDC 2014 (2014)
18. Lin, Z., Kahng, M., Sabrin, K.Md., et al.: MMap: fast billion-scale graph computation on a pc via memory mapping. In: Big Data, pp. 159–164 (2014)
19. Zhang, Y., et al.: CGraph: a correlations-aware approach for efficient concurrent iterative graph processing. In: ATC 2018 (2018)
20. Zhu, X., Han, W., Chen, W.: GridGraph: large-scale graph processing on a single machine using 2-level hierarchical partitioning. In: ATC 2015 (2015)

Game-Based Multi-MD with QoS Computation Offloading for Mobile Edge Computing of Limited Computation Capacity

Junyan Hu[1,2], Chubo Liu[1,2(✉)], Kenli Li[1,2], and Keqin Li[1,2,3(✉)]

[1] College of Computer Science and Electronic Engineering, Hunan University,
Changsha 410082, Hunan, China
{junyanhu,liuchubo,lkl,likq}@hnu.edu.cn
[2] National Supercomputing Center in Changsha, Changsha 410082, Hunan, China
[3] Department of Computer Science, State University of New York,
New Paltz, NY 12561, USA
lik@newpaltz.edu

Abstract. Mobile edge computing (MEC) is becoming a promising paradigm of providing cloud computing capabilities to the edge network, which can serve mobile devices (MDs) with computation-intensive and delay-sensitive tasks. Facing with high requirements of many MDs, it's essential for MEC with limited computation capacity to serve more MDs with QoS. For each mobile device, it is also desirable to have a low energy consumption with an expected deadline. To solve above problems, we propose a Game-based Computation Offloading (GCO) algorithm, which includes the task offloading profile and the transmission power controlling with the method of non-cooperative game. Our mechanism maximizes the number of served MDs with deadline, as well as minimizing the energy consumption of each MD whose task is executed on MEC. Specifically, Given the allocation of transmission power, a Greedy-Pruning algorithm is proposed to determine the number of tasks executed on MEC. Besides, each MD adopts his/her transmission power controlling strategy to compete the computation resource of MEC or minimize the energy consumption. A game model for illustrating the problem of task offloading is formulated to find a proper transmission power for each task and is proved the existence of Nash equilibrium solution. Experiments are simulated to evaluate the proposed algorithm in terms of effectiveness evaluation.

Keywords: Mobile edge computing · Nash equilibrium ·
Non-cooperative game theory · Task offloading · Power controlling

1 Introduction

Nowadays, Mobile Devices (MDs) are indispensable part in our daily life [1,2]. With the popularity of smart MDs, many new computation-intensive and

© IFIP International Federation for Information Processing 2019
Published by Springer Nature Switzerland AG 2019
X. Tang et al. (Eds.): NPC 2019, LNCS 11783, pp. 16–27, 2019.
https://doi.org/10.1007/978-3-030-30709-7_2

delay-sensitive applications have higher demands on quality of service (QoS) [3]. However, the limited resources of MDs e.g., battery, computation capacity, cannot meet their own needs. Therefore, how to meet the high QoS requirements of multiple MDs with low energy consumption is an urgent problem to be solved.

Mobile edge computing (MEC) provides high-bandwidth, high-computing resources for nearby MDs to meet the high QoS demands for computation-intensive and latency-sensitive applications via edge network [4,5]. For a multi-device MEC system with multiple parallel computation tasks requiring computing resources, MEC can be viewed as a small cloud with limited resources (processing speed, CPU cycle). Facing with resource requests from numerous devices, MEC should propose a resource allocation strategy that maximizes the number of served MDs with QoS requirements. For each MD, it has an expected value of delay, and on this basis, it is desirable to have a minimum energy consumption. The transmission rate and the received computation resource of each MD are affected by other MDs. Thus, if there are many devices that offload their tasks, the QoS experience of each MD will be deduced. In order to compete the resource for CPU cycle, a suitable transmission power controlling strategy mechanism for each MD should be proposed.

The remainder of the paper is organized as follows. In Sect. 2, we introduce the related work. Section 3 describes the system model and presents the problem that needs to be solved. In Sect. 4, we consider the problem as a non-cooperative game and propose Algorithm GCO to compute the Nash equilibrium solution. In Sect. 5, extensive experiments results indicate the feasibility of our algorithms. We conclude the works of this paper in Sect. 6.

2   Related Work

Task offloading for user requirements in MEC has been studied by many scholars and most of studies are analyzed from computational offloading, latency, storage, and energy efficiency. [6–8] are considered from optimizing the energy consumption of users. In [6], Chen et al. computed the energy harvesting for MEC by using Lyaponuv Optimization method. Besides, some works and models considered from guaranteeing the deadline or minimizing average delay [9–12]. Fan et al. proposed an application aware workload scheduling mechanism for IoT based on MEC to minimize the average delay of application resource requests in [10]. [11] solved the problem of minimizing delay by using the method of one-dimensional search. And then in [13] and [14], Zhang and Chen et al. considered the proportional overhead on power consumption and latency. In addition, [15–17] optimized the transmission to achieve the offloading balance in the MEC by controlling the transmission power. In [15], Rodrigues et al. proposed a workload balance strategy for cloudlets to minimize the cost by using Transmission Power Control (TPC). In [16], Mao et al. minimized the weighted sum of the execution delay and energy consumption by optimizing the transmission. Different from above all, our work considers not only from the view of serving the maximum number of MDs with deadline constraint, but also from the perspective of each MD's minimum energy consumption.

Game theory plays an increasingly important method in MEC [18–21]. In [18], Chen *et al.* analyzed the multi-task offloading problem for MEC under the condition of multi-channel from the view of game theory. In [21], by using the theory of Minority Games, Ranadheera *et al.* proposed a novel distributed server activation mechanism for computation offloading which guaranteed energy-efficient activation of servers as well as satisfaction of users quality-of-experience (QoE) requirements in terms of latency. Heuristically, our work introduces an adaptive transmission power mechanism in the competing process for limited-computation resources provided by MEC. We formulate a non-cooperative game-based mechanism for MEC's offloading decision making and MDs' power control.

3   System Model

We denote $\mathcal{N} = \{1, 2, \ldots, N\}$ as the set of N MDs, each of which has computation-intensive and time-sensitive task to be completed. Let τ_n be the task of n, and the requirement of MD τ_n can be denoted as a tuple (c_n, d_n, T_n), where c_n denotes the total number of required CPU cycles, d_n denotes the size of the input task data, and T_n denotes the expected time required to complete task τ_n. The task can be computed either locally on the mobile device or remotely executed on MEC via computation offloading. Therefore, we denote the decision profile $X = \{x_1, x_2, \ldots, x_N\}$ as the set of indicator function for N MDs, where $x_n \in \{0, 1\}$. If the task of MD n is computed on MEC $\mathcal{S}$, $x_n = 1$, otherwise, $x_n = 0$. Besides, we denote J as the set of mobile devices, where $J = \{n | x_n = 1\}$. Here we consider the computational capacity of MEC $\mathcal{S}$, denoted as $\mathcal{C}$, is limited. If a MD prepares to offload his task to MEC $\mathcal{S}$, the energy and time consumption of communication and computation are considered.

3.1   Communication Model

If MD n offloads task τ_n to remotely edge execution, the input data should be transmitted to MEC servers of $\mathcal{S}$. Given the decision profile X and J, the communication rate of MD n $(n \in J)$ via the wireless channel can be denoted as

$$r_n(X, \mathcal{P}) = B \log_2(1 + \frac{p_n G_n}{\eta_0 + \Sigma_{i \in J \setminus \{n\}} p_i G_i}). \tag{1}$$

Here B is the channel bandwidth, and for simplicity, we only consider one channel. $\mathcal{P} = \{p_1, p_2, \ldots, p_N\}$ is the transmission power profile of all MDs and each p_n can be chosen from the internal $[\underline{p_n}, \overline{p}_n]$. Further, G_n, related to the environment and the distance, denotes the channel gain between MD n and MEC $\mathcal{S}$ and η_0 is the background noise power. In (1), let $I_n = \Sigma_{i \in J \setminus \{n\}} p_i G_i$ be the sum of interference from other MDs who belong to set J. Note that the transmission rate can be affected not only the transmission power of itself but also the MDs which offload tasks to MEC $\mathcal{S}$.

3.2   Computation Model

If task τ_n of mobile device n is offloaded to MEC $\mathcal{S}$ to execute, i.e., $x_n = 1$, the completion time will contain communication time and computation time. We define the completion time as

$$t_{n,off} = \frac{d_n}{r_n(X,\mathcal{P})} + \frac{c_n}{f_n} = \frac{d_n}{B \log_2(1 + \frac{p_n G_n}{\eta_0 + \Sigma_{i \in J \backslash \{n\}} p_i G_i})} + \frac{c_n}{f_n}, \qquad (2)$$

where f_n is the computation capability (i.e., CPU cycles per second) assigned to MD n by the MEC $\mathcal{S}$. Therefore, the energy consumption can be denoted as

$$E_{n,off}(X,\mathcal{P}) = \frac{p_n d_n}{B \log_2(1 + \frac{p_n G_n}{\eta_0 + \Sigma_{i \in J \backslash \{n\}} p_i G_i})}. \qquad (3)$$

3.3   MEC's Resource Allocation Strategy

From the perspective of MEC $\mathcal{S}$ with limited resource, serving as many MDs as possible is its primary goal. We consider distributed resource allocation for MDs, and model it as $\max_{X} |J|$ with the constraints $t_n \leq T_n, n \in J$ and $\sum_{n \in J} f_n \leq \mathcal{C}$, where $|\cdot|$ is the number of elements in set $\cdot$.

Theorem 1. *The issue* $\max_{X} |J|$ *that maximizes the number of tasks with QoS executed on MEC is NP-hard.*

Algorithm 1. Greedy-Pruning algorithm

Require: $\mathcal{N}, \mathcal{P}, G, B, \mathcal{C}$.
Ensure: $J, f_n(J)$.
1: $J \leftarrow \mathcal{N}, J_1 \leftarrow \{\emptyset\}, J_2 \leftarrow \{\emptyset\}$;
2: Calculate each $f'_n(J)$ $(n \in J)$ based on Eq. (4);
3: **while** $(\sum_{k \in J} f'_k(J) > \mathcal{C})$ **do**
4:     $J_1 \leftarrow J_1 \bigcup \{\arg\min_{i} \sum_{j \in J \backslash \{i\}} f'_j(J \backslash \{i\})\}$;
5:     **while** $(J_1 \neq J_2)$ **do**
6:       $J_2 \leftarrow J_1$;
7:       **for** $(l \in J_2)$ **do**
8:         $J_1 \leftarrow (J_1 \backslash \{l\}) \bigcup$
                $\{\arg\min_{i} \sum_{j \in \mathbb{C}_{\mathcal{N}}^{J_1 \backslash \{l\}} \backslash \{i\}} f'_j(\mathbb{C}_{\mathcal{N}}^{J_1 \backslash \{l\}} \backslash \{l\})\}$;
9:   $J \leftarrow \mathcal{N} \backslash J_1$;
10: **return** $J, f_n(J)$.

In order to solve the problem $\max_{X} |J|$, we propose a Greedy-Pruning algorithm (Algorithm 1). Let $f'_n(J)$ be the critical point of computation capability that MD n needs.

$$f_n(J) \geq \frac{c_n}{T_n - \frac{d_n}{\gamma_n(X,\mathcal{P})}} = f_n'(J). \tag{4}$$

Assuming $\sum_{n\in\mathcal{N}} f_n' \leq \mathcal{C}$, then there is $J = \mathcal{N}$. Otherwise, MEC $\mathcal{S}$ needs to filter out some MDs to maximize the number of beneficial MDs with QoS. In Algorithm 1, J is the set of MDs to be selected, and J_1 is the set of MDs to be filtered out. In the outer *while* loop of the line 3–9, once $\sum_{k\in J} f_k'(J) > \mathcal{C}$, an appropriate MD will be added to J_1 to check whether the condition $\sum_{k\in\widetilde{J}} f_k'(\widetilde{J}) \leq \mathcal{C}$ is satisfied, where $\widetilde{J}$ is the updated J. In each round of preparation to remove a MD to J_1, we use $\min \sum_{j\in\widetilde{J}} f_j'(\widetilde{J})$ as the objective function. But removing MD i in J that minimizes $\sum_{j\in J\backslash\{i\}} f_k'(J\backslash\{i\})$ directly does not guarantee that updated J is globally optimal. If there is always

$$l = \arg\min_i \sum_{j\in(J\cup\{l\})\backslash\{i\}} f_j'((J\cup\{l\})\backslash\{i\}). \tag{5}$$

for any MD $(l \in \mathcal{N}\backslash J)$, J is optimal.

3.4   Power Control Strategy of Mobile Device

In this section, we explore that how to minimize each MD's energy consumption within the expected delay range. As can be seen from Eq. (2), $t_{n,off}$ decreases as p_n increases. Given $\mathcal{F}$ and the expected time T_n required to complete the task τ_n, $t_{n,off} \leq T_n$ can be introduced as follows

$$p_n \geq (2^{\frac{d_n}{(T_n - \frac{c_n}{f_n})B}} - 1)(\frac{\eta_0 + I_n}{G_n}) = p_n'. \tag{6}$$

We denote p_n' as the critical power of MD n. If $p_n' > \bar{p}_n$, MD n will not choose to execute his task τ_n on MEC. We assume $p_n' \leq \bar{p}_n$ and consider the energy consumption of MD n in the internal $[\max\{p_n', \underline{p}_n\}, \bar{p}_n]$.

In each round, MD n, who does not execute his task τ_n on the MEC, can increase p_n to provide his own competitiveness. This leads to two outcomes: removing one of the other MDs in J or adding to the set J directly.

Removing one of the other MDs in J: Increasing p_n to p_n^1 and satisfying the conditions

$$\arg\min_k \sum_{j\in J_3} f_j'(J_3\backslash\{k\}) \neq n, (J_3 = J\cup\{n\}),$$

$$\min_k \sum_{j\in J_3} f_j'(J_3\backslash\{k\}) \leq \mathcal{C}. \tag{7}$$

Adding to the set J directly: Increasing p_n to p_n^2 and satisfying the condition

$$\sum_{j\in J\cup\{n\}} f_j'(J\cup\{n\}) = \mathcal{C}. \tag{8}$$

Considering the energy consumption and $p_n \leq \overline{p}_n$, we define $\widetilde{p}_n = \min\{p_n^1, p_n^2, \overline{p}_n\}$, where $\widetilde{p}_n$ is the updated p_n in next round. We denote $\widetilde{\mathcal{P}} = (\widetilde{p}_1, \widetilde{p}_2, \ldots, \widetilde{p}_N)$. We propose a Binary search algorithm ($\mathcal{C}alculate\_\widetilde{\mathcal{P}}(\cdot)$) to update p_n.

4 Game Formulation and Analyses

4.1 Game Formulation

Let $\mathcal{P}_{-n} = (p_1, \cdots, p_{n-1}, p_{n+1}, \cdots, p_N)$ be the transmission power profile of all MDs except MD n. Let $\mathcal{P}_n$ be the set of power and decision making strategies for MD n, i.e., $p_n \in \mathcal{P}_n$. Given other MDs' transmission power $\mathcal{P}_{-n}$, MD n would like to select a proper decision p_n to compete the computation resource of MEC $\mathcal{S}$ and minimize his own energy consumption, under the condition of satisfying QoS. The objective function of MD n can be written as follows $\min E_n(X, \mathcal{P})$. The strategy set of MEC $\mathcal{S}$ is $\mathcal{X}$ and his objective function is maximizing the beneficial number of MDs $|J|$. Then, the multi-device computation offloading game can be represented as G, where $G = \{(\mathcal{P}_n)_{n \in \mathcal{N}}, \mathcal{X}; (E_n)_{n \in \mathcal{N}}, |J|\}$.

Algorithm 2. $\mathcal{C}alculate\_\widetilde{\mathcal{P}}(\cdot)$

Require: $\mathcal{N}, \mathcal{P}, G, B, \mathcal{C}, J, \varepsilon$.
Ensure: $\widetilde{\mathcal{P}}$.
  1: **for** $n \in J$ **do**
  2:     $\widetilde{p}_n = p_n$;
  3: **for** $n \in \mathcal{N} \backslash J$ **do**
  4:     $l_1 p_n \leftarrow p_n, r_1 p_n \leftarrow \overline{p}_n$;
  5:     $l_2 p_n \leftarrow p_n, r_2 p_n \leftarrow \overline{p}_n$;
  6:     **while** $(|r_1 p_n - l_1 p_n > \varepsilon|)$ **do**
  7:         $mid_1 \leftarrow \frac{l_1 p_n + r_1 p_n}{2}$;
  8:         **if** Conditions in Eq. (7) are satisfied **then**
  9:             $r_1 p_n \leftarrow mid_1$;
 10:         **else**
 11:             $l_1 p_n \leftarrow mid_1$;
 12:     $p_n^1 \leftarrow r_1 p_n$;
 13:     **while** $(|r_2 p_n - l_2 p_n > \varepsilon|)$ **do**
 14:         $mid_2 \leftarrow \frac{l_2 p_n + r_2 p_n}{2}$;
 15:         **if** Condition in Eq. (8) is satisfied **then**
 16:             $r_2 p_n \leftarrow mid_2$;
 17:         **else**
 18:             $l_2 p_n \leftarrow mid_2$;
 19:     $p_n^2 \leftarrow r_2 p_n$;
 20:     $\widetilde{p}_n = \min\{p_n^1, p_n^2, \overline{p}_n\}$;
 21: **return** $\widetilde{\mathcal{P}}$.

For all MDs, $\mathcal{P}^* = \{p_1^*, \ldots, p_N^*\}$ is the optimal countermeasure strategy. That is to say, for MD n and any $p_n \in \mathcal{P}_n$, there is $E_n(p_n, \mathcal{P}_{-n}^*) \geq E_n(p_n^*, \mathcal{P}_{-n}^*)$. For MEC $\mathcal{S}$ and any $X = (x_1, x_2, \ldots, x_N)$, $|J(X^*)| \geq |J(X)|$.

4.2   Nash Equilibrium Existence Analysis

Theorem 1. *Given* $\mathcal{N}$, G, B, C, *and* $p_n \geq \max\{p'_n, \underline{p_n}\}$, *non-cooperative game strategies for N MDs and MEC* $\mathcal{S}$ $\mathcal{M} = (\mathcal{N}, \{\mathcal{P}_n\}_{n \in \mathcal{N}}, \{E_{n,off}\}; \mathcal{S}, \mathcal{X}, |J|)$ *have a Nash equilibrium* $\langle \mathcal{P}^*, X^* \rangle$, $(p^* \in \mathcal{P}_n, X^* \in \mathcal{X})$.

Proof. We easily know that $\frac{\partial E_{n,off}(X,\mathcal{P})}{\partial p_n} > 0$ $(p_n > 0)$. Based on Eq. (1) we can obtain that

$$\frac{\partial^2 r_n(X,\mathcal{P})}{\partial p_n^2} = -\frac{BG_n^2}{\ln 2(\eta_0 + I_n + p_n G_n)^2}. \tag{9}$$

$E_{n,off}(X,\mathcal{P})$ is taken the second derivative with respect to p_n, and it yields that

$$\frac{\partial^2 E_{n,off}}{\partial p_n^2} = \frac{d_n B^2 G_n}{(\eta_0 + I_n)\mu r_n^3 \ln 2}[(-1 - \frac{1}{\mu})\log_2 \mu + \frac{2}{\ln 2}(1 - \frac{1}{\mu})], \tag{10}$$

where $\mu = \frac{\eta_0 + I_n + p_n G_n}{\eta_0 + I_n}$ and $\mu > 1$.

Let function $g(x) = (-1 - \frac{1}{x})\log_2 x + \frac{2}{\ln 2}(1 - \frac{1}{x})$. We analyse function $g(x)$, and its derivative for x is

$$g'(x) = \frac{-x + \ln 2 \log_2 x + 1}{x^2 \ln 2}. \tag{11}$$

Let function $s(x) = -x + \ln 2 \log_2 x$. When $x \geq 1$, $s(x)$ is monotonically decreasing, and $s(x) \leq s(1) = 0$. Therefore, when $x \geq 1$, $g'(x) < 0$, $g(x)$ is monotonically decreasing, and $g(x) \leq g(1) = 0$. Because $\mu > 1$, the second derivative of $E_{n,off}(X,\mathcal{P})$ with respect to p_n is always less than 0, i.e., $\frac{\partial^2 E_{n,off}}{\partial p_n^2} \leq 0$ $(p_n \geq \max\{p'_n, \underline{p_n}\})$. Based on $\frac{\partial E_{n,off}(X,\mathcal{P})}{\partial p_n} > 0$ and the power variable of each MD is a closed interval, $E_{n,off}(X,\mathcal{P})$ takes the minimal value when $p_n = \max\{p'_n, \underline{p_n}\}$. Thus, $p_n^* = \max\{p'_n, \underline{p_n}\}$, and for any $p_n \geq p_n^*$, there always is $E_n(p_n, \mathcal{P}^*_{-n}) \geq E_n(p_n^*, \mathcal{P}^*_{-n})$.

For MEC $\mathcal{S}$, J^* is the first set in Algorithm Greedy-pruning that satisfies the following conditions: (1) $\sum_{k \in J^*} f'_k(J^*) \leq C$; (2) for any MD $l \in \mathcal{N} \backslash J^*$, there is always

$$l = \arg\min_i \sum_{j \in (J^* \cup \{l\}) \backslash \{i\}} f'_j((J^* \cup \{l\}) \backslash \{i\}).$$

Then, the maximum number of beneficial MDs with QoS will no longer decrease. Therefore, for any offloading scheduling profile $X \in \mathcal{X}$ satisfying the conditions $t_n \leq T_n, n \in J$ and $\sum_{n \in J} f_n \leq C$, there always will be $|J(X)| = \sum_{n \in \mathcal{N}} x_n \leq |J(X^*)| = \sum_{n \in \mathcal{N}} x_n^*$.

4.3   Nash Equilibrium Solution Computation

We propose a Game-based Computation Offloading (GCO) Algorithm 3 to find the equilibrium solution.

Algorithm 3. Game-based Computation Offloading (GCO)

Require: $\mathcal{N}, \mathcal{P}, \overline{\mathcal{P}}, G, B, \mathcal{C}, \varepsilon, \delta$.
Ensure: $\mathcal{P}, J, \overline{X}$.
  1: $\mathcal{N}(0) \leftarrow \mathcal{N}$;
  2: $s \leftarrow 1$;
  3: $p_{\mathbf{n}}(0) \leftarrow \underline{\mathcal{P}}(\mathcal{N}(s-1))$;
  4: $\langle J(0), f_{\mathbf{n}}(J(0)) \rangle \leftarrow \mathrm{GP}(\mathcal{N}(0), p_{\mathbf{n}}(0), G, B, \mathcal{C})$;
  5: $t \leftarrow 0$;
  6: **while** $|P(t+1) - P(t)| < \delta$ **do**
  7:     $p_{\mathbf{n}}(t+1) \leftarrow Calculate\_\widetilde{\mathcal{P}}(\mathcal{N}(t), p_{\mathbf{n}}(t), G, B, \mathcal{C}, J(t), \varepsilon)$;
  8:     $\langle J(t+1), f_{\mathbf{n}}(J(t+1)) \rangle \leftarrow \mathrm{GP}(\mathcal{N}(s), p_{\mathbf{n}}(t+1), G, B, \mathcal{C})$;
  9:     $t \leftarrow t+1$;
 10: $J \leftarrow J(t)$;
 11: $\mathcal{N}(s) \leftarrow \mathcal{N}(s-1)$;
 12: **while** $(\mathcal{N}(s) \neq \mathcal{N}(s-1))$ **do**
 13:     $s \leftarrow s + 1$;
 14:     *loop* steps 3 to 11;
 15: **return** $\mathcal{P}, J, X$.

---

5   Simulations

5.1   Simulation Settings

We evaluate the system performance of the proposed GCO based on the interaction of MEC $\mathcal{S}$ and multiple mobile devices in this section. We consider 50
MDs in this system. The size of the input task data d_n of each MD n is randomly selected from the interval $(0, 2]$ MB and the total number of required
CPU cycles $c_n = d_n \cdot w_n$, where w_n is the workload requirements of task τ_n
($w_n \in [100, 500]$ cycles/bit). Similarly, the expected Time T_n of MD n also
follows a uniform distribution with $(0, 3]$s. The minimum transmission power
p_n is 100 mW, and the maximum value is randomly selected from the interval
$[1000, 3000]$ mW. We consider MEC $\mathcal{S}$ has a coverage range of 50 m. The computational capacity $\mathcal{C}$ of MEC $\mathcal{S}$ is 1GHz. The bandwidth $B = 10$ MHz and
the background noise power $\eta_0 = -100$ dBm. Based on the wireless interference
model for urban cellular radio environment, the channel gain $G_n = dis_n^{\alpha}$, where
dis_n is the distance between MD n and the MEC $\mathcal{S}$ and $\alpha = -4$ is the path loss
factor.

5.2   Convergence of Algorithm GCO

Figures 1 and 2 illustrate the convergence process of transmission power for each
MD by executing our proposed GCO algorithm. With the number of iterations
increasing, the transmission power of each MD is increasing and then the curve
reaches to a stable value. During the process of computing, some MDs will
withdraw the resource competition if the transmission power is higher than their
accepted maximum value, i.e., $p_i > \overline{p}_i$. Figure 2 is the transmission power curve
of MDs who cancel to compete the computation resource of MEC $\mathcal{S}$. Moreover,
we can know that the transmission power can be obtained after 6 iterations,
which shows high efficiency of our proposed algorithm.

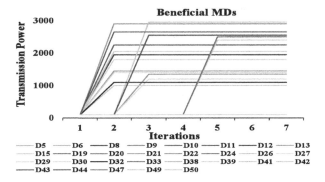

Fig. 1. Change of transmission power and beneficial MDs in the iterative process.

Figures 3 and 4 is a curve of the number of MDs who obtain computing resources provided by MEC $\mathcal{S}$ and a bar graph of the average energy consumption during the iterative process, respectively. At the beginning, each MD's transmission power is set as the initial value, i.e., the minimum value. The number of MDs with QoS served by the MEC $\mathcal{S}$ with limited computing resources is 23 and the average energy consumption is about 70. Each MD increases its transmission power to complete for the computing resources of MEC $\mathcal{S}$, which causes the average energy consumption to rise during the iteration, as shown in Fig. 4. In Fig. 3, after several rounds of mutual negotiation between MDs, the number of MDs who can use the computing resources provided by MEC $\mathcal{S}$ gradually increases and maintains stable at the value 29 as the number of iterations increases.

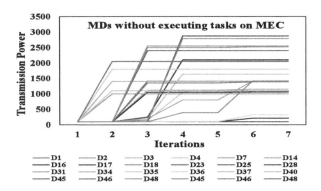

Fig. 2. Change of transmission power and non-beneficial MDs in the iterative process.

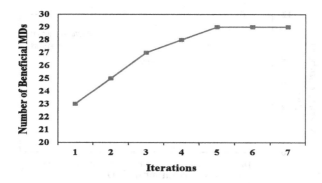

Fig. 3. The process change of number of beneficial MDs.

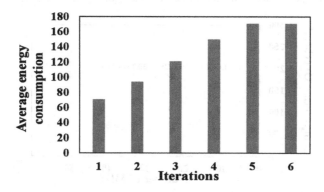

Fig. 4. The process change of average energy consumption.

5.3   Performance Evaluation

The performance of GCO algorithm is evaluated from two respects: the number of iterations and the execution time. The variable is the number of MDs N, which increases by 10 from 10 to 50. For each N, we repeat the experiment many times. The experimental results are shown in Figs. 5 and 6.

Figures 5 and 6 show the number curve of iterations and iterative time curve of Algorithm GCO as the number of MDs increases, respectively. The blue line is an average curve in each figure. In Fig. 5, the general trend of the curve increases linearly and slowly. Besides, even if the number of MDs is 50, the average number of iterations is very small. In Fig. 6, as the scale of MDs increases, the computation overhead curve increases in a polynomial. The red dashed line is the trend line of the computation overhead curve, which is a second order polynomial. The fitting degree of the trend line and the time curve is 0.9908. When the number of MDs reaches 50, the average overhead is 225 ms, which is rapid and shows the high efficiency of our proposed algorithm.

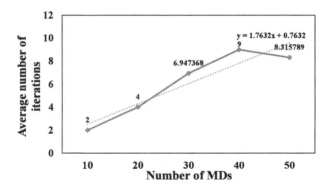

Fig. 5. Average iterative times of different scales of MDs. (Color figure online)

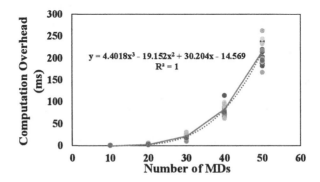

Fig. 6. Computation Overhead of different scales of MDs. (Color figure online)

6   Conclusions

Our study focuses on the task offloading problem of one MEC and multiple MDs with delay deadlines. From the perspective of non-cooperative game theoretical method, the number of served MDs with delay deadline and the energy consumption of all tasks executed on MEC $\mathcal{S}$ are alternately optimized. We prove the existence of Nash equilibrium solution and propose GCO algorithm to solve it. Besides, the convergence of the algorithm is also analyzed. Extensive simulated experiments results validate and show the feasibility of our proposed method.

Acknowledgments. The research was partially funded by the National Key R&D Program of China (Grant No. 2018YFB1003401), the Program of National Natural Science Foundation of China (Grant No. 61751204).

References

1. Abbas, N., Zhang, Y., Taherkordi, A., Skeie, T.: Mobile edge computing: a survey. IEEE Internet Things J. **5**(1), 450–465 (2018)

2. Porambage, P., Okwuibe, J., Liyanage, M., Taleb, T., Ylianttila, M.: Survey on multi-access edge computing for internet of things realization. IEEE Commun. Surv. Tutor. **20**, 2961–2991 (2018)
3. Ning, Z., Wang, X., Huang, J.: Mobile edge computing-enabled 5G vehicular networks: toward the integration of communication and computing. IEEE Veh. Technol. Mag. **14**, 54–61 (2018)
4. Kai, W., Hao, Y., Wei, Q., Min, G.: Enabling collaborative edge computing for software defined vehicular networks. IEEE Netw. **32**, 112–117 (2018)
5. Guo, H., Liu, J.: Collaborative computation offloading for multiaccess edge computing over fibercwireless networks. IEEE Trans. Veh. Technol. **67**(5), 4514–4526 (2018)
6. Chen, W., Dong, W., Li, K.: Multi-user multi-task computation offloading in green mobile edge cloud computing. IEEE Trans. Serv. Comput. **99**, 1 (2018)
7. Yang, L., Zhang, H., Ming, L., Guo, J., Hong, J.: Mobile edge computing empowered energy efficient task offloading in 5G. IEEE Trans. Veh. Technol. **67**, 6398–6409 (2018)
8. Feng, W., et al.: Joint offloading and computing optimization in wireless powered mobile-edge computing systems. IEEE Trans. Wirel. Commun. **17**(3), 1784–1797 (2017)
9. Min, C., Hao, Y.: Task offloading for mobile edge computing in software defined ultra-dense network. IEEE J. Sel. Areas Commun. **36**(3), 587–597 (2018)
10. Qiang, F., Ansari, N.: Application aware workload allocation for edge computing based IoT. IEEE Internet Things J. **5**(3), 2146–2153 (2018)
11. Liu, J., Mao, Y., Zhang, J., Letaief, K.B.: Delay-optimal computation task scheduling for mobile-edge computing systems. In: IEEE International Symposium on Information Theory, April 2016
12. Xiang, S., Ansari, N.: Latency aware workload offloading in the cloudlet network. IEEE Commun. Lett. **21**(7), 1481–1484 (2017)
13. Jiao, Z., et al.: Energy-latency trade-off for energy-aware offloading in mobile edge computing networks. IEEE Internet Things J. **5**, 2633–2645 (2018)
14. Chen, X., Jiao, L., Li, W., Fu, X.: Efficient multi-user computation offloading for mobile-edge cloud computing. IEEE/ACM Trans. Netw. **24**, 2795–2808 (2016)
15. Rodrigues, T.G., Suto, K., Nishiyama, H., Kato, N., Temma, K.: Cloudlets activation scheme for scalable mobile edge computing with transmission power control and virtual machine migration. IEEE Trans. Comput. **67**, 1287–1300 (2018)
16. Mao, Y., Zhang, J., Letaief, K.B.: Joint task offloading scheduling and transmit power allocation for mobile-edge computing systems. In: Wireless Communications and Networking Conference (2017)
17. Tao, X., Ota, K., Dong, M., Qi, H., Li, K.: Performance guaranteed computation offloading for mobile-edge cloud computing. IEEE Wirel. Commun. Lett. **6**(6), 774–777 (2017)
18. Xu, C., Lei, J., Li, W., Fu, X.: Efficient multi-user computation offloading for mobile-edge cloud computing. IEEE/ACM Trans. Netw. **24**(5), 2795–2808 (2016)
19. Hu, X., Wong, K.K., Yang, K.: Wireless powered cooperation-assisted mobile edge computing. IEEE Trans. Wirel. Commun. **17**(4), 2375–2388 (2018)
20. Li, K.: A game theoretic approach to computation offloading strategy optimization for non-cooperative users in mobile edge computing. IEEE Trans. Sustain. Comput. **99**, 1 (2018)
21. Ranadheera, S., Maghsudi, S., Hossain, E.: Computation offloading and activation of mobile edge computing servers: a minority game. IEEE Wirel. Commun. Lett. **7**, 688–691 (2018)

NOC and Networks

KLSAT: An Application Mapping Algorithm Based on Kernighan–Lin Partition and Simulated Annealing for a Specific WK-Recursive NoC Architecture

XiaoJun Wang[1,2](✉) ⓘ, Feng Shi[1](✉) ⓘ, and Hong Zhang[2](✉) ⓘ

[1] Beijing Institute of Technology, Beijing 100081, China
wxjred9915@163.com, bitsf@bit.edu.cn
[2] Henan University of Economics and Law, Zhengzhou 450046, Henan, China
gracezxkl@126.com

Abstract. Application mapping is a critical phase in NoC design because of the running time, the network latency and the power consumption. In order to reduce these problems of applications running on multicore architecture, we propose a novel application mapping algorithm, called KLSAT mapping algorithm. It is used for the triplet-based architecture (TriBA) topology which is WK-recursive based networks well conform to a modular design due to the properties of regularity and scalability. The KLSAT mapping algorithm exploits the advantage of both the Kernighan–Lin partitioning algorithm and simulated annealing algorithm to reduce the overall power consumption and network latency. Compared to the random mapping algorithm, the experiment results reveal that the solutions generated by the proposed mapping algorithm reduce average power consumption and network latency by 6.4%, 12.2% in mapping 27 cores and 29.5%, 26.7% in mapping 81 cores respectively.

Keywords: WK-recursive network · Kernighan–Lin algorithm · Simulated annealing algorithm · Application mapping · Network-on-Chip

1 Introduction and Motivation

On-chip communication plays one of the crucial roles in multicore architecture topology design. Network-on-Chip (NoC) has been proposed to reduce the power consumption and the network latency limitations of bus-based on-chip multicore architecture [1,2]. There are several factors affecting the NoC performance, such as the network topology, the routing algorithm, application mapping. So the network-on-chip (NoC) topology design is an important factor in the on-chip multicore architecture. Our team proposed the triplet-based multicore architecture (TriBA) on-chip network is a kind of the multicore WK-recursive network [3,4], which has several advantages such as scalability, regularity, locality and hierarchy.

© IFIP International Federation for Information Processing 2019
Published by Springer Nature Switzerland AG 2019
X. Tang et al. (Eds.): NPC 2019, LNCS 11783, pp. 31–42, 2019.
https://doi.org/10.1007/978-3-030-30709-7_3

Definition 1: Given a WK-recursive NoC topology with N^L ($L \geq 0$) cores (in here N = 3), the core's ID number is encoded in the sequence $a_L a_{L-1} a_{L-2} \cdots a_1 a_0$, where $a_i \in 1, 2, \cdots, N$ ($0 \leq i \leq L-1$) which contains the level number and the core number after partition at $level_i$ and the value of a_i means the position of the level number. The Fig. 1 shows TriBA NoC topology as L = 1, 2.

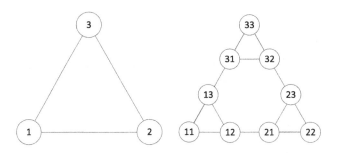

Fig. 1. TriBA multicore network topology with L = 1, 2

TriBA network topology has smaller degree, bisection width, smaller network diameter and less number of total links than 2DMesh topology with the same number of cores. It indicates any of TriBA's cores can spend less time to send a data package to other cores. Meanwhile, TriBA has a less total links which means capacity of low power consumption. The researches [5,6] elaborate that on-chip multicore processors such as the power consumption of Terascale and MIT RAW with respect to the whole on-chip power are 30%, 40% respectively. So another crucial challenge in NoC is how associate the IP cores implementing tasks of an application to reduce the power consumption. This is application mapping algorithm which is a crucial design decision to improve the performance of the overall multicore architecture at an early design phase. In this paper, we propose a mapping heuristic algorithm (KLSAT mapping algorithm) that is based on Kernighan–Lin (KL) algorithm, simulated annealing (SA) algorithm and the WK-recursive multicore architecture TriBA to reduce the overall power consumption and network latency. KL algorithm can reduce the fact network communication cost by placing frequently communicating cores closely. SA algorithm is a kind of mapping algorithm for exploiting optimization and searching solutions that originates from the annealing in engineering. TriBA [7] (Triplet-based architecture) is a novelty WK-recursive on-chip multicore architecture with the characteristic of scalability and locality.

2   Related Work

Several previous works have been proposed to use specially designed application mapping algorithms, for example Kernighan-Lin partitioning algorithm and simulated annealing algorithm, to improve the different NoC architectures performance or reduce power consumption and network latency.

Sahu et al. proposed several mapping algorithms which extends the basic Kernighan-Lin bi-partitioning algorithm to enhance the static and dynamic performances of three different NoC architectures [8]. Authors explored the opportunities in optimizing application mapping based on Kernighan-Lin algorithms for express channel-based on-chip network [9]. Manna et al. presented a KL bi-partitioning based approach to perform mapping the core graph of an application onto 2DMesh-based NoC architecture [10]. In [11], the authors proposed an application mapping algorithm for the mesh-of-tree network topology. Authors represented core mapping procedure based on the Kernighan-Lin graph bi-partitioning algorithm to select Through-Silicon-Via positions [12].

However, the KL mapping algorithm has its limitations and the resulted mappings generated by the KL algorithm may not be global optimal solution. It differs from KL algorithm, the SA algorithm has been observed to perform better application mappings. SA is one heuristic algorithm that has been used in a set of previous works for solving the application mapping problems [13–21]. Compared with KL mapping algorithm, the significant strength of SA is the ability of finding the global optimum solution.

Hu and Marculescu first used the SA algorithm in application mapping problem to evaluate the Branch and Bound application mapping algorithm on 2DMesh NoC [13]. In [14], the authors proposed algorithms using the simulated annealing and tabu search with communication-weighted model for obtaining low energy. The authors proposed an application mapping technique based on particle swarm optimization combined with simulated annealing for comparison of the performance of Zmesh with that of other NoC topologies [15]. In [16], the authors proposed two heuristics mapping algorithm based on the simulated annealing method for solving the capacitated version of the location-routing problem. The authors [17] used SA algorithm with two functions to map application onto multiprocessor system-on-chip (MPSoC). Bo et al. [18] proposed SA algorithm by using the Nelder-Mead simplex method for selecting a set of parameters applied. Tosun et al. [19] presented a mapping algorithm based on simulated annealing for energy- and communication-aware mapping problems of mesh-based NoC architecture. In [20], the authors proposed a heuristic algorithm CHMAP to solve the application mapping problem on the mesh topology to reduce energy consumption. In [21], an optimized mapping algorithm based on simulated annealing, which allocates tasks that have big communication volume to adjacent places on the mesh, was proposed for reducing the energy consumption of applications running on multicore architecture.

Based on the above mentioned reasons, the novelty of our proposed KLSAT mapping algorithm employs the advantages of KL algorithm and SA algorithm for mapping application onto TriBA multicore architecture. Firstly, we use with a Kernighan-Lin tri-partitioning algorithm which idea is come from the reference [22]. The modified Kernighan-Lin tri-partitioning algorithm which fits for the triplet-based characteristic of TriBA ensures the communication value among cores in the same partition is maximum value and the communication value among cores between partitions is minimum value. Secondly, we employ a

SA algorithm to find the final optimal mapping. To the best of our knowledge, the KLSAT mapping algorithm is the first work that employs the modified KL algorithm and SA algorithm onto TriBA, which satisfies the performance requirement of the application mapping and minimizes the average power consumption and network latency. Our experimental results show that, compared to the random mapping algorithm, the KLSAT mapping algorithm reduces the average power consumption and network latency by 6.4%, 12.2% in mapping 27 cores and 29.5%, 26.7% in mapping 81 cores respectively.

3   Problem Formulations

In this section we focus on minimizing the power consumption associated to the application mapping.

3.1   Power Consumption Model

Ye et al. [23] proposed a power consumption model for evaluating the power consumption of switch fabrics in network routers. For the on-chip multicore architecture, however, links between nodes should also be included in the power consumption model. So Hu and Marculescu [24] proposed a modified power consumption model for the on-chip multicore architecture. By evaluating the difference of the power consumption of various components on-chip multicore architecture, Hu and Marculescu found that the power consumed by buffering and internal wires is negligible compared with switch and link. Thus, the power consumption model can be reduced to:

$$E_{bit} = E_{Sbit} + E_{Lbit} \tag{1}$$

where E_{Sbit} and E_{Lbit} represent the energy consumed by switch and link respectively. So, the power consumption of sending one bit from node i to node j can be ex-pressed as following:

$$E_{bit}^{i,j} = n_{hops} \times E_{Sbit} + (n_{hops} - 1) \times E_{Lbit} \tag{2}$$

where n_{hops} is the number of routers the bit passes on its way along a path from node i to node j.

So the total power consumption of the NoC is the sum of weight value of all edges as following:

$$E_{total} = \sum_{i,j}^{E} \sum_{bit} E_{bit}^{i,j} \tag{3}$$

3.2   Definition of Application Mapping

The goal of application mapping algorithms is to assign a given task to a specific core in the NoC to match the certain requirement such as minimizing the network latency and power consumption.

Definition 2: The task core graph is a weighted edge graph, C(V, E). A vertex $v_i \in V$ represents a task and the weighted edge $e_{i,j} \in E$ represents the communication bandwidth between the cores v_i and v_j. $Comm_{i,j}$ denotes the weighted value of edge $e_{i,j}$, which indicates the bandwidth constraints of the communication from vertex v_i to vertex v_j.

Definition 3: The NoC topology graph is a multicore interconnects architecture graph T(U, F). A vertex $u_i \in U$ represents a node in multicore NoC topology and the directed edge $f_{i,j} \in F$ indicates a physical link for directed communicating between the vertices u_i and u_j. $Bw_{i,j}$ denotes the weighted value of the edge $f_{i,j}$, which shows the available communication bandwidth across the edge $f_{i,j}$.

The application mapping algorithm can be formulated as the following one-to-one mapping function:

Mapping algorithm: given a task core graph C(V, E) and the NoC topology graph T(U, F), find the function:

map: $V \rightarrow U$, such that, $map(v_i) = u_j$, $\forall\, v_i \in V$, $\exists\, u_j \in U$, $|\,V\,| \leq |\,U\,|$

$\forall\, v_i \in V$, $map(v_i) = U$

$\forall\, v_i \neq v_j$, $map(v_i) \neq map(v_j)$

Number(V)≤Number(U)

Minimam(E_{total}).

4   The Proposed KLSAT Mapping Algorithm

In this section, we present the proposed KLSAT mapping algorithm which includes the Kernighan-Lin partitioning algorithm and simulated annealing algorithm to minimize the overall communication cost among all of cores. The goal of KL partitioning algorithm is to partition a task graph into subsets recursively and get the minimum value of the communication costs between the subsets. So we use the KL partitioning algorithm to obtain the first stage optimal solution as the initial solution as the input of the next stage SA algorithm. The simulated annealing algorithm is an effective global optimization algorithm which simulates the physical annealing process of solid and solves large scale combinatorial optimization problems. Along with the Metropolis acceptance criterion is introduced to the optimization process, the result of the simulated annealing achieves an approximate global optimal solution. So, we apply the simulated annealing algorithm and obtain the optimal mapping solutions at the second stage.

The KL partitioning algorithm is applied to recursively partition the core graph. Firstly, all cores are in one partition group at level-0. At level-1, there are three partition subsets, naming partition number 1, 2 and 3, each partition containing one third the nodes of the core graph. At level-2, nine partitions are generated (three each from partition-1, partition-2 and partition-3 of level-1) having partition number 11, 12, 13, 21, 22, 23, 31, 32 and 33. This continues until there are 3 cores left in each partition for TriBA. Because the initial partitioning determines the KL algorithm partitioning results, in this paper, this algorithm runs several times for the best result with different randomly generated initial partitions which is used for subsequent mapping and iterative improvement.

Figure 2 shows an example with N = 27 and how the IP-sets are merged. By merging three IP-sets, it finds the best contact between boundaries.

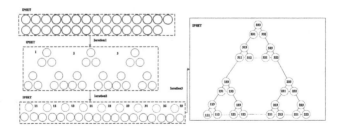

Fig. 2. An example of trinomial merging iteration (N = 27)

Algorithm 1. KL _Tri-Partitioning(C)

Input: Core graph C=(V, E)
Output: Partition number of each core at each level of partitioning
  if $|V| \leq 3$ **then**
    return
  end if
  best _tri-partition = NULL
  best _cost = ∞
  for $i = 0$ to L **do**
    tri-partition = KL_Tri(C)
    if cost(tri-partition$l$ best_cost **then**
      best_cost = cost (tri-partition)
      best_tri-partition = tri-partition
    end if
  end for
  Generate graphs C1, C2 and C3 based on best_tri-partition
  KL _ Tri-Partitioning(C1)
  KL _ Tri-Partitioning(C2)
  KL _ Tri-Partitioning(C3)

Now the next stage, each of these 3-core subsets is assigned to the appropriate basic unit of the multicore architecture TriBA, L is the level of TriBA and the number of cores is 3^L. Although these 3-core subsets are attached to the nearby basic unit arbitrarily, it is still great opportunity to resolve an optimization solution by the proposed KLSAT mapping algorithm.

KLSAT Mapping Algorithm:

When the temperature initialization of the system is completed, the KLSAT mapping algorithm executes two nested loops. After the external loop with KL partition algorithm reaching the global minima, the internal loop refines and finds the optimal local solution. The number of external loop iterations is limited to U^2 as suggested in [14]. The internal loop randomly selects two nodes in a

L-level subset and swaps them to determine a new solution. Then the algorithm calculates whether the new solution is better than the old solution. If it is, the new solution replaces the current solution. Otherwise, the algorithm automatically generates a random variable γ ($0 \leq \gamma \leq 1$), and compares with the acceptance probability function $(-\Delta P)/$Temperature. If the value of the function result is higher than γ, the new solution is accepted. The acceptance probability is high at high temperatures. However, with the temperature of the system lowing, the acceptance probability decreases. We limit the iteration of the internal loop to L^2 consecutive rejects and the Temperature is more than 0.01. When each internal loop completed, temperature of the system decreases and the algorithm starts a new loop accepting the new solution as our initial solution for the next iteration.

Algorithm 2. Algorithm KLSAT mapping Task mapping algorithm based on simulated annealing

---

Input: Core Graph C= (V, E), Topology Graph T=(U, F), U=3^L, M=$400L^2$
  Partition_ID (partition number) for each core at each level of KL_Tripartition algorithm and simulated annealing algorithm
Output: Addressing number of each core in term of Q(level, C) S = KL_Tripartion(C)
  P=KL_Tri(C) S_best = S P_best = P Temperature = 1000L
  for i=0 to U^2 do
    R = 0
    while R<L^2 and Temperature>0.01 do
      S'=neighbor(S)
      P'=KL-Tri(S')
      Δ P=P–P'
      Generates a random variable γ
      if ΔP$\leq$0 or $\gamma \leq$ e(-ΔP)/Temperature then
        P = P'
        R = 0
      else
        R++
      end if
      if R=0 and PlP$_{best}$ then
        S_{best} = S
        P_{best} = P
      end if
    end while
    Decrement Temperature
  end for
  Q=MAPPING(C)
  return Q

We produce a mapping by using MAPPING (G) algorithm. At each level of tri-partitioning, we assign a partition number 1, 2 and 3 to each subset by turn. These numbers have been utilized in the address assignment process in the MAPPING (G) algorithm. In the mapping algorithm, these 3-core subsets are assigned according to the output results generated by KLSAT mapping algorithm. After the mapping algorithm completed, each core has an assigned (level number, subset number) to identify its mapping position on the on-chip multicore TriBA.

At last the KLSAT mapping algorithm completed, we obtain the global optimal solution. All of task cores are mapped onto the corresponding position of TriBA multicore architecture.

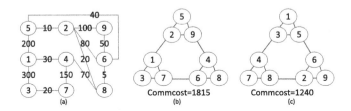

Fig. 3. An example for KLSAT mapping algorithm ((a) an example task graph, (b) communication cost of random mapping, (c) communication cost of KLSAT mapping)

In Fig. 3, we present an example of our KLSAT mapping algorithm. Figure 3(a) shows an example of task graph with communication weighted between nodes. Figure 3(b) and (c) shows communication cost with random mapping and KLSAT mapping. The communication cost of mapping with random mapping is calculated as Commcost = 1815. And with KLSAT mapping, the communication cost becomes Commcost = 1240, which is accepted as the new solution. The KLSAT mapping algorithm continues executing the iteration process until the predefined terminated condition value is reached.

5   Experimentation and Results

5.1   Simulator and Benchmarks

In this paper, we used Gem5 as our simulator to evaluate the KLSAT mapping algorithm, which is widely used as a configurable architecture simulator for multicore on-chip architecture-related research. In Gem5, the Orion [25] model is used to evaluate the power consumption of the various NoC topologies. Meanwhile, the benchmarks of PARSEC [26] are used in the following experiments. We use the WK-recursive NoC TriBA topology as the NoC topology, which is a regular topology with better NoC topology characteristics such as smaller network diameter, less total links and lower node degree than the 2DMesh topology. We compare the KLSAT mapping algorithm with several other algorithms on the TriBA NoC architectures: (1) BL_TriBA (the baseline): which maps the tasks onto the TriBA NoC topology randomly; (2) KL_TriBA: KL mapping algorithm on the TriBA structure; (3) SA_TriBA: which is the conventional simulated annealing algorithm on TriBA NoC structure; (4) KLSAT: our proposed mapping algorithm on TriBA NoC structure.

5.2   Results and Analysis

Based on the previous research experience, we set the initial parameters of the algorithm as follows: $M = 4000$, $temperature_0 = 5000$, terminated temperature $\varepsilon = 0.01$. We implement the algorithm in Matlab R2013b environment. Host CPU is Intel Core i7 3.40 GHz, 8 GB memory and operating system is Windows 7. Host has 8 processor cores and the sizes of the target machine are 27 and 81.

The network latency of TriBA multicore architecture normalized to the baseline case shows in Figs. 4 and 5. Due to the various communications characteristics of these benchmarks, the network latency of experimental results varies significantly. For 27 cores of TriBA, compared to the baseline case, KL_TriBA, SA_TriBA and KLSAT mapping algorithm decrease the network latency by the average 2.9, 6.1 and 12.2% respectively. In the experimental result of TriBA with 81 cores, the differences between four mapping algorithms are more significant because the communication overheads among cores are dramatically increased. The KLSAT mapping algorithm decreases the network latency by an average of 26.7% compared to the baseline as shown in Fig. 5.

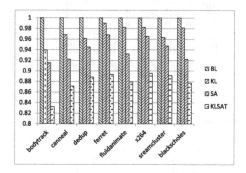

Fig. 4. Network latency of the four algorithms with 27 cores

Figure 6 shows the power consumption of TriBA with 27 cores. The power consumption is normalized to the BL_TriBA random mapping algorithm. As shown in the Fig. 6, the BL_TriBA random mapping algorithm consumes the highest power consumption while the KLSAT mapping algorithm has the least power, with an average of 6.4% than the random mapping. Figure 7 shows the experimental results of TriBA's power consumption with 81 cores. For 81 cores of TriBA, compared to the baseline case, KL_TriBA, SA_TriBA and KLSAT mapping algorithm decrease the power consumption by the average 12.0, 23.1 and 29.5% respectively. In this experimental result, the power savings of KLSAT mapping algorithm in Fig. 7 is more significant than that in the 27 cores of TriBA architecture in Fig. 6. Overall, KLSAT mapping algorithm saves power consumption by an average of 29.5% compared to the baseline and achieves better performance compared to KL_TriBA and SA_TriBA.

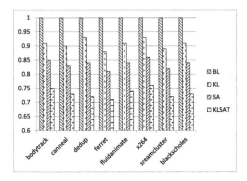

Fig. 5. Network latency of the four algorithms with 81 cores

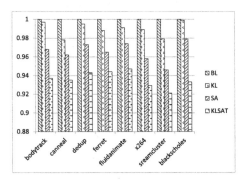

Fig. 6. Power consumption of the four algorithms with 27 cores

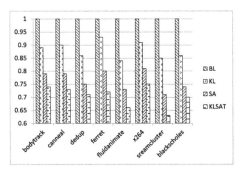

Fig. 7. Power consumption of the four algorithms with 81 cores

The reason is that KLSAT mapping algorithm has a smaller chance to get trapped in local optimum than random mapping algorithm because we add KL partition as the initial solution in the KLSAT mapping algorithm. Because KL partition algorithm combines the triplet-based characteristic of TriBA to make more communication transfer among three cores which have the characteristic of local full interconnect flavor. In consequence, the solution generated by the KLSAT mapping algorithm has less network communication cost and lower power consumption than the other mapping algorithms.

In general, the KLSAT mapping algorithm sharply decreases the number of iterations, the power consumption and network latency, compared with the random mapping algorithm. Our proposed KLSAT algorithm achieves better performance than both the KL algorithm and the simulated annealing algorithm.

6   Conclusion

One of the important research fields on NoC is the design of the application mapping algorithms. Several different mapping algorithms have been presented to reduce network latency, lower power consumption, satisfy bandwidth constraint or minimize on-chip area and so on. This paper focused on a new mapping algorithm based on KL partition algorithm and the simulated annealing algorithm in order to generate better performance in application mapping problems. We designed and implemented an application mapping algorithm on multicore architecture TriBA for performance simulation based on KL partition algorithm and simulated annealing, and verified the KLSAT mapping algorithm by experiments. Our experimental results show that the algorithm has significant reduction in the number of iterations, the network latency and the power consumption. It also shows that the algorithm can solve the large-scale problem.

References

1. Dally, W., Towles, B.: Route packets, not wires: on-chip interconnection networks. In: DAC 2001, pp. 684–689, June 2001
2. Benini, L., Micheli, G.: Networks on chips: a new SoC paradigm. IEEE Trans. Comput. **35**(1), 70–78 (2002)
3. Farahabady, M., Sarbazi-Azad, H.: The WK-recursive pyramid: an efficient network topology. In: PAAN 2005, pp. 6–11, December 2005
4. Wang, Y., Juan, S.: Hamiltonicity of the basic WK-recursive pyramid with and without faulty nodes. J. Theor. Comput. Sci. **562**(C), 542–556 (2015)
5. Taylor, M.B., Lee, W., Miller, J., et al.: Evaluation of the raw microprocessor: an exposed-wire-delay architecture for ILP and streams. ACM SIGARCH Comput. Archit. News **32**(2), 2–13 (2004)
6. Hoskote, Y., Vangal, S., et al.: A 5-GHz mesh interconnect for a teraflops processor. IEEE Micro **27**(5), 51–61 (2007)
7. Shi, F., Ji, W., et al.: A triplet-based computer architecture supporting parallel object computing. In: IEEE ASSAP 2007, pp. 192–197, July 2007
8. Sahu, P., Manna, N., Shah, N., Chattopadhyay, S.: Extending Kernighan-Lin partitioning heuristic for application mapping onto network-on-chip. J. Syst. Archit. **60**(7), 562–578 (2014)
9. Zhu, D., Chen, L., Yue, S., Pedram, M.: Application mapping for express channel-based networks-on-chip. In: DATE 2014, pp. 1–6, March 2014

10. Manna, K., Choubey, V., et al.: Thermal variance-aware application mapping for mesh based network-on-chip design using Kernighan-Lin partitioning. In: PDGC 2014, pp. 274–279. IEEE, December 2014

11. Fang, J., Yu, L., et al.: KLGA: an application mapping algorithm for mesh-of-tree (MoT) architecture in network-on-chip design. J. Supercomputing **71**(11), 4056–4071 (2015)

12. Manna, K., Teja, V., Chattopadhyay, S., et al.: TSV placement and core mapping for 3D mesh based network-on-chip design using extended Kernighan-Lin Partitioning. In: VLSI 2015, pp. 392–397. IEEE, July 2015

13. Hu, J., Marculescu, R.: Energy- and performance-aware mapping for regular NoC architectures. IEEE Trans. Comput. Aided Des. Integr. Circuits Syst. **24**(4), 551–562 (2005)

14. Marcon, C., Moreno, E., Calazans, L., Moraes, F.: Comparison of network-on-chip mapping algorithms targeting low energy consumption. IET Comput. Digit. Tech. **2**(6), 471–482 (2008)

15. Prasad, N., Mukherjee, P., Chattopadhyay, S., et al.: Design and evaluation of ZMesh topology for on-chip interconnection networks. J. Parall. Distrib. Comput. **113**(2), 17–36 (2018)

16. Dong, Z., Yang, B., Hu, P., et al.: An efficient global energy optimization approach for robust 3D plane segmentation of point clouds. ISPRS J. Photogra. Remote Sens. **137**(1), 112–133 (2018)

17. Orsila, H., Salminen, E., Timo, D.: Best practices for simulated annealing in multiprocessor task distribution problems. Tech. **4**(2), 197–198 (2008)

18. Yang, B., Liang, G., et al.: Parameter-optimized simulated annealing for application mapping on networks-on-chip. In: LIO 2012, pp. 307–322 (2012)

19. Tosun, S., Ozturk, O., Ozkan, E., Ozen, M.: Application mapping algorithms for mesh-based network-on-chip architectures. J. Supercomputing **71**(3), 995–1017 (2015)

20. Cheng, C., Chen, W.: Application mapping onto mesh-based network-on-chip using constructive heuristic algorithms. J. Supercomputing **72**(11), 1–14 (2016)

21. Zhong, L., Sheng, J., et al.: An optimized mapping algorithm based on simulated annealing for regular NoC architecture. In: ASIC 2011, pp. 389–392. IEEE, October 2011

22. Larsson, T., Jesper, F.: Direct graph k-partitioning with a Kernighan-Lin like heuristic. Oper. Res. Lett. **34**(6), 621–629 (2006)

23. Ye, T., Benini, L., Micheli, G.: Analysis of power consumption on switch fabrics in network routers. In: DAC 2002, pp. 524–529, June 2002

24. Hu, J., Marculescu, R.: Energy-aware mapping for tile-based NoC architectures under performance constraints. In: ASPDAC 2003, pp. 233–239. IEEE (2003)

25. Kahng, A., Li, B., Peh, L., Samadi, K.: ORION 2.0: a power-area simulator for interconnection networks. IEEE Trans. Very Large Scale Integr. Syst. **20**(1), 191–196 (2012)

26. Bienia, C., Li, K.: PARSEC 2.0: a new benchmark suite for chip-multiprocessors. In: AWMBS 2009, pp. 1–9. IEEE (2009)

Modeling and Analysis of the Latency-Based Congestion Control Algorithm DX

Wanchun Jiang[✉], Lijuan Peng, Chang Ruan, Jia Wu, and Jianxin Wang

School of Computer Science and Engineering, Center South University,
Changsha 410083, Hunan, China
{jiangwc,ruanchang,jxwang}@csu.edu.cn

Abstract. Nowadays, low latency has become one of the primary goals of congestion control in data center networks. To achieve low latency, many congestion control algorithms have been proposed, wherein DX is the first latency-based one. Specifically, DX tackles the accurate latency measurement problem, reduces the flow completion time and outperforms the *de facto* DCTCP algorithm significantly in term of median queueing delay. Although the advantages of DX have been confirmed by experimental results, the behaviors of DX have not been fully revealed. Accordingly, some drawbacks of DX under special environment are unexplored. Therefore, in this paper, we conduct fluid-flow analysis over DX, deducing sufficient condition for the stability of DX and revealing the behaviors of DX. Analytical results uncover two problems of DX: (1) it has poor throughput when either the base RTT is very large or the number of flows is relatively small; (2) it suffers from large queueing delay when either the base RTT is relatively small or the number of flows is very large. These results are instructive to the improvement and deployment of DX. Simulation results based on NS-3 verify our analytical results.

Keywords: Congestion control · Fluid-flow analysis · Stability · Latency

1 Introduction

Nowadays, low latency becomes one of the primary goals of designing congestion control algorithms for the data center network. To achieve low latency, accurate and fine-grained feedback signals are needed to represent the degree of congestion. Recently, many congestion control algorithms have been proposed [1,5,10,12]. Generally speaking, most of them employ the following feedback

Supported by the Projects of Hunan Province Science and Technology Plan in China under Grant No. 2016JC2009.

X. Tang et al. (Eds.): NPC 2019, LNCS 11783, pp. 43–55, 2019.
https://doi.org/10.1007/978-3-030-30709-7_4

signals: the packet loss, the explicit in-network feedback like ECN, and the latency-based feedback. Compared to the other two signals, latency-based feedback signals have the following advantages. The endpoint can detect the fine-grained degree of congestion, or even estimate the switch queue-size [10,12] by measuring the Round Trip Time (RTT) and the base RTT. Moreover, the in-network support is never required.

However, the latency-based feedback signal is difficult to be measured accurately [1]. This is because most kernel implementations can only track RTTs at the granularity of $1ms$ [9], while the RTT is only a few hundreds of microseconds in the data center network. Recently, DX, the latency-based congestion control algorithm proposed for data center networks, tackles the measurement problem of RTT and has good performance. By setting its operating point close to zero, DX can reduce the flow completion time and outperform the *de facto* DCTCP algorithm significantly in term of median queueing delay.

In this paper, we model and analyze the DX algorithm because (1) DX is the up-to-date latency-based congestion control algorithm while other state-of-art algorithms such as ExpressPass and NDP are not. As a latency-based algorithm, DX has very good performance, which are validated by experimental results in [6]. (2) Although some advantages of DX have been confirmed by experimental results, the behaviors of DX have not been explored theoretically. (3) Moreover, existing analytical work on congestion control cannot be applied to the window-based latency-based DX algorithm. In details, we first model DX with the fluid-flow method and linearize the fluid-flow model such that the Nyquist stability criterion [6] can be applied to the model. Subsequently, we deduce the sufficient condition for the stability of DX. In this way, the influence of some parameters, such as the number of flows and the RTT, on the stability of DX can be exhibited. Moreover, we theoretically uncover a special behavior of DX under the condition of a large number of flows and small RTT. Finally, we implement DX in NS-3 simulator to confirm our analytical results.

In total, our analytical results mainly reveal two problems of DX. (1) DX has poor throughput when the DX system is unstable when either the base RTT is very large or the number of flows is relatively small. (2) DX suffers from large queueing delay when either the base RTT is relatively small or the number of flows is very large. Under these conditions, DX enters into the special stable state. It implies that DX should not be employed under these kinds of environments. We believe these results are instructive to the improvement and deployment of DX in practical data center network.

2   Background and Related Work

In this section, we first introduce the DX algorithm in brief, and then present the related work on the theoretical analysis of congestion control algorithms.

2.1   The DX Algorithm

DX is a window-based congestion control algorithm, which uses the latency-based feedback signal to determine the congestion window should be increased or decreased. Similar to TCP, its congestion avoidance algorithm follows the Additive Increase Multiplicative Decrease (AIMD) style. The DX algorithm is characterized by dropping the queue size down to zero quickly as soon as it observes congestion. In the following, we introduce the DX algorithm in detail.

The main DX algorithm is composed of two parts: one is measuring the latency accurately, the other one is a congestion control algorithm for adjusting the congestion window. For accurately measure queueing delay, [10] exhibits sources of measurement errors and their magnitude and their elimination technique.

The congestion control algorithm of DX works as follows. In each RTT, DX measures the queueing delay, which is the difference between the base RTT and a sample RTT. If the queueing delay is not 0, DX considers the network is congested. Otherwise, DX considers that there is no congestion. Mathematically, the window adaption algorithm of DX is as follows:

$$W(t+1) = \begin{cases} W(t) + 1, & if \quad Q(t) = 0, \\ W(t)(1 - \frac{Q(t)}{U(t)}), & if \quad Q(t) > 0, \end{cases} \tag{1}$$

where $W(t)$ is the window size at time t, $Q(t)$ represents the average queueing delay measured by DX in current RTT. $U(t)$ is a self-updated coefficient.

$$U(t) = \frac{R_0 \cdot W(t)}{W(t) - 1}, \tag{2}$$

where R_0 is the base RTT. The self-updated coefficient $U(t)$ is deduced in consideration of high utilization and the number of flows in the network.

According to Eq. (1), DX decreases the congestion window as soon as it detects the network congestion according to $Q(t)$. Therefore, DX keeps the near-zero queueing delay.

2.2   Related Work

Although there are many theoretical works on congestion control algorithms, such as those in [7,11,13], we focus on those works analyzing the state-of-art congestion control algorithms for data center networks in this paper.

Analysis on Non-latency-Based Algorithms. DCTCP [1] is a famous congestion control algorithm using ECN. In [2], Alizadeh et al. develop a fluid-flow model of DCTCP and analyze its stability by the Bode Stability Criterion [6]. The analysis insights guide the configurations of design parameters like the threshold. DCQCN is the latest protocol which outperforms DCTCP in terms of reducing the flow completion time. In [14], the authors analyze its stability condition using the same method as DCTCP.

All these algorithms for data center works are based on non-latency congestion signals, while DX adopts the latency-based feedback signal. Therefore, the theoretical analysis of these algorithms cannot be directly applied to DX.

Analysis on Latency-Based Algorithms. TIMELY [12] is an end-to-end, rate-based congestion control algorithm that uses changes in RTT as a congestion signal. In [14], the author finds that TIMELY has no unique fixed point. To analyze the stability of TIMELY, they modify the algorithm. Its stability condition is analyzed through the Nyquist Stability Criterion [6].

Similar to TIMELY, DX is also a latency-based transport protocol. Different from TIMELY, DX is a window-based algorithm and adjusts the congestion window according to the queueing delay. In [10], authors show that DX exhibits very good performance by extensive experiments. However, to the best of our knowledge, there is no theoretical work on the window-based latency-based DX up to now, which motivates us to perform this investigation.

3   Analysis of DX

In this section, we first build a fluid-flow model for the DX algorithm and then analyze its stability based on its linearized version.

3.1   Modeling

Considering the oversubscribed link and the applications like MapReduce [4], we assume that the sources are homogeneous and flows arrive according to the Poisson process, the same as [2,3,8]. In other words, we assume that all sources have identical sending rates and RTTs, and the RTT equals to τ seconds.

Suppose that N sources share a single link of capacity C. Let $W(t)$ denote the congestion window, R_0 represent the fixed base RTT, and $Q(t)$ be the queueing delay. Let p denote the probability of $Q(t) > 0$. Although in practice, the probability p is time-varying. We find that p is close to a constant in the stable state, as shown in the simulation results under the condition of varying p in Sect. 4. Therefore, we assume that p is constant for the simplicity of analysis. With this assumption, we plug the Eq. (2) into Eq. (1), and can model the DX algorithm as follows by using the method of [11].

$$\frac{dW(t)}{dt} = \frac{seg * (1-p)}{R_0 + Q(t-\tau)} - \frac{Q(t-\tau)(W(t) - seg)}{R_0(R_0 + Q(t-\tau))}p, \tag{3}$$

$$\frac{dQ(t)}{dt} = \begin{cases} \frac{NW(t)}{C(R_0+Q(t-\tau))} - 1 & if \quad Q(t) > 0, \\ max\{0, \frac{NW(t)}{C(R_0+Q(t-\tau))} - 1\} & if \quad Q(t) = 0. \end{cases} \tag{4}$$

The Eq. (3) describes the dynamic evolution of the window size $W(t)$. The Eq. (4) models the evolution of the queueing delay $Q(t)$.

3.2 Stability Analysis

We analyze the stability of DX based on its fluid-flow model (3) and (4). Assume that the equilibrium point of DX is (W_0, Q_0). At the equilibrium point, we have $\dot{W}(t) = 0$ and $\dot{Q}(t) = 0$. Referring to Eqs. (3) and (4), we have

$$seg * R_0(1 - p) = pQ_0(W_0 - seg). \tag{5}$$

$$NW_0 = C(R_0 + Q_0). \tag{6}$$

Substituting Eq. (6) into Eq. (5), we can get the following expression of Q_0

$$Q_0 = \frac{p(N * seg - CR_0) + \sqrt{\Delta}}{2Cp}, \tag{7}$$

where

$$\Delta = (CpR_0 - pN * seg)^2 + 4CpNR_0 * seg * (1 - p). \tag{8}$$

Next, we will linearize the fluid-flow model around the equilibrium point (W_0, Q_0) to obtain

$$\begin{aligned} \delta\dot{W} &= a_1\delta W + a_2\delta Q(t - \tau), \\ \delta\dot{Q} &= b_1\delta W + b_2\delta Q(t - \tau), \end{aligned} \tag{9}$$

where

$$\begin{aligned} \delta W &\doteq W - W_0, \\ \delta Q &\doteq Q - Q_0, \end{aligned} \tag{10}$$

and

$$\begin{aligned} a_1 &= -\frac{Q_0 p}{R_0(R_0 + Q_0)}, & a_2 &= \frac{2p*seg - pW_0 - seg}{(Q_0 + R_0)^2}, \\ b_1 &= \frac{N}{C(R_0 + Q_0)}, & b_2 &= -\frac{NW_0}{C(R_0 + Q_0)^2}. \end{aligned} \tag{11}$$

To obtain the characteristic equation, we compute the Laplace transform of (9). Then we can obtain the transfer function of the linear time-delayed system

$$G(s) = e^{-s\tau}\frac{a_1b_2 - a_2b_1 - b_2s}{s(s - a_1)}. \tag{12}$$

Then, we apply the Bode Stability Criteria [6] to the transfer function (12). Specifically, define the frequency characteristic function $G(j\omega) = G(s)|_{s=j\omega}$ of the system, we have

$$G(j\omega) = A(\omega)e^{j\varphi(\omega)}, \tag{13}$$

where

$$|A(\omega)|^2 = \frac{b_2^2[\omega^2 + (a_1 - \frac{a_2b_1}{b_2})^2]}{\omega^2(\omega^2 + a_1^2)}, \tag{14}$$

$$\varphi(\omega) = -\frac{\pi}{2} - \omega\tau + \arctan\frac{\omega}{a_1} + \arctan\frac{\omega b_2}{a_1b_2 - a_2b_1}, \tag{15}$$

where $A(\omega)$ is amplitude - frequency characteristics and $\varphi(\omega)$ is phase-frequency characteristic. Assume that ω_c is the cross-over frequency which makes $L(\omega_c) = 0$, i.e., $A(\omega_c) = 1$. From Eq. (14), we have

$$\omega_c = \sqrt{\frac{b_2^2 - a_1^2 + \sqrt{(a_1^2 - b_2^2)^2 + 4(a_1 b_2 - a_2 b_1)^2}}{2}}. \tag{16}$$

Note that $\varphi(0) = -\frac{\pi}{2}$. According to Bode Stability Criteria [6], the DX system is stable when $\varphi(\omega_c) > -\pi$, i.e., we have the following theorem in summary.

Theorem 1. *The DX system is stable if the delay satisfies*

$$\tau < \frac{1}{\omega_c}(arctan\frac{\omega_c}{a_1} + arctan\frac{\omega_c b_2}{a_1 b_2 - a_2 b_1} + \frac{\pi}{2}), \tag{17}$$

where ω_c is defined in (16), and a_1, b_1, a_2 and b_2 are defined in (11).

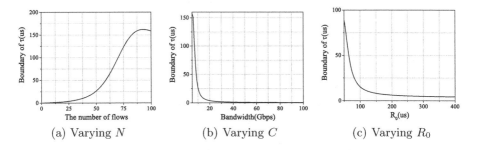

(a) Varying N (b) Varying C (c) Varying R_0

Fig. 1. The variation of the boundary of τ with different N, C, R_0.

Theorem 1 implies that the stability of the DX system holds just when τ is limited. The boundary of τ is associated with both the bottleneck bandwidth C and the number of flows N. In fact, according to Eq. (17), the boundary of τ decreases when either the bandwidth C increases or the number of flows decreases. In order to verify the result, we assume that the value of p is 0.95, the bandwidth C is 10 Gbps, the number of flows is 50, the packet size seg is 1500 and the base RTT R_0 is 80 μs by default. Figure 1 shows the variation of the boundary of τ with different N, C, R_0 respectively. In Fig. 1(a), when N is small, the boundary of τ is small and accordingly Theorem 1 is probably not satisfied. Consider this condition, we do not know whether the DX system is stable. When DX becomes unstable, it will suffer from large queue-size oscillation and poor link utilization. However, when N is large, Theorem 1 is satisfied, i.e., the DX system is stable. In Fig. 1(b) and (c), when C or R_0 changes, similar results can be obtained according to Theorem 1. This is also why the evaluation of DX in [10] always shows good performance.

In total, Theorem 1 reveals the problem that DX may become unstable and have poor throughput when either the base RTT is very large or the number of flows is relatively small.

3.3   A Special Stable State

When we conduct the stability analysis of the DX algorithm, we do not consider the limitation on the congestion window size. In fact, the window size of the DX cannot be less than a segment in real networks. When there are too many flows, i.e., when $\frac{N*seg}{R_0} > C$, the aggregated sending rate of all flows are always larger than the bandwidth C. As a result, Q would be always greater than 0. Meanwhile, the congestion window of every flow is already at the minimum value 1 and cannot be decreased again. In other words, although the queueing delay is still greater than 0 in this scenario, the window size cannot be adjusted by the congestion control algorithm.

To obtain the stable point in this situation, $W(t)$ is kept invariant and its value is always a segment, which can be plugged into the Eq. (4). We can get the new model.

$$\frac{dW(t)}{dt} = 0,$$

$$\frac{dQ(t)}{dt} = \begin{cases} \frac{N*seg}{C(R_0+Q(t-\tau))} - 1 & if \quad Q(t) > 0, \\ max\{0, \frac{N*seg}{C(R_0+Q(t-\tau))} - 1\} & if \quad Q(t) = 0, \end{cases}$$

we can get the fixed point (W^*, Q^*) as follows.

$$W^* = seg,$$

$$Q^* = \frac{N * seg - CR_0}{C}.$$

We find that the system is absolutely stable when $N \geq \frac{CR_0}{seg}$, and this special stable state is different from the stable state under $N < \frac{CR_0}{seg}$. In the stable state, the queueing delay will always drop to zero when $N < \frac{CR_0}{seg}$, so the stable state in this case still has the jitter. But if $N \geq \frac{CR_0}{seg}$, the window size does not change and the queueing delay will increase with the increasing number of flows. We can summarize this phenomenon as the following theorem.

Theorem 2. *When the condition $N \geq \frac{CR_0}{seg}$ is satisfied, the DX system enters a special stable state where*

(1) The system is stable;
(2) The congestion window of every flow is unchanged with size 1;
(3) The link is fully utilized.

Obviously, the queueing delay would increase under this case. In other words, Theorem 2 reveals the problem that DX would suffer from large queueing delay when either the base RTT is relatively small or the number of flows is very large.

4   Evaluation

In this section, we validate our theoretical analysis by NS-3 simulations. First, we evaluate the accuracy of our model by comparing the numerical solution of the model conclusion by Matlab 2014a with NS-3 simulation results. Subsequently, we validate our assumption about the probability p by simulations. Next, we examine the conclusion on the special stable state in Theorem 2. Finally, the theoretical conclusion in Theorem 1 is validated by several experiments with the changing parameter.

We use a many-to-one network topology with 10 Gbps link capacity in our experiments. The switch buffer is set to be 256 KB. To validate the stability of a system, we use the metric of the link utilization. If a system is stable, the link utilization keeps a high level since the queue length at switch cannot be zero. We also show the queueing delay and queue size in a few experiments.

Note that in all experiments, we do not explore all values exhaustively for a parameter due to practical consideration. Specifically, the concurrent number of flows, which occupy the link fully, can not surpass the number of ports of a switch (often less than 96). The commonly deployed maximum bandwidth is not greater than 40 Gbps in data center networks, and the base RTT is less than $500\,\mu s$ [1].

Table 1. Probability of decreasing windows

N RTT	10	20	30	40	50	60	70	80	90	100
$80\mu s$	0.79599	0.887787	0.934742	0.956067	0.973239	0.982844	0.995288	0.999902	0.999896	0.999891
$200\mu s$	0.782627	0.835583	0.869653	0.898591	0.914323	0.922147	0.933485	0.944501	0.953232	0.960271
$400\mu s$	0.743014	0.827145	0.876389	0.881062	0.889631	0.890585	0.898752	0.901909	0.913419	0.920201

4.1   Model Validation

Although we model the DX system in Sect. 3, how well the model can match the behavior of practical DX is yet unknown. We answer this question by comparing the queue length obtained by the model with that by running with the NS-3 code of DX. Before that, we first check the assumption that the probability of decreasing windows or $Q(t) > 0$, i.e., p, is constant in the stable state.

We select the scenario where the system enters a stable state and a special stable state, and test the change of p with N ranging from 10 to 100 when the base RTT (R_0) is $80\,\mu s$, $200\,\mu s$, $400\,\mu s$, as shown in Table 1. According to Theorem 1, we know that when R_0 is $80\,\mu s$, $200\,\mu s$, or $400\,\mu s$, the system stability conditions are $N > 30$, $N > 50$ or $N > 140$, respectively. Meanwhile, if the R_0 is $80\,\mu s$ and N is greater than 70, the system is in a special stable state. According to our measurement of p, all values of p are greater than 0.9 when the system is stable. When the system enters a special stable state, the value of p is even greater than 0.99. Using the average value 0.95, p represents those values in the two states basically. This is the reason why we set p as a constant.

Next, we examine the accuracy of our whole model. Figure 2(a) and (b) are respectively the evolution of the queue length under the condition of $N = 50, R_0 = 20\,\mu s$, where DX is in the special stable state, and $N = 50$, $R_0 = 80\,\mu s$, where the behaviors of DX are described by Eqs. (12) and (13). The results of the fluid-flow model are close to the simulation results of NS-3. Therefore, the accuracy of our model for DX is good.

4.2 The Special Stable State

Through the stability analysis in Sect. 3.3, there is a special stable state under the condition of a large number of flows or small base RTT, according to $N < \frac{CR_0}{seg}$ in Theorem 2. When the DX system enters the special stable state, the utilization can even achieve 99.9% and the window size of each flow keeps 1. In this scenario, we will verify this conclusion.

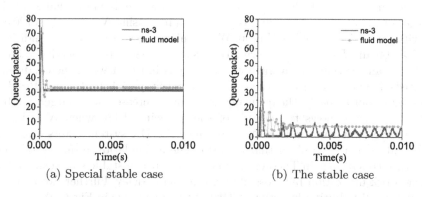

(a) Special stable case   (b) The stable case

Fig. 2. Comparison of the numerical results of fluid flow model with NS-3 simulation.

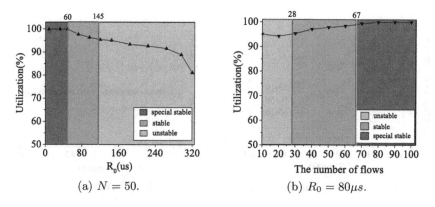

(a) $N = 50$.   (b) $R_0 = 80\mu s$.

Fig. 3. The three states of DX.

We first set the number of flows to be 50 and the bottleneck bandwidth 10 Gbps. Figure 3(a) shows the three states of DX including the special stable, the stable and the unstable states with varying R_0. In the special stable state, the link utilization is 99.9%. We observe that the transition from the stable state to the unstable state is smooth. In fact, the boundary between these two states is not absolute. This is because of we model and analysis DX with some assumptions, like homogeneous sources. In this case, the τ calculated according to Theorem 1, corresponding to the boundary line, is not the absolute upper bound of maintaining the stable state of DX.

Second, we set the base RTT to be $80\,\mu$ s and test the link utilization with varying N. When the number of flows exceeds the threshold (67 in Fig. 3(b)), the system enters the special stable state. Although the window sizes of these flows should be reduced due to the queueing delay, there is a limit on the window sizes, which cannot be lower than 1. As a result, the injected traffic may be greater than the bandwidth delay product, resulting in the queue at the switch cannot be drained up and high utilization.

Next, we inspect the special stable state further by taking deep study into the experiment detail. In Fig. 4(a) and (b), we show the dynamic change of the average congestion window ($cwnd$) of flows with increasing N when R_0 is 120 μs, and with increasing R_0 when N is 50. We calculate the corresponding conditions are $N \geq 100$ and $R_0 \leq 60$ for entering the special stable state, respectively. From Fig. 4, we can see that the average window size is indeed 1 when the conditions are satisfied, which means that the system enters the special stable state. Besides, according to our analysis, the queueing delay may increase with a larger number of flows. Figure 5 shows the change of the queueing delay when N is larger than 70. We omit the result of $N < 70$ since the DX system enters a special stable state when $N \geq 67$ in this scenario. These simulation results verify our theoretical conclusions in Theorem 2, that is, the special stable state can lead to high network utilization but possible high queueing delay. Further, we plot the Cumulative Distribution Function (CDF) of the queue size in Fig. 6. When N is

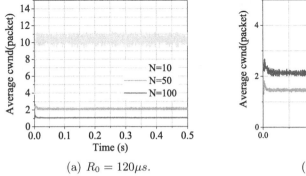
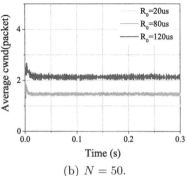

(a) $R_0 = 120\mu s$.   (b) $N = 50$.

Fig. 4. Comparison of the size of the average congestion window.

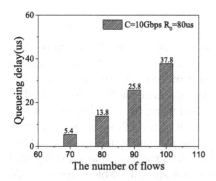

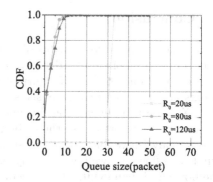

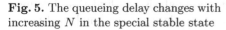

Fig. 5. The queueing delay changes with increasing N in the special stable state

Fig. 6. The CDF of the queue size with the different base RTT

fixed, the system will enter the special stable state for smaller R_0, resulting in that DX has the larger queue size or queueing delay for small R_0. In this figure, the queue size is constantly larger than 30 when $R_0 = 20\,\mu s$.

4.3 Stability Criterion

According to the analysis in Sect. 3 and Theorem 1, the system stability is affected by the number of flows N, R_0 and C. Further, the larger N or the smaller the R_0 or the smaller the C, the more stable the system. To verify this conclusion, we just change one parameter and keep other parameters invariant to investigate its sole influence on the stability of DX in our simulations.

Varying R_0 In this test, we fix the network parameter N as 50 and vary the base RTT R_0 from $20\,\mu s$, $120\,\mu s$ to $320\,\mu s$. According to Theorem 1, we calculate the upper bound of R_0 for keeping the DX system stable as $145\,\mu s$. We observe that the larger R_0 is, the lower the link utilization is, which ranges from 99.91%, 96.11% to 89.1%. The low link utilization means that the system becomes more unstable. This is consistent with the theoretical result.

Varying N In this test, we vary N from 10, 50 to 100 with fixed R_0 $120\,\mu s$. The link utilization increases from 94.81%, 96.11% to 98.53% when N becomes larger and larger. Our theoretical conclusion is that when N is larger than 42, DX is stable according to Theorem 1. From the increase of the link utilization, our theoretical analysis is basically correct.

Varying C In this test, the bottleneck bandwidth C is changed from 1 Gbps, 10 Gbps to 40 Gbps. We set N as 50 and the base RTT R_0 as $120\,\mu s$. In particular, the link utilization decreases from 99.86%, 96.11% to 86.44%. When C is 40 Gbps, the utilization is lowest, which means that the system suffers from unstable. This confirms the theoretical analysis that the larger bandwidth will lead to the instability of the system in Sect. 3.

5   Conclusion

In this paper, we perform a theoretical analysis of DX, which is the up-to-date latency-based algorithm in data center network and has a better performance than the well-known DCTCP. Current investigations on DX are based on experiments and its theoretical analysis is spare. We establish the fluid-flow model of the DX system. By linearizing the fluid model and using the stability criterion of the linear system, we derive the stability condition of the DX system. According to our analysis, we found that the stability of the system is proportional to the number of flows, as well as inversely proportional to the propagation delay and the bottleneck bandwidth. In particular, there is a special stable state when N is too large or RTT is too small. Through the analysis, we find that DX has poor throughput when either the base RTT is very large or the number of flows is relatively small. Besides, DX suffers from large queueing delay when either the base RTT is relatively small or the number of flows is very large. Finally, we verify the conclusion in the NS-3 simulation. Our analysis takes a step forward for understanding DX deeply and can be helpful to deploy DX in the data center network or design new latency-based protocols built on DX.

References

1. Alizadeh, M., et al.: Data center TCP (DCTCP). ACM SIGCOMM Comput. Commun. Rev. **40**, 63–74 (2010)
2. Alizadeh, M., Javanmard, A., Prabhakar, B.: Analysis of DCTCP: stability, convergence, and fairness. In: Proceedings of the ACM SIGMETRICS Joint International Conference on Measurement and Modeling of Computer Systems, pp. 73–84. ACM (2011)
3. Alizadeh, M., Kabbani, A., Atikoglu, B., Prabhakar, B.: Stability analysis of QCN: the averaging principle. In: Proceedings of the ACM SIGMETRICS Joint International Conference on Measurement and Modeling of Computer Systems, pp. 49–60. ACM (2011)
4. Dean, J., Ghemawat, S.: MapReduce: simplified data processing on large clusters. Commun. ACM **51**(1), 107–113 (2008)
5. Gao, P.X., Narayan, A., Kumar, G., Agarwal, R., Ratnasamy, S., Shenker, S.: pHost: distributed near-optimal datacenter transport over commodity network fabric. In: ACM Conference on Emerging Networking Experiments & Technologies (2015)
6. Golnaraghi, F., Kuo, B.: Automatic control systems. Complex Variables **2**, 1–1 (2010)
7. Hollot, C.V., Misra, V., Towsley, D., Gong, W.: Analysis and design of controllers for AQM routers supporting TCP flows. IEEE Trans. Autom. Control **47**(6), 945–959 (2002)
8. Jiang, W., Ren, F., Shu, R., Wu, Y., Lin, C.: Sliding mode congestion control for data center ethernet networks. IEEE Trans. Comput. **64**(9), 2675–2690 (2015)
9. Lee, C., Park, C.: Accurate latency-based congestion feedback for datacenters. In: USENIX ATC, pp. 403–415 (2015)
10. Lee, C., Park, C., Jang, K., Moon, S., Han, D.: DX: latency-based congestion control for datacenters. IEEE/ACM Trans. Networking **25**(1), 335–348 (2017)

11. Misra, V., Gong, W.B., Towsley, D.: Fluid-based analysis of a network of AQM routers supporting TCP flows with an application to red. ACM SIGCOMM Comput. Commun. Rev. **30**, 151–160 (2000)
12. Mittal, R., et al.: TIMELY: RTT-based congestion control for the datacenter. ACM SIGCOMM Comput. Commun. Rev. **45**, 537–550 (2015)
13. Srikant, R.: The Mathematics of Internet Congestion Control. Springer, New York (2012)
14. Zhu, Y., Ghobadi, M., Misra, V., Padhye, J.: ECN or Delay: lessons learnt from analysis of DCQCN and TIMELY. In: Proceedings of the 12th International on Conference on Emerging Networking Experiments and Technologies, pp. 313–327. ACM (2016)

Distributed Quality-Aware Resource Allocation for Video Transmission in Wireless Networks

Chao He[1,2], Zhidong Xie[1,2(✉)], and Chang Tian[1]

[1] College of Communications Engineering,
Army Engineering University of PLA, Nanjing 210007, China
xzd313@163.com
[2] National Innovation Institute of Defense Technology,
Academy of Military Sciences of PLA, Beijing 100071, China

Abstract. The rapid development of wireless networks makes it more convenient for people to enjoy high quality multimedia. However, video applications are throughput-demanding, and relatively, radio resource always seems insufficient. Hence, a distributed algorithm is designed in this paper to allocate the limited wireless resource among multiple users for video streaming. In order to specify multimedia service from other ordinary data transmission, the QoE-oriented utility function is considered first. Then, a potential game model is formulated and all the video receivers can update their rate strategies with very little information exchange. By this kind of updating, the bandwidth allocation could be achieved intelligently. The algorithm converges to a set of correlated equilibria. Numeric simulation results indicate that it brings remarkable benefits to both the resource provider and the video users.

Keywords: Distributed algorithm · Resource allocation · QoE · Potential game

1 Introduction

The massive layout of different wireless networks makes handheld devices more and more pervasive. Meanwhile, High Definition (HD) and Ultra High Definition (UHD) multimedia gradually bring people high-grade visual experience. When delivered in wireless network, high definition videos need more available bandwidth. Although they have been greatly compressed by video coding algorithms, such as H.265/HEVC, wireless networks still can not afford the burdens when users become abundant. Thus, it's very crucial to properly allocate the limited bandwidth resource to different video terminals.

This work was funded by the Project of Natural Science Foundations of China (No. 91738201 and 61401507).

Some literatures solved the rate allocation problems by improving the Quality of Service (QoS) and formulating them as optimization problems [1,7,18]. Meanwhile, we can find that game theory has been widely used in tackling with these issues and gotten very good effects. [17] established a two-level game framework, with an evolutionary game for underlying service and a differential game for upper bandwidth selection. [9] proposed a Stackelberg dynamic game model to get the optimal allocated resources. [4] modeled resource competition as the process of replicator dynamics and formulated a decentralized way to deal with task offloading. However, some of them were not about video transmission, or some of the utility functions could not effectively describe the Quality of Experience (QoE) of video application. Some others [3,5,11,13] studied the QoE and solved the allocation problems of handing off and interface selection, etc. Compared with game theory, they often followed a traditional optimization-based approach. [16] formulated the issue as a cooperative bargaining problem of game theory. It also took both the QoE and fairness into account. But they need a proxy server to allocate the bandwidth collaboratively. Thus, the robust of the system may mainly depend on the server which could be invaded and influenced easily.

In this paper, we propose a distributed rate allocation framework and construct a potential game model so that the bandwidth can be allocated in a reasonable and efficient way. The total utility of all users is our primary consideration and we fully consider the users' experience. After several iteration, the algorithm will converge to correlated equilibria and each receiver will choose and keep a proper transmission rate. The rest of this paper is organized as follows. In Sect. 2, we describe the system in detail and discuss our preliminary goal. Also, the utility function is introduced here. In Sect. 3, we model the problem as a potential game and prove the existence of correlated equilibrium. Section 4 follows the way of regret-matching to solve the model. The experimental results and some discussions are settled in Sect. 5. And we draw the conclusions in Sect. 6.

2   System Model and Utility Function

2.1   System Model

Figure 1 presents the typical scenario we discuss. There are N Unmanned Aerial Vehicles (UAV) flying in the coverage of the same access point (AP) and all of them are equipped with Wireless Cameras (WC). Thus, the UAV platforms have the functions of recording and compressing videos, and then, they send the encoded videos to the Video Processing Center (VPC) via the wireless network. All videos are finally edited there together for all kinds of commercial purposes. Because the AP is owned by a Network Service Provider (NSP), as the receiver of the communication, the VPC should lease the wireless channels from the NSP. In order to gain good total video experience, the limited wireless resource should be allocated properly among the UAVs.

Suppose $Cband$ is the total constant throughput provided by the AP. r_i denotes the channel rate for the i-th WC (WC_i), which varies from the minimum

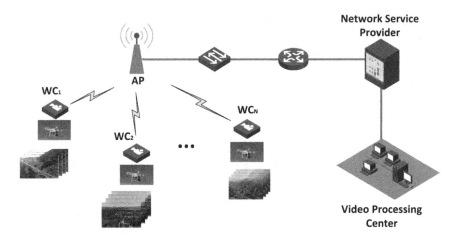

Fig. 1. Video transmission network

rate constraint $R_i^{\min}$ to the maximum one $R_i^{\max}$, $i = 1, 2, ..., N$. $R = [r_1, r_2, ..., r_N]$ denotes the rate vector of the N WCs and the utility function vector is $U = [u_1, u_2, ..., u_N]$, correspondingly. In order to send back high quality video, each WC intends to magnify their own utilities:

$$
\begin{aligned}
\max \quad & u_i \\
s.t. \quad & R_i^{\min} \leq r_i \leq R_i^{\max} \\
& 0 \leq \sum_{i=1}^{N} r_i \leq Cband
\end{aligned}
\tag{1}
$$

2.2   QoE-Based Utility

In ordinary data transmission, the utility is always formulated as the function of QoS. However, users and service providers focus more on the QoE. Peak Signal to Noise Ratio (PSNR) is a typical video quality assessment (VQA) metric based on the error statistics of pixel domain. Some studies regarded the mapping relation between PSNR and Mean Opinion Score (MOS) as a straight line [8,15]. But figures in [10,14] show that the mapping is closer to a sigmoid function. Anyhow, an effective expression of PSNR could properly describe the variation trend of video MOS. Furthermore, PSNR has the lowest complexity among many evaluation methods [10], which makes it more convenient to use in real-time services. In this paper, we adopt the formula in [2] and the utility of video of WC_i can be expressed as

$$
\begin{aligned}
u_i &= PSNR(r_i) - P(r_i) \\
&= a + b \cdot \sqrt{\frac{r_i}{c}} (1 - \frac{c}{r_i}) - P(r_i)
\end{aligned}
\tag{2}
$$

where $P(r_i) = \theta \cdot \frac{r_i}{Cband}$ is the price the VPC pays for the channel leasing of WC_i and θ is the price constant.

3   Potential Game Based Resource Allocation

From the view of individual, the WC_i intends to maximize its channel rate so that its QoE will be favorable. However, it's not allowed too much channel occupation, because the whole bandwidth the NSP can provide is limited and the VPC has to concentrate on maximizing the total utility rather than the single ones. This makes the problem complex. In this paper, we regard it as decentralized process and model the problem as a potential game. The game is denoted as $G = [\Omega, \{R_i\}_{i \in \Omega}, \{U_i\}_{i \in \Omega}]$, where $\Omega = \{1, 2, ..., N\}$ is the set of players. R_i represents the strategy of the i-th player and U_i is the utility correspondingly. It's hoped that, by very little information exchange, everyone can choose proper strategy and no one will break the equilibrium.

In this section, we construct a potential function as: $I = \sum_{i=1}^{N} u_i$ and the problem can be presented as:

$$\max \quad I = \sum_{i=1}^{N} \{a + b \cdot \sqrt{\frac{r_i}{c}}(1 - \frac{c}{r_i})\} - P(\sum_{i=1}^{N} r_i)$$

$$s.t. \quad R_i^{\min} \le r_i \le R_i^{\max} \tag{3}$$

$$0 \le \sum_{i=1}^{N} r_i \le Cband$$

where $P(\sum_{i=1}^{N} r_i) = \theta \cdot \frac{\sum_{i=1}^{N} r_i}{Cband}$ means the price the VPC pays for all the N channel leasing. The more the bandwidth is occupied in the AP, the more the VPC should pay for it, which will cut its benefits in another way. On the other hand, NSP can also effectively guarantee the quality of the network and avoid congestion by changing the price.

Proposition 1. *Each WC selfishly switches its rate strategy will lead to a near optimal solution to the whole utility of the VPC and the game G we propose is a standard potential game.*

Proof. We set another rate strategy for WC_i as r_i'. Because the potential function I is the sum of the single utilities, we can get

$$\Delta I = I(r_i) - I(r_i') = u(r_i) - u(r_i') = \Delta u \tag{4}$$

The variation of the potential function equals to the difference value of the utility function. According to [12], we know that game G is a standard potential game.

Proposition 2. *The correlated equilibrium uniquely exists in this model.*

Proof. From Eq. (3), we can get the partial derivative of the potential function

$$\frac{\partial I(r_i)}{\partial r_i} = \frac{b}{2\sqrt{cr_i}} + \frac{b\sqrt{c}}{2\sqrt{r_i}^3} - \frac{\theta}{Cband} \tag{5}$$

As $\frac{b}{2\sqrt{cr_i}} + \frac{b\sqrt{c}}{2\sqrt{r_i}^3} > 0$ and $\frac{\theta}{Cband} > 0$, there exists $r_i = \tilde{r}_i$, which makes Eq. (5) identically equal to zero.

Then, we get the second-order partial derivative of Eq. (3) at $r_i = \tilde{r}_i$

$$\left.\frac{\partial^2 I(r_i)}{\partial r_i^2}\right|_{r_i = \tilde{r}_i} = -\frac{b}{4\sqrt{c}\sqrt{\tilde{r}_i}^3} - \frac{3b\sqrt{c}}{4\sqrt{\tilde{r}_i}^5} \tag{6}$$

Because $\frac{b}{4\sqrt{c}\sqrt{\tilde{r}_i}^3} > 0$ and $\frac{3b\sqrt{c}}{4\sqrt{\tilde{r}_i}^5} > 0$, we can get $\left.\frac{\partial^2 I(r_i)}{\partial r_i^2}\right|_{r_i = \tilde{r}_i} < 0$. The results show that, if WC_i doesn't choose the data rate $\tilde{r}_i$, a higher total utility will not be obtained.

4   Resource Allocation Algorithm

Literature [6] discussed a simple adaptive procedure leading to correlated equilibrium and provided the way of "regret-matching". According to this method, and also on the basis of the game model we build above, a distributed algorithm can be obtained to reach correlated equilibrium as follow:

Initialization: At the initial time when $t = 1$, each WC can get the minimum rate of the video to start its strategy. As a matter of fact, the rate can be selected arbitrarily within the range. Meanwhile, a strategy space $\{R\_space\}$ is formulated.

Iterative Update Process:

Strategy Update: At the time $t \geq 2$, each WC calculates the utility of the current strategy r_i and the utility for choosing another strategy r_i'. The average difference between r_i and r_i' needs to be calculated as:

$$L_i^t(r_i, r_i') = \frac{\lambda}{t} L_i^\lambda(r_i, r_i') + \frac{1}{t}[u_i^t(r_i') - u_i^t(r_i)] \tag{7}$$

where λ denotes for time and $\lambda \leq t$. Then $R_i^t(r_i, r_i') = \max\{L_i^t(r_i, r_i'), 0\}$ and it's a measure of "regretting" [6].

Strategy Decision: Suppose r_i is chosen by player i at time t. Then, at time $t+1$, the strategy will be reconsidered and it will follow the probability distribution:

$$\begin{cases} \pi_i^{t+1}(r_i') = \frac{1}{\mu} R_i^t(r_i, r_i') & \forall r_i' \neq r_i \\ \pi_i^{t+1}(r_i) = 1 - \sum_{r_i' \neq r_i} \pi_i^{t+1}(r_i') \end{cases} \quad (8)$$

where $\mu > 0$ is large enough. According to the distribution, we can choose a more proper strategy within the space $\{R\_space\}$ who has a higher possibility. After multiple iterations, the results won't be changed and the equilibrium will be achieved.

5 Simulation Results and Analyses

We conduct some simulations to evaluate the scheme we propose. It's assumed that 3 UAVs fly in the AP's coverage area as Fig. 1. They shoot and record independently and send back 3 different compressed videos, Carphone, Coastguard and Football to the VPC. The minimum and maximum rates are shown in Table 1 and their different styles are also listed.

Table 1. Parameters of different videos

	$R_i^{\min}(kb/s)$	$R_i^{\max}(kb/s)$	Style
Carphone	20.2554	322.0153	Medium motion and smooth scene
Coastguard	28.4987	878.8011	Medium motion and complex scene
Football	286.311	1720	Fast or complex motion

Figure 2 plots the real-time video transmission rates and the total utilities of the three when the channel adopts two different bandwidth values. We can find that when $Cband = 20\,Mbps$, which means the channel can offer sufficient bandwidth, each video can be encoded and transmitted at the maximum of their rates. The faster and the complexer the videos are, the more resource they will occupy. While the resource is insufficient, $Cband = 2\,Mbps$, video Carphone and Coastguard should decrease their rates correspondingly, so that the whole utility can still maintain at a proper level. The utility of our distributed algorithm is very close to the optimal one who takes the global information exchanges. From Fig. 2 we also find that after about 20 iterations, all the curves become smooth and steady, which means the system converges to the equilibrium in a very short time by our scheme.

The total bandwidth of the channel can obviously affect the results of resource allocation. Figure 3 shows the final results at different values of $Cband$. From the aspect of different video styles, we can find that the slight insufficience of bandwidth will first influence the videos which are fast and complex, while the medium and smooth videos are affected relatively less. When the channel condition becomes much worse, less than 1Mbps, all the videos have to reduce their rates. Correspondingly, the total utility increases along with the total bandwidth and the results of our scheme are very close to the optimal solutions at different rates.

Not only the bandwidth of the channel, but also the price parameter θ can influence the results. In Fig. 4, we vary the price factor θ and keep the total channel bandwidth at 3Mbps. When θ increases, the rate of Football decreases obviously. When it's greater than 3.5, Coastguard's rate also reduces. From the total utility curves, we can find the decrease, too. Thus, the NSP, as the resource provider, can easily control both the bandwidth allocation and the robust of the network by adjusting the price parameter.

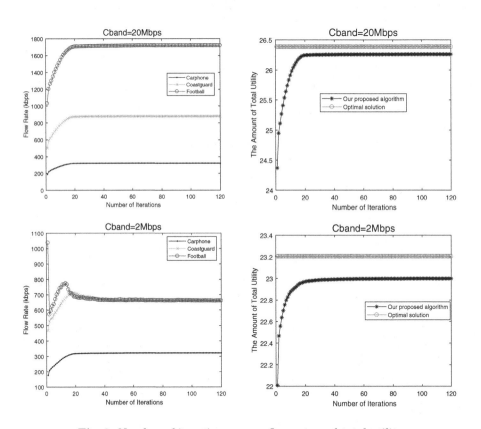

Fig. 2. Number of iterations versus flow rate and total utility

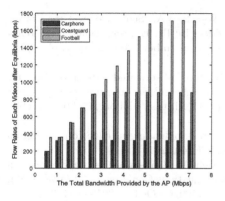
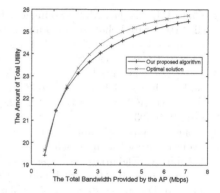

Fig. 3. The influence of total bandwidth $Cband$

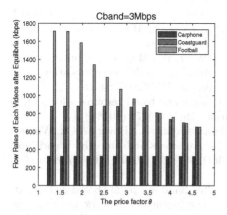
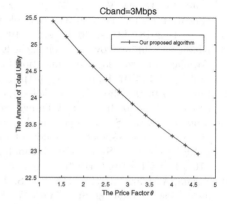

Fig. 4. The influence of price parameter θ

6   Conclusion

Both the dramatic growth of mobile users and the improvement of video quality has made the bandwidth competition of wireless network much fiercer. In this paper, a distributed algorithm was designed to allocate the limited resource among multiple users in order to gain a better total QoE utility of all videos. A model based on potential game theory was constructed and a distributed algorithm was leveraged to solve it. The results could rapidly converge to a set of correlated equilibria. Simulation implied that the proposed strategy could provide a favorable way to solve the resource allocation problem for video transmission.

References

1. Bai, X., Li, Q., Tang, Y.: A low-complexity resource allocation algorithm for indoor visible light communication ultra-dense networks. Appl. Sci. **9**(7), 1391 (2019)

2. Choi, L.U., Ivrlac, M.T., Steinbach, E., Nossek, J.A.: Sequence-level models for distortion-rate behaviour of compressed video. In: IEEE International Conference on Image Processing 2005, vol. 2, pp. II-486, September 2005. https://doi.org/10.1109/ICIP.2005.1530098

3. Deng, Z., Liu, Y., Liu, J., Zhou, X., Ci, S.: QoE-oriented rate allocation for multipath high-definition video streaming over heterogeneous wireless access networks. IEEE Syst. J. **11**(4), 2524–2535 (2017). https://doi.org/10.1109/JSYST.2015.2430893

4. Dong, C., Wen, W.: Joint optimization for task offloading in edge computing: an evolutionary game approach. Sensors **19**(3), E740 (2019)

5. Elgabli, A., Elghariani, A., Aggarwal, V., Bell, M.: QoE-aware resource allocation for small cells. In: 2018 IEEE Global Communications Conference (GLOBECOM), pp. 1–6, December 2018. https://doi.org/10.1109/GLOCOM.2018.8647828

6. Hart, S., Mas-Colell, A.: A simple adaptive procedure leading to correlated equilibrium. Econometrica **68**(5), 1127–1150 (2000)

7. Jiang, Q., Leung, V.C.M., Tang, H., Xi, H.: QoS-guaranteed adaptive bandwidth allocation for mobile multiuser scalable video streaming. IEEE Wireless Commun. Lett., 1 (2018). https://doi.org/10.1109/LWC.2018.2889078

8. Khan, S., Duhovnikov, S., Steinbach, E., Kellerer, W.: MoS-based multiuser multiapplication cross-layer optimization for mobile multimedia communication. Adv. MultiMedia **2007**(1), 6 (2007). https://doi.org/10.1155/2007/94918

9. Liu, B., Xu, H., Zhou, X.: Stackelberg dynamic game-based resource allocation in threat defense for Internet of Things. Sensors **18**(11), 4074 (2018)

10. Moorthy, A.K., Seshadrinathan, K., Soundararajan, R., Bovik, A.C.: Wireless video quality assessment: a study of subjective scores and objective algorithms. IEEE Trans. Circ. Syst. Video Technol. **20**(4), 587–599 (2010). https://doi.org/10.1109/TCSVT.2010.2041829

11. Sarma, A., Chakraborty, S., Nandi, S.: Deciding handover points based on context-aware load balancing in a WiFi-WiMAX heterogeneous network environment. IEEE Trans. Veh. Technol. **65**(1), 348–357 (2016). https://doi.org/10.1109/TVT.2015.2394371

12. Scutari, G., Barbarossa, S., Palomar, D.P.: Potential games: a framework for vector power control problems with coupled constraints. In: 2006 IEEE International Conference on Acoustics Speech and Signal Processing Proceedings, vol. 4, p. IV, May 2006. https://doi.org/10.1109/ICASSP.2006.1660950

13. Senouci, M.A., Souihi, S., Hoceini, S., Mellouk, A.: QoE-based network interface selection for heterogeneous wireless networks: a survey and e-health case proposal. In: 2016 IEEE Wireless Communications and Networking Conference, pp. 1–6, April 2016. https://doi.org/10.1109/WCNC.2016.7564979

14. Seshadrinathan, K., Soundararajan, R., Bovik, A.C., Cormack, L.K.: Study of subjective and objective quality assessment of video. IEEE Trans. Image Process. **19**(6), 1427–1441 (2010). https://doi.org/10.1109/TIP.2010.2042111

15. Thakolsri, S., Kellerer, W., Steinbach, E.: QoE-based cross-layer optimization of wireless video with unperceivable temporal video quality fluctuation. In: 2011 IEEE International Conference on Communications (ICC), pp. 1–6, June 2011. https://doi.org/10.1109/icc.2011.5963296

16. Yuan, H., Wei, X., Yang, F., Xiao, J., Kwong, S.: Cooperative bargaining game-based multiuser bandwidth allocation for dynamic adaptive streaming over HTTP. IEEE Trans. Multimedia **20**(1), 183–197 (2018). https://doi.org/10.1109/TMM.2017.2724850

17. Zhu, K., Niyato, D., Wang, P.: Optimal bandwidth allocation with dynamic service selection in heterogeneous wireless networks. In: 2010 IEEE Global Telecommunications Conference GLOBECOM 2010, pp. 1–5, December 2010. https://doi.org/10.1109/GLOCOM.2010.5683238
18. Zhu, L., Zhan, C., Hu, H.: Transmission rate allocation for reliable video transmission in aerial vehicle networks. In: 2018 14th International Wireless Communications Mobile Computing Conference (IWCMC), pp. 30–35, June 2018. https://doi.org/10.1109/IWCMC.2018.8450454

Neural Networks

PRTSM: Hardware Data Arrangement Mechanisms for Convolutional Layer Computation on the Systolic Array

Shuquan Wang[✉], Lei Wang, Shiming Li, Tian Shuo, Shasha Guo,
Ziyang Kang, Shuzheng Zhang, and Weixia Xu

National University of Defense Technology, Changsha, China
wangshuquan3@163.com

Abstract. The systolic array is an array of processing units which share
the inner data flow. Since the 2D systolic array fits the operation of multiplication and accumulation (MAC) naturally, there are many groups
which use the systolic array to accelerate the computation of DNN (Deep
Neural Network). However, the performance of the systolic array is limited by the data bandwidth. Some groups solve this problem with the
method of loop tiling and care little about the pixel reuse potential of the
convolutional layer. In this paper, we propose a novel method of PRTSM
(Pixels Reuse with Time and Spatial Multiplexing) which reuses the pixels of the input feature map with time and spatial multiplexing. With it,
we can significantly reduce the pressure of bandwidth and save the time
of data preparing for convolutional layers on the systolic array. We propose three algorithms for this method and implement the corresponding
hardware mechanisms on Xilinx FPGA XCVU440. Experiments show
that our hardware mechanisms can reduce at least 72.03% of the off-
chip traffic. The mechanisms proposed by this paper can reach a peak
performance of 64.034 GOPS with a frequency of 167 MHz.

Keywords: DNN · FPGA · Systolic array ·
Hardware data arrangement

1 Introduction

The systolic array is an array of processing units which share the inner data flow.
With the development of DNN acceleration technique, many groups use the 2D
systolic array to improve the efficiency of processing element (PE) [1–3]. However, the performance of the systolic array is limited by the data communication
bandwidth. Here, we give an example of the systolic array with 8×8 PEs (pixel
width: 16-bit, weight width: 16-bit). Each time, both pixels and weights should

---

Supported by organization x.

be fetched. If the work frequency is 100 MHz, the systolic array needs 200 Gbps ($8 \times 8 \times 16 \times 2 \times 100$) to make a full use of the PEs. However, the best performance of DDR4 is 6.4 Gbps. Concerning the fact we typically use much more PEs and design a higher work frequency, the gap of speed will be getting bigger. That increases the pressure of bandwidth. On the other hand, the operation of convolution is a data-intensive operation. There are many overlapped parts between different convolutional operations. We can not store all the data on the hardware accelerator because of the limited on-chip store space. Facing this, many groups prefer to use the tiling method to reduce the pressures of bandwidth and on-chip storage. However, this tiling method ignores the potential of pixels reuse in the convolutional layer. On the other hand, the systolic array fetches the pixels one by one. The pixels of the input feature map have to be reordered so that they can fulfill the timing requirement of the systolic array. This procedure is called data arrangement. We typically use software to reorder the pixels on the host. However, this leads to a lot of additional data traffic.

Here, we give an example of 6×6 input feature map with a convolutional kernel of 3×3 and set the pixel to be 16-bit. The stride of convolution is 1. The total data we need is 576 bit ($6 \times 6 \times 16$). However, the systolic array processes pixels in sequence. We need to reorder these pixels based on the convolutional operation. The final data we transform in the data channel is 2304 bit ($(6 - 3 + 1) \times (6 - 3 + 1) \times 3 \times 3 \times 16$). Most of the data come from the overlapped parts of different convolution operations. The method of tiling has no idea to reduce this additional off-chip traffic. Facing this, we propose a new method PRTSM (Pixels Reuse with Time and Spatial Multiplexing) and reorder the pixels on-chip. With it, we can reduce the off-chip traffic significantly. In our scheme, the hardware accelerator fetches one input feature map every time. We reorder these pixels to build up the final input sequence.

So far, there are a few of works focusing on the data arrangement optimization for DNN hardware accelerator [6,7], especially for the systolic array. Google's TPU [7] shows a scheme of WS but does not give many details about how the intermediate data is set up for the computation of the next layer. [6] proposed a method of data mapping for CNN accelerator but didn't show the procedure of data arrangement. Concerning this, we propose three algorithms to reorder the pixels so that these pixels can be used by the systolic array directly. The main contributions of this paper are as follows.

– We propose three reordering algorithms which reuse the pixels of the input feature map to found the input data flows for convolutional layers on the systolic array;
– We implement the corresponding hardware mechanisms with Xilinx FPGA XCVU440. These mechanisms can accomplish the task of pixels reordering.

Experiments show that the algorithms we propose can reuse at least 71% of overlapping pixels. The corresponding mechanisms can reduce at least 72.03% of off-chip traffic with a power of less than 5.623 W. The best throughput of our mechanisms is 64.034 GOPS with a working frequency of 167 MHz. Note, since this work is a follow-up of SNN (Spiking Neural Network) accelerator, we denote the pixel with 1-bit. Our mechanisms can be configured in other representation schemes (for example 16-bit). The baselines are a straightforward algorithm we proposed and [14].

2   Related Works

The research of using the systolic array to accelerate DNN is getting popular. Bao et al. [3] proposed a reconfigurable macro-pipelined systolic accelerator architecture and implemented 32-PE accelerator on Xilinx ML605. Samajdar et al. [1] proposed a software simulator called SCALE-Sim to explore the micro-architectural features and parameters configuration optimization. Zhang et al. [2] analyzed the impact of permanent faults on the systolic array based neural network accelerator. They proposed two strategies to improve the fault-tolerant rate of DNN accelerator with a negligible drop in classification accuracy. So far, most of works focused on the functional design and performance optimization [8–10]. They paid a little attention to the optimization of overlapped pixels reuse for DNN on the systolic array. Some groups try to arrange the computation order efficiently. Qiu et al. [11] proposed a scheme which focused on the loop operations. Chen et al. [12] try to reduce off-chip traffic with the help of input feature map compression. Their scheme is limited by the number of zeros. The others try to mine the data reuse across the layers. Alwani et al. [13] proposed fused-layer CNN to reuse the intermediate output feature map tiles. Their scheme needs a double feature map buffer for input-output feature map exchanging. Azizimazreah et al. [14] proposed a scheme of shortcut mining which reuses the input feature maps in the residual networks. They decoupled the logical-physical banks and made a difficulty on the control logic design. Many works are focusing on the area of hardware accelerator optimization. However, a few of them try to reuse the pixels in the inner part of the convolutional layer.

3   Background and Preliminaries

3.1   Tiling and Optimization

The systolic array fetches pixels in sequence and computes the dot product with pixels (activations) and weights. All the pixels must be ordered in the right way so that they can calculate with the corresponding weights. The input feature map has to be reordered. However, the systolic array doesn't have this function. We should reorder the input feature map with software. This leads to a blow-up of pixels, which aggravates the problem of bandwidth limitation. Facing this, many groups choose to split the feature map into many tiles. Each time, the systolic array processes one tile. However, this method destroys the potential of pixels reusing in the convolutional layer. Here, we define the pixels which belong to the same convolution kernel area as one kernel block. There are many overlapped parts between different kernel blocks (see Fig. 1). These pixels can

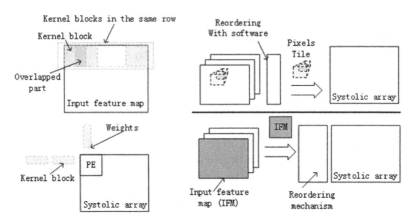

Fig. 1. The left figure shows the detail of pixels reordering and the right shows the principle of our design. Unlike the tiling method (shown in the right-up subfigure), we fetch one input feature map rather than one tile and reorder the pixels on-chip.

be reused. We divide the situations of pixels reusing in two forms, i.e., horizontal and vertical. We can reuse columns of pixels in the horizontal direction and rows in the vertical direction. Here, we can not use the classic tiling method.

We propose a new scheme which has two steps. First, we change the manner of data fetching. Each time, the systolic array fetches one input feature map completely. For the first convolutional layer of DNN, one input feature map corresponds to one output feature map. The change of data fetching manner leads to a little waste of on-chip storage. As for the convolutional layers in the middle of DNN, many feature maps correspond to one output feature map. We should store the intermediate results on-chip which waste a lot of storage space. The input feature map should be split. Unlike the classic tiling method, we only split one input feature map into smaller feature maps. These smaller feature maps still have the potential of pixels reuse. Second, we should reorder the input feature map on-chip. We need a hardware mechanism to reorder these pixels to fulfill the timing requirement of MAC computation. We decide to reorder the pixels in a kernel block form. The motivation of this paper is reordering pixels as soon as possible and reuses pixels if possible.

3.2   Unfold Data Arrangement

The baseline of our data arrangement scheme is a straightforward algorithm which is called unfold data arrangement (see Algorithm 1). For discussion convenience, we introduce Eq. (1) [15], where F denotes the size of feature map is $F \times F$, F' represents the size of output feature map is $F' \times F'$, Z represents the zero padding and S denotes the stride. Here, we use three numbers $(F/S/K)$ to denote the parameters of the feature map. We denote the number of pixels in one feature map with N_{pixel} which equals $F \times F$ and the number of kernel blocks with N_{kernel} which equals $F' \times F'$. i denotes the index of current pixel and j denotes the index of current kernel block. We use $KernelB$ to denote the kenel block.

$$F' = \frac{F - K + Z}{S} + 1 \tag{1}$$

Algorithm 1. Unfold Data Arrangement

 1: set $i = 0; j = 0$;
 2: **repeat**
 3:   **repeat**
 4:     **if** $Pixel_i$ *belongs to* $KernelB_j$ **then**
 5:       $Buffer\ Pixel_i$;
 6:       $i = i + 1$;
 7:     **end if**
 8:   **until** $i > N_{pixel}$
 9:   $Pop\ KernelB_j$ *into the systolic array*;
10:   $j = j + 1$;
11: **until** $j > N_{kernel}$

We assume the input feature map has been stored on-chip. The size of the reordering buffer equals the size of kernel block, i.e., $K \times K$. The reordering buffer fetches the pixels one by one. If the pixel belongs to the current kernel block, it will be buffered. When the reordering buffer gets all the pixels of the current kernel block, it pops these pixels into the systolic array. The time complexity of this algorithm is $O(mn)$ where n equals $(F')^2$ and m equals K^2. The space complexity is $O(m)$.

The hardware mechanism of the unfold data arrangement algorithm has three modules, i.e., RAM, reordering buffer, and fetching address generator. The fetching address generator generates the fetching addresses so that the reordering buffer can fetch the pixels in a given order. Since we store the input feature map in one RAM, the operation of pixels reordering is limited by the reading channel number of RAM. Here, we get two variants, i.e., the unfold data arrangement mechanism with one reading channel ($UnFoldR1$) and the mechanism with two reading channels ($UnFoldR2$). Besides, this algorithm has two weaknesses. First, it works in a serial manner, which means it processes the pixels one by one. Second, it reuses little of the pixels. Since there are overlapped parts between different kernel blocks, we should fetch these pixels again and again. To solve these problems, we propose three improved algorithms.

4   Data Arrangement Algorithm

4.1   Fold Data Arrangement

To reuse as many pixels as we can, we propose the fold data arrangement algorithm (see Algorithm 2). We fetch all the pixels of the input feature map once

---

Algorithm 2. Fold Data Arrangement

1: set $i = 0; j = 0$;
2: **repeat**
3:     *Fetch $Pixel_i$ into the $Buffers$ based on the arrange information*;
4:     **if** $KernelB_j$ gets all the pixels **then**
5:         *All buffers pop the pixels in their first banks i.e. pop $KernelB_j$*;
6:         *The pixels in the rest banks move to their former banks one by one*;
7:         $j = j + 1$;
8:     **end if**
9:     $i = i + 1$;
10: **until** $j > N_{kernel}$

---

and reorder them with the help of the arrange information. Here, we divide the kernel block into many places. Each place corresponds to one pixel. The arrange information is a piece of information which denotes the places where the pixel is needed. We use one buffer to store all the pixels of the same place in different kernel blocks. The number of buffers equals $K \times K$. If the current pixel corresponds to the place of one kernel block, we buffer it in the corresponding buffer. Each column of buffer banks corresponds to one kernel block. We reorder several kernel block in parallel. When one kernel block gets all the pixels, we pop them into the systolic array. The time complexity is $O(n)$ while the space complexity is $O(m \lg n)$. The hardware implementation of the fold data arrangement algorithm is shown (see Fig. 2). This mechanism has five parts, i.e., fetching address generator, RAM, switch logic, reordering buffer, and the arrange information related part. The switch logic is used to push the pixel into the right buffer. The decision is made with the help of the arrange information. Since the arrange information can be generated by software or hardware and stored on-chip or off-chip, we get three variants. The first variant is we use software to generate the arrange information and store them on-chip ($FoldS$). The second is we fetch the arrange information and pixels from the off-chip in pair ($FoldN$). The third is we use hardware module to generate the arrange information ($FoldH$). The reordering buffer has K^2 small buffers. Note, the bank number of each buffer is decreased. Each buffer has a bank index pointer to denote the current bank. When a new pixel is arranged into the current buffer, the buffer stores it in the current bank. When one kernel block gets all the pixels, all the buffers pop their first banks. The rest pixels are passed to their former banks one by one.

4.2 Half-Fold Data Arrangement Variant 1

To reorder the pixels as soon as possible, we propose the half-fold data arrangement algorithm. We divide the situations of pixels reusing in two forms, i.e., horizontal and vertical. It means we reuse columns of pixels in the horizontal direction and rows in the vertical direction. Here, we get two variants. The first variant is that we only reuse pixels in the horizontal direction (see Algorithm 3).

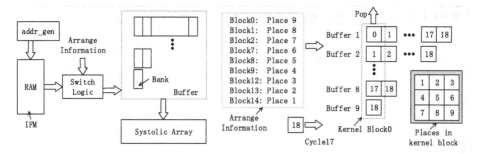

Fig. 2. The detail of fold data arrangement mechanism. The left figure shows the architecture and the right shows the state of reordering buffer (input feature: 6×6, kernel: 3×3, stride: 1).

We organize all the kernel blocks in the same row into one group. All the groups are reordered in parallel. Each time, we fetch one pixel. When we found the first kernel block, the procedure is similar to the fold data arrangement algorithm. However, when the kernel block gets all the pixels, we copy these pixels rather than pop them to the systolic array. When we found the next kernel block, we reuse some pixels which belong to the previous kernel block. The time complexity is $O(\lg n)$ while the space complexity is $O(m \lg n)$. Another variant is we reuse pixels in both the horizontal and vertical direction, which is discussed in the next section.

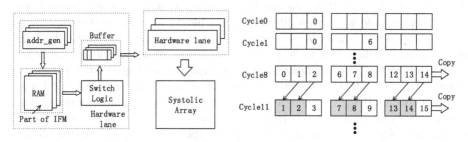

Fig. 3. The detail of half-fold data arrangement variant 1 mechanism (HalfFoldV1).

The hardware implementation of the fold data arrangement algorithm is shown (see Fig. 3). Here, we define a series of hardware modules as one hardware lane. One hardware lane processes one group. Each hardware lane has four parts, i.e., fetching address generators, RAMs, Buffers, and switch logic. We use K small buffer to found reordering buffer. Each small buffer has K banks. Each small buffer corresponds to one RAM and fetching address generator. We reuse the pixels with internal data moving in the small buffer. The internal data moving stride of the pixels in the small buffer is based on S.

Algorithm 3. Half-Fold Data Arrangement Variant 1

1: set $i = 0; j = 0$;
2: Organize all the kernel blocks in the same row into one group.
3: #Process each group in parallel.
4: *Reorder pixels to found the $KernelB_1$ as same as Unfold Data Arrangement Algorithm;*
5: **if** $j > 0$ **then**
6:     **repeat**
7:         **repeat**
8:             **if** $Pixel_i$ *belongs to* $KernelB_j$ *and it's a new pixel;* **then**
9:                 *Buffer* $Pixel_i$;
10:                $i = i + 1$;
11:            **end if**
12:        **until** $i > N_{pixel}$
13:        Reuse some pixels of the former kernel block with internal data moving;
14:        **if** $KernelB_j$ *gets all the pixels;* **then**
15:            *Copy* $KernelB_j$ *into the systolic array;*
16:        **end if**
17:        $j = j + 1$;
18:    **until** $j > N_{kernel}$
19: **end if**

4.3 Half-Fold Data Arrangement Variant 2

Variant 2 is shown (see Algorithm 4). Unlike the Variant 1, we organize a series of rows of kernel blocks into one group called one big group. Then, we process them in parallel. Here, we denote the number of big groups with G. One hardware lane processes one big group. There are some changes. First, we use pixels block to speed up the procedure of kernel block founding. In Variant 1, we fetch the pixels one by one. Here, we can fetch the pixels in block form. Second, we reuse the pixels block between different kernel blocks rows. Since there are some overlapping parts in the vertical direction, we reuse these pixels with the help of the history buffer. When we process a new row of kernel blocks, we reuse some pixels in the previous row. The time complexity is $O(\lg n)$ while the space complexity is $O(m \lg n)$. The hardware lane is also different (see Fig. 4). First, we use a specific RAM to store the pixels block. Second, we carefully design the switch logic to make a switch between pixel input and pixels block input. Third, the small buffer has been designed to support the operation of updating all the banks one time. Besides, we use a history buffer to buffer the pixels block of the former rows.

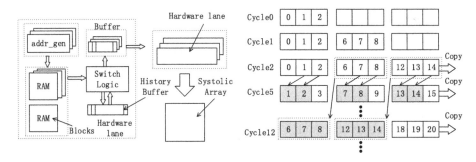

Fig. 4. The detail of half-fold data arrangement variant 2 mechanism (HalfFoldV2).

Algorithm 4. Half-Fold Data Arrangement Variant 2

1: set $i = 0; j = 0;$
2: Organize several groups (defined in Variant 1) into one big group.
3: #Process each big group in parallel.
4: *Found $KernelB_1$ similar to the Variant 1. Each time, fetch one pixels block.*
5: *Found the rest kernel blocks of the first row like Variant 1.*
6: *#The rest rows of kernel blocks.*
7: **if** *start a new row of kernel blocks* **then**
8:    **repeat**
9:        *Reuse some pixels blocks of the former kernel blocks row;*
10:       *Reuse some pixels of the former kernel block;*
11:       *Fetch new pixels to found the $KernelB_j$;*
12:       *Copy $KernelB_j$ into the systolic array;*
13:       $j = j + 1;$
14:    **until** $j > N_{kernel}$
15: **end if**

---

5 Experimental Setup and Result

We implement all the data arrangement mechanisms at the RTL level with Verilog. We use Vivado 2016.4 for synthesizing and choose FPGA XCVU440 for implementation. For the concerning of performance exploration, we implement all the data arrangement mechanisms in a parameter configurable manner. Since this work is the follow-up of SNN accelerator, the pixel width is 1 bit. Our mechanisms can be configured in other representation schemes (for example 16-bit). The baselines are $UnFoldR1$, $UnFoldR2$ and [14].

5.1 Power Efficiency and Hardware Consumption

The detail of power consumption is shown in Table 2. As it shows, $HalfFoldV1$ wastes the most with a power consumption of 5.62 W. The detail of power efficiency is shown (see Fig. 5). For different feature maps, fold data arrangement mechanisms work the best while $HalfFoldV1$ works the worst. Though it works worse comparing to the others (except $HalfFoldV1$), $HalfFoldV2$ works better than $HalfFoldV1$.

Since 229/2/7 (one layer of ResNet152) is the biggest among all the convolutional layers, which leads to the biggest hardware consumption, we choose it to be the evaluation target. G is $F'/2$, which leads to the biggest hardware consumption among all the settings of $HalfFoldV2$. As it shows in Table 1, $HalfFoldV1$ wastes the most of hardware resource while $UnFoldR1$ wastes the least. Since we don't store the pixels and arrange information on-chip, $FoldN$ needs little RAM. We find $HalfFoldV1$ wastes the most of RAM while $HalfFoldV2$ gets a 14× reduction. We also notice that $HalfFoldV2$ needs the most of LUT, which is caused by the complexity of switch logic circuit.

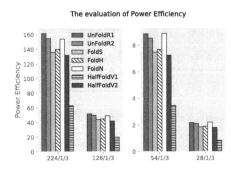

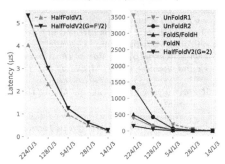

Fig. 5. The detail of power efficiency.   **Fig. 6.** The detail of reordering latency.

Table 1. Hardware resource consumption (229/2/7).

Mechanism	UnFoldR1	UnFoldR2	FoldS	FoldH	FoldN	HalfFoldV1	HalfFoldV2
LUT	**195**	4803	430	238	125	17421	**20543**
LUTRAM	**0**	4096	9953	0	0	0	**488**
FF	**145**	270	0	9721	9651	20235	**13479**
BRAM	**2**	0	90	2	0	392	**28**
DSP	**1**	2	0	0	0	0	**0**

5.2   Latency and Data Reuse Rate

The latency of one input feature map reordering is defined with $Cycles/Frequency$. The total cycles needed for one feature map and the working frequency is shown in Table 1. As it shows, $HalfFoldV1$, $HalfFoldV2$ and $UnFoldR2$ get the best working frequency of 167 MHz while $FoldS$ and $FoldH$ get the worst. The detail of reordering latency is shown (see Fig. 6). $HalfFoldV1$ works the best. As for $HalfFoldV2$, it depends on the parameter G. When we set the G to be $F'/2$, we get the best latency of $HalfFoldV2$. When we set the G to be 2, we get the worst latency of $HalfFoldV2$. All the latencys of $HalfFoldV2$ are better than the others (except $HalfFoldV1$).

Table 2. Total cycles, Frequency and Power.

Mechanism	Cycles	Frequency (MHz)	Power
UnFoldR1/UnFoldR2	$K^2(F')^2, K^2(F')^2/2$	125/167	**2.68/2.77**
FoldS/H/N	F^2	**100/100/125**	2.94/2.7/2.71
HalfFoldV1	FK	167	**5.62**
HalfFoldV2	$((F'/G-1)S+K)(F-K+1)$	167	3.03

The data reuse rate R is defined with Eq. (2), where D_{rest} denotes the overlapping pixels which the mechanism can not reuse, $D_{overlap}$ denotes the total overlapping pixels which equals $(F')^2 K^2 - F^2$. The detail of data reuse rate is shown (see Table 3). We make a comparison between $HalfFoldV1$ and $HalfFoldV2$ (see Fig. 7). When G equals F', the data reuse rate of $HalfFoldV2$ equals the one of $HalfFoldV1$. When G equals 1, $HalfFoldV2$ turns to be a form of fold data arrangement mechanism, and its data reuse rate becomes 100%.

$$R = (D_{overlap} - D_{rest})/D_{overlap} \times 100\% \qquad (2)$$

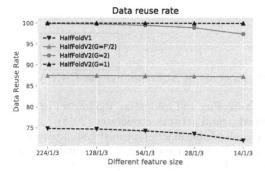

Table 3. Data reuse rate.

Mechanism	Data reuse rate
UnFoldR1/2	0%
FoldS/H/N	100%
HalfFoldV1	$\frac{(F')^2 K^2 - F^2 - (F' K - F)F}{(F')^2 K^2 - F^2}$
HalfFoldV2	$\frac{(F')^2 K^2 - F^2 - (G-1)(K-S)F}{(F')^2 K^2 - F^2}$

Fig. 7. Comparation of $HalfFoldV1/V2$.

5.3 Comparison with State-of-the-Art

Here, we set the pixel width to be 16 bit (as same as [13,14]). We use the layer of 229/2/7 for evaluation. We can reduce 100% overlapping pixels with $FoldS$, $FoldH$ and $FoldN$, 78.8% with $HalfFoldV1$ and 89.6% with $HalfFoldV2(G = 2)$. If we treat the total data to be the off-traffic and concern convolutional layers only, we can reduce 91.5% of the off-chip traffic with $FoldS$, $FoldH$ and $FoldN$, 72.03% with $HalfFoldV1$ and 73.39% with $HalfFoldV2$. All of them are better than [13] with a reduction of 26% and [14] with 43%. We can get at least 1.67× improvement. We compare the working frequency with [14]. $HalfFoldV1$ and $HalfFoldV2$ work with a frequency of 167 MHz, while [14] with 150 MHz. Here, one operation denotes placing one pixel to the corresponding place of kernel block. We can get a throughput of 51.14 GOPS ($HalfFoldV2$) and 64.03 GOPS ($HalfFoldV1$). Though they are lower than [14], our throughput is limited by the number of reordering operations in one convolutional layer. We also compare the on-chip RAM resource consumption with [14]. Our mechanisms waste at most 392 slices of BRAM while [14] needs 3210 BRAM. Our mechanisms waste a lower scale of RAM resource.

6 Conclusions

With the development of DNN research, the systolic array based accelerator is becoming more and more popular. The input data arrangement for convolutional layers on the systolic array turns to be a problem. Concerning this, we propose three algorithms and implement all the hardware mechanisms on FPGA XCVU440. We finally find that the $HalfFoldV2$ mechanism gets a good balance between reordering speed and hardware resource consumption. It can work in a frequency of 167 MHz and reach the peak performance of 51.14 GOPS while reducing at least 73.39% of off-chip traffic. We can get a better throughput of 64.03 GOPS ($HalfFoldV1$). In the future, we will make further research on pixels reusing.

References

1. Samajdar, A., Zhu, Y., Whatmough, P., et al.: SCALE-Sim: Systolic CNN Accelerator (2018)
2. Zhang, J., Gu, T., Basu, K., et al.: Analyzing and mitigating the impact of permanent faults on a systolic array based neural network accelerator (2018)
3. Bao, W., Jiang, J., Fu, Y., et al.: A reconfigurable macro-pipelined systolic accelerator architecture. In: 2011 International Conference on Field-Programmable Technology, FPT 2011, New Delhi, India, 12–14 December 2011. IEEE (2011)
4. Chen, Y.-H., Krishna, T., Emer, J., Sze, V.: Eyeriss: an energy-efficient reconfigurable accelerator for deep convolutional neural networks. In: International Solid-State Circuits Conference, Ser. ISSCC (2016)
5. Sze, V., Chen, Y.H., Yang, T.J., et al.: Efficient processing of deep neural networks: a tutorial and survey. Proc. IEEE **105**(12), 2295–2329 (2017)
6. Du, Z., Fasthuber, R., Chen, T., et al.: ShiDianNao: shifting vision processing closer to the sensor. In: ACM/IEEE International Symposium on Computer Architecture (2015)
7. In-Datacenter Performance Analysis of a Tensor Processing Unit (2017)
8. Razip, M.I.M., Junid, S.A.M.A., Halim, A.K., et al.: Sequence alignment using systolic array for an accelerator. In: Power Engineering and Optimization Conference. IEEE (2014)
9. Razip, M.I.M., Al Junid, S.A.M., Halim, A.K., et al.: Sequence alignment using systolic array for an accelerator (2014)
10. Ito, M.: A power-efficient FPGA accelerator: systolic array with cache-coherent interface for pair-HMM algorithm. In: Low-Power and High-Speed Chips (2016)
11. Qiu, J., et al.: Going deeper with embedded FPGA platform for convolutional neural network. In: Proceedings of the 24th ACM/SIGDA International Symposium on Field-Programmable Gate Arrays (2016)
12. Chen, Y., Emer, J., Sze, V.: Eyeriss: a spatial architecture for energy-efficient dataflow for convolutional neural networks. In: Proceedings of 43rd Annual International Symposium on Computer Architecture (2016)
13. Alwani, M., Chen, H., Ferdman, M., Milder, P.: Fused-layer CNN accelerators. In: Proceedings of the 49th Annual IEEE/ACM International Symposium on Microarchitecture (2016)

14. Azizimazreah, A., Chen, L.: Shortcut mining: exploiting cross-layer shortcut reuse in DCNN accelerators. In: 2019 IEEE International Symposium on High-Performance Computer Architecture
15. Ma, Y., Kim, M., Cao, Y., Vrudhula, S., Seo, J.: End-to-end scalable FPGA accelerator for deep residual networks. In: IEEE International Symposium on Circuits and Systems (2017)

PParabel: Parallel Partitioned Label Trees for Extreme Classification

Jiaqi Lu[1], Jun Zheng[2], and Wenxin Hu[2(✉)]

[1] School of Computer Science and Software Engineering, East China Normal University, 3663 Zhong Shan Rd. N., Shanghai, China
[2] The Computer Center, East China Normal University, 3663 Zhong Shan Rd. N., Shanghai, China
`wxhu@cc.ecnu.edu.cn`

Abstract. Extreme classification consists of extreme multi-class or multi-label predictions, whose objective is to learn classifiers that can label each data point with the most relevant labels. Recently, some approaches such as 1-vs-all method have been proposed to accomplish the task. However, their training time is linear with the number of classes, which makes them unrealistic in real-world applications such as text and image tagging. In this work, we are motivated to present a two-stage thread-level parallelism which is based on Partitioned Label Trees for Extreme Classification (Parabel). Our method is able to train the tree nodes in different parallel ways according to their number of labels. We compare our algorithm with recent state-of-the-art approach on some publicly available real-world datasets which have up to 670,000 labels. The experimental results demonstrate that our algorithm achieves the shortest training time.

Keywords: Extreme multi-label classification ·
Thread-level parallelism · OpenMP

1 Introduction

Extreme classification was coined by John Langford[1] and Manik Varma[2] in 2013. It is the emerging research field in machine learning which solves classification problems in presence of a large number of categories (which are also called classes or labels) [8]. And the number of these categories is often more than 10^5. To be specific, extreme classification consists of extreme multi-class (only one label is correct) or multi-label predictions (more than one label is relevant to the given item).

In this work, we focus on extreme multi-label classification where the label set has dimensionality of the order of hundreds of thousands or even millions,

---

[1] http://hunch.net/~jl/.
[2] http://manikvarma.org/.

© IFIP International Federation for Information Processing 2019
Published by Springer Nature Switzerland AG 2019
X. Tang et al. (Eds.): NPC 2019, LNCS 11783, pp. 82–92, 2019.
https://doi.org/10.1007/978-3-030-30709-7_7

because this task has been of more and more significance in real-world applications such as text tagging. The goal in extreme multi-label classification is to learn a classifier which can annotate a new instance with relevant labels from the extremely large label set. Take web tagging as an example, the pages in Wikipedia are all tagged with several relevant labels. Extreme multi-label classification can be used to learn a classifier to automatically label new pages by training on the existing pages. Furthermore, extreme multi-label classification can effectively address machine learning problems in web-scale data mining, such as recommendation systems and ad landing pages' queries [1,10,11]. Due to its capability for dealing with web-scale data, extreme multi-label classification has attracted more and more attention in recent years.

The popular approaches to extreme multi-label classification can be divided into two categories, namely 1-vs-all approaches [2,9,12,13] and tree-based approaches [5,6,10,11]. 1-vs-all approaches train a classifier for each label and they usually take months to train on large datasets on a standard desktop [11]. It is intolerable since extreme multi-label classification has been applied in real-world applications such as recommendation systems and ad landing pages' queries which are required to quickly predict the labels of items and give users an immediate answer. To overcome this, DiSMEC [2] and PPDSparse [12] take advantage of distributed systems and partition the training jobs on several computing nodes. Although it is effective, the cost of hardware is heavy. Taking dataset, WikiLSHTC-325K[3], as an example, it has 1,778,351 training instances and 325,056 categories. On this dataset, DiSMEC needs 3 h train on 1000 cores. While for PPDSparse, it takes much shorter training time (i.e. 353 s on 100 cores). If we reduce the hardware cost and train on a single core, tree-based approaches can train much faster. For example, PfastreXML [5] only needs 7.42 h on a single core relative to 749 h for DiSMEC. However, tree-based approaches have not been parallel to accelerate the training process. So is Parabel which is the fastest 1-vs-all approach built with tree structure [11]. To overcome this, we analyze the data independence between nodes and propose PParabel method to accelerate the training process which is the fastest method on one core [11].

Our contributions are shown as follows:

- We analyze the hierarchy of Parabel and find that each label only exists in one node on the same level which means nodes on the same level have data independence. With data independence, we can make the training process parallel at each level.
- We parallelize the training process in two stages. In the first stage, we parallelize the training process of nodes on the same level. In the second stage, we parallelize the k-means in nodes with OpenMP according to the number of labels in nodes.
- We conduct our training process in a thread-level parallelism way and apply OpenMP to accelerate our training. We can enable PParabel work on standard desktops to minimize hardware costs.

---

[3] http://manikvarma.org/downloads/XC/XMLRepository.html.

- We shorten the training time from one day to just one hour without appending more machines.

The rest of the paper is organized as follows. Section 2 introduces the existing approaches to extreme multi-label classification. Section 3 describes the detail of our proposed PParabel method. Section 4 reports our experiments and we will analyze the results. At the end of this paper, we conclude our work and indicate future directions.

2   Related Work

The existing approaches to solving the extreme multi-label classification task can be divided into four categories, namely 1-vs-all approaches [2,9,12,13], label-embedding approaches [3,4,14], tree-based approaches [5,6,10,11] and deep learning based approaches [7].

1-vs-all approaches: 1-vs-all approaches train a separate classifier per label on the whole dataset. This kind of approach leads to training time linear in the number of classes. Therefore, when it comes to big datasets, the training costs can be heavy [11]. But this kind of method ignores the relevance between labels which makes each label independent and easy to be parallelized. DiSMEC [2] and PPDSparse [12] took advantage of the irrelevance between labels and scaled the training progress in large-scale distributed settings. Despite it can make the method easy to be parallelized, the cost of hardware is heavy.

Label-embedding approaches: Label-embedding approaches make the assumption that label matrix is low-rank. Therefore, it can be projected into low dimensional space. In this way, effective number of labels can be reduced. However, since the training points follow a power-law distribution, it will lead to low accuracy [2]. Moreover, embedding approaches need long time for training and prediction even on small embedding dimensions, let alone large datasets.

To overcome these limitations, SLEEC [3] was proposed. It learned local embedding instead of global embedding. To be specific, it used kNN method to preserve nearest neighbors in the label space. However, SLEEC ignores to model the label structure. [15] proposed a deep embedding method for extreme multi-label classification to overcome this. The deep embedding method uses label graph to depict the label structure. In the label graph, an edge exists if the two labels are active at the same sample. With the label graph established, DeepWalk method is used to make word2vec representation for all nodes in the graph. Then the distance between features and labels can be computed and all the training points can be clustered.

Tree-based approaches: Tree-based approaches usually have two types: decision trees and label trees. FastXML [10] is a state-of-the-art classifier for extreme multi-label classification. It recursively partitioned a parent's feature space between its children. To learn the hierarchy, FastXML optimizes the normalized Discounted Cumulative Gain (nDCG).

Another popular tree-based approach is PfastreXML [5]. The algorithm replaces the nDCG loss with its propensity scored variant. It also assigns higher rewards for accurate tail label predictions. In this way, it can improve tail label prediction which is the most challenging factor of extreme multi-label classification.

Unlike the above two methods, Parabel [11] learned a few balanced label hierarchies. The root node of each hierarchy contains the whole set of labels. Each label tree recursively partitions the nodes into two balanced nodes until the number of labels in the leaf nodes is smaller than a threshold. When it comes to leaf nodes, a classifier will be learned for each label. It can be conducted on a single core with the shortest training time while matching its prediction accuracy with other methods. Although Parabel learned balanced label trees in a parallel way, it didn't optimize the node partition process in a parallel way to save time.

Deep learning based approaches: Deep learning based approach is a new way for extreme multi-label classification. Although it has achieved great success in other areas, it has not been applied to extreme classification until 2017. The first attempt is XML-CNN [7]. It utilizes the CNN model to learn a rich number of feature representations. Unlike the traditional CNN model, XML-CNN adopts a dynamic max pooling scheme to get more than one feature. Therefore, it can capture more fine-grained features.

Nowadays, the most popular approach is 1-vs-all approach since we can make log-time training and prediction. But how can we reduce the hardware cost and accelerate the training process is still a problem. Therefore, we are motivated to propose our solution in the next section.

3   Methodology: Parallel Partitioned Label Trees (PParabel)

Our method is designed to accelerate the node partition process parallel. There are two main components in our method, label trees and idle threads. Label trees are used for training the model. Internal nodes in label trees are processed parallel in idle threads. Figure 1 shows the main structure of PParabel. The proposed method is described in the algorithmic format in Algorithm 1. Detailed information is shown as follows. As we all know, every node in the tree which is learnt by Parabel is partitioned into two groups, and not a single label can

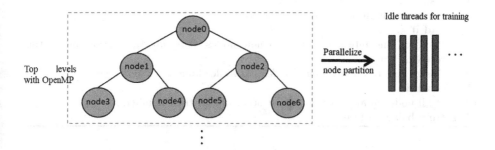

Fig. 1. The main structure of PParabel.

be appeared in two groups. In other words, the split of different nodes is independent. Therefore, we can use its independence to make Parabel parallel. Each node is carried on a single thread.

For each label tree, we load feature matrix $X = x_1, x_2, \ldots, x_n$ and label matrix $Y = y_1, y_2, \ldots, y_n$. Then we represent the label representation in the way Parabel did. We average the feature vectors of instances which are positive to the label as the representation of label. With all labels represented, we put all label representations in the root node and start partitioning. We parallelize the partition process in a two-stage way which will be discussed in Sects. 3.1 and 3.2. For each node partition, we apply k-means to split the node into two nodes. We first randomly choose two label representations as centroids. After that, we calculate the distance between centroids and labels. For nodes on top levels, we parallelize the calculating distance with OpenMP. If it is not converged, we calculate the new centroids and repeat the k-means clustering. When the k-means clustering is converged, we split the labels into two nodes according to the resulting clusters. All these child nodes will be sent to idle threads to further split. The partition process will not stop until the number of labels in any leaf nodes is smaller than a threshold.

We also implement a two-stage parallelization which executes different strategies according to different nodes on different levels. In the following, we will elaborate more details about the two-stage parallelization. The first stage parallelization is applied to all nodes. The second is applied to nodes on top levels.

Algorithm 1. PParabel - Parallel Partitioned Label Trees for Extreme Classification

Input: Feature matrix $\mathbf{X}$, label matrix $\mathbf{Y}$
Output: Balanced tree with the number of leaf nodes' labels smaller than a threshold

1: Load single copy of feature matrix $\mathbf{X}=x_1,x_2,...,x_n$ and label matrix $\mathbf{Y}=y_1,y_2,...,y_n$
2: Compute the label representation by averaging the feature vectors of instances which are positive to the label.
3: **while** the number of labels in any leaf node is larger than the threshold **do**
4:     **if** node is on top levels **then**
5:         Calculate the distance between the cluster centroids and labels with OpenMP.
6:     **else**
7:         Execute K-means in its own thread
8:     **end if**
9:     Partition the internal node into new nodes $node_1,node_2,...$ according to the resulting clusters.
10:     Assign the new nodes $node_1,node_2,...$ to idle threads
11: **end while**
12: Sort all nodes in an ascending order according to their numbers.
13: **return** balanced tree

3.1   First-Stage Parallelization

In this stage, training process of each node is carried on a single thread. And we parallelize the training process of nodes on the same level. The notion behind is that each node is split into two nodes with totally different labels. When it comes to splitting these two child nodes, they do not affect each other. In other words, siblings do not have data dependency.

3.2   Second-Stage Parallelization

Since each tree node is halved, the training time of child nodes will also be halved. In other words, the time child nodes take for training should be half of the time their parent takes. We make the analysis to demonstrate this and we will discuss this in Sect. 4.3. To maximize the usage of threads and speedup the training process, we parallelize the k-means, which is used to split the parent node into two parts, for the nodes on top layers with OpenMP. Here, we set the first five layers as top layer. Since calculating the distance between labels and cluster centroids is the most time consuming step in k-means, we parallelize this process with OpenMP. For the rest nodes which are on the sixth or after sixth layers, since their label sets are not large enough and there is no idle thread, they do not need to parallelize the k-means.

4   Experiments

4.1   Dataset Description

Table 1. Dataset Statistics

Dataset	Number of train points	Number of test points	Label dimensionality	Feature dimensionality
EURLex-4K	15,539	3,809	3,993	5,000
WikiLSHTC-325K	1,778,351	587,084	325,056	1,617,899
Wiki-500K	1,813,391	783,743	501,070	2,381,304
Amazon-670K	490,449	153,025	670,091	135,909

We carry out experiments on publicly available datasets from the Extreme Classification repository[4]. The detailed information of these datasets is shown in Table 1. All these datasets are processed from their original sources such as Wikipedia and Amazon. To figure out the effectiveness of the algorithm on different scale dataset, we choose one small dataset (EURLex-4K with 3,993 labels) and three large scale datasets (WikiLSHTC-325K, Wiki-500K and Amazon-670K) which include hundreds of thousands labels along with million train points.

---

[4] http://manikvarma.org/downloads/XC/XMLRepository.html.

4.2   Evaluation Metrics

We use precision at k and speedup as the metrics for comparison. Precision at k is a commonly used metrics in extreme multi-label classification to show the classification accuracy. And speedup is used to show the effectiveness of the parallelization. For a predicted score vector $\hat{y} \in R^L$ and the ground truth label vector $y \in \{0,1\}^L$, the precision at k is defined as:

$$P@k := \frac{1}{k} \sum_{l \in rank_k(\hat{y})} y_l \qquad (1)$$

The speedup is defined as:

$$S = \frac{T_s}{T_p} \qquad (2)$$

where T_s is the time that the experiment takes in a serial way and T_p is the time that the experiment takes in a parallel way. Higher value of S means more effectiveness of the algorithm.

4.3   Results

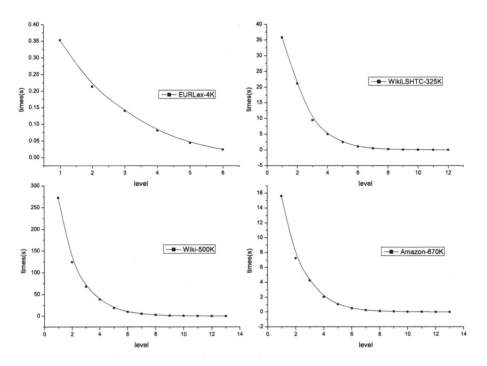

Fig. 2. Average training time of each tree level for four datasets.

Figure 2 shows the average split time of each level. As we can see, the average split time of second-layer nodes is about half of the root node. The average split time of third-layer nodes is about half of the second-layer nodes. So are the forth layer and fifth layer. When it comes to layers after fifth layer, the split time is almost the same. The reason is that the original label set has been divided into more than 32 parts and the k-means process in these nodes can be quickly converged. As we can see, k-means process is the most time consuming step in top level nodes. When it comes to other levels, a large number of node partition is the most time consuming step. Therefore, we can parallelize the k-means process with OpenMP on top levels to accelerate the training.

All experiments are run on two Intel Xeon E5-2620 v4 2.10 GHz CPUs. Each CPU has 8 physical cores. There are no hyper threads per core. While the proposed method is based on Parabel, the precision@k and speedup are calculated between the Parabel, PParabel and fastXML. For Parabel and PParabel, the number of balanced trees trained is three and two algorithms use squared hinge loss. But fastXML trains fifty trees in order to achieve high accuracy. Table 2 shows the results on extreme classification datasets. It turns out that the prediction accuracy of PParabel is almost the same as Parabel. It is much better than fastXML which trains much more trees to increase the accuracy. Since we

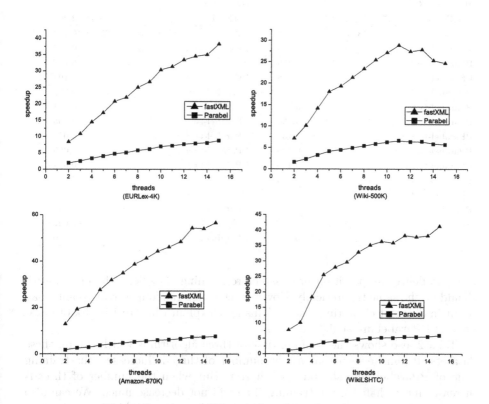

Fig. 3. Speedup of different threads on four datasets.

Table 2. Results on extreme classification datasets.

Method	P1(%)	P3(%)	P5(%)	Training time (hr)	Method	P1(%)	P3(%)	P5(%)	Training time (hr)
EURLex-4K					*Wiki-500K*				
fastXML	71.36	59.90	50.39	0.0590	fastXML	49.27	33.30	25.63	27.72
Parabel	82.25	68.71	57.53	0.0136	Parabel	68.52	48.42	38.55	6.37
PParabel-t=2	81.31	68.33	57.04	0.0070	PParabel-t=2	67.97	48.83	38.02	3.88
PParabel-t=3				0.0054	PParabel-t=3				2.75
PParabel-t=4				0.0041	PParabel-t=4				1.96
PParabel-t=5				0.0034	PParabel-t=5				1.53
PParabel-t=6				0.0029	PParabel-t=6				1.43
PParabel-t=7				0.0027	PParabel-t=7				1.30
PParabel-t=8				0.0024	PParabel-t=8				1.19
PParabel-t=9				0.0022	PParabel-t=9				1.09
PParabel-t=10				0.0019	PParabel-t=10				1.02
PParabel-t=11				0.0019	PParabel-t=11				0.96
PParabel-t=12				0.0018	PParabel-t=12				1.01
PParabel-t=13				0.0017	PParabel-t=13				0.995
PParabel-t=14				0.016	PParabel-t=14				1.09
PParabel-t=15				0.0015	PParabel-t=15				1.12
WikiLSHTC-325K					*Amazon-670K*				
fastXML	49.75	33.10	24.45	4.556	fastXML	36.99	33.28	30.53	2.263
Parabel	65.04	43.23	32.05	0.651	Parabel	44.90	39.81	35.99	0.302
PParabel-t=2	64.08	42.54	31.46	0.585	PParabel-t=2	44.38	39.28	35.44	0.174
PParabel-t=3				0.458	PParabel-t=3				0.117
PParabel-t=4				0.248	PParabel-t=4				0.109
PParabel-t=5				0.178	PParabel-t=5				0.082
PParabel-t=6				0.163	PParabel-t=6				0.071
PParabel-t=7				0.154	PParabel-t=7				0.065
PParabel-t=8				0.138	PParabel-t=8				0.059
PParabel-t=9				0.129	PParabel-t=9				0.055
PParabel-t=10				0.125	PParabel-t=10				0.051
PParabel-t=11				0.126	PParabel-t=11				0.049
PParabel-t=12				0.119	PParabel-t=12				0.047
PParabel-t=13				0.120	PParabel-t=13				0.0426
PParabel-t=14				0.119	PParabel-t=14				0.0428
PParabel-t=15				0.110	PParabel-t=15				0.040

just parallelize the partition process before learning classifiers, the precision@k should be the same theoretically. However, choosing random starting points may result in different clustering results. This may explain why precision@k of Parabel and PParabel are slightly different.

Table 2 also shows the training time of three algorithms which run on these datasets. It can be seen that as the number of threads increases, the training time of PParabel gets shorter and shorter. But when the number of threads increases more than 10, the training time will not decrease much. We can also reduce the training time from 27 h to 1 h just using one machine. It is great to shorten the training time while using much fewer machines.

Figure 3 shows the speedup of different threads on four datasets. The number of threads for PParabel varies from 2 to 15. Both Parabel and fastXML are run with a single thread. We set the maximum threads 15 since we want to make sure that every thread can work on different processor separately which make parallelization happen. PParabel uses multi-threads, while Parabel and fastXML use just one thread. And all these experiments are run in this situation.

The maximum speedup of Parabel is around 9 which is achieved in EURLex-4K dataset. And the maximum speedup of fastXML is around 57 which is achieved in Amazon-670K dataset. As it can be seen, the speedup gain per thread is getting down with the number of threads increasing. The reason is that to protect the data consistency, we need to block other completed threads until the current thread finishes writing. When the number of threads increases, the chance of being blocked is getting bigger and bigger which wastes a lot of time to communicate. Therefore, the speedup gain per thread is getting down with the number of threads increasing. But the speedup is achieved almost linearly with the threads number increasing. In this consideration, to maximize the performance, the optimal number of threads is around 15.

5   Conclusion

In this paper, we have discussed the hardware cost and training time of four typical kinds of approaches to extreme multi-label classification. In order to reduce the hard-ware cost and speedup the training process, we have proposed PParabel algorithm based on Parabel. Our main contribution is employing a two-stage thread-level parallelism. Moreover, we analyze the data independence of nodes on the same level to make sure the training process can be successfully parallelized. The experiment results show that our method is successful to accelerate the training process. All our experiments are conducted on a standard desktop. However, the speedup is achieved almost linearly with the thread number increasing. In the future work, we will study more sufficient approaches in thread level.

Acknowledgement. We thank all viewers who provided the thoughtful and constructive comments on this paper. The third author is the corresponding author. We are grateful to Dr. Manik Varma and his group for their preprocessed datasets. We also thank ECNU Public Platform for Innovation (001) for their equipment to carry out our experiments.

References

1. Agrawal, R., Gupta, A., Prabhu, Y., Varma, M.: Multi-label learning with millions of labels: recommending advertiser bid phrases for web pages. In: Proceedings of the 22nd International Conference on World Wide Web (WWW), pp. 13–24. ACM, New York (2013)

2. Babbar, R., Schölkopf, B.: DiSMEC: distributed sparse machines for extreme multi-label classification. In: Proceedings of the Tenth ACM International Conference on Web Search and Data Mining (WSDM), pp. 721–729. ACM, New York (2017)

3. Bhatia, K., Jain, H., Kar, P., Varma, M., Jain, P.: Sparse local embeddings for extreme multi-label classification. In: Proceedings of the 28th International Conference on Neural Information Processing Systems (NIPS), vol. 1, pp. 730–738. MIT Press Cambridge, MA (2015)

4. Choromanska, A.E., Langford, J.: Logarithmic time online multi-class prediction. In: Proceedings of the 28th International Conference on Neural Information Processing Systems (NIPS), vol. 1, pp. 55–63. MIT Press Cambridge, MA (2015)

5. Jain, H., Prabhu, Y., Varma, M.: Extreme multi-label loss functions for recommendation, tagging, ranking & other missing label applications. In: Proceedings of the 22nd ACM SIGKDD International Conference on Knowledge Discovery and Data Mining (KDD), pp. 935–944. ACM, New York (2016)

6. Jasinska, K., Dembczynski, K., Busa-Fekete, R., Pfannschmidt, K., Klerx, T., Hullermeier, E.: Extreme F-measure maximization using sparse probability estimates. In: Proceedings of the 33rd International Conference on Machine Learning(ICML), vol. 48, pp. 1435–1444 (2016)

7. Liu, J., Chang, W.C., Wu, Y., Yang, Y.: Deep learning for extreme multi-label text classification. In: Proceedings of the 40th International ACM SIGIR Conference on Research and Development in Information Retrieval (SIGIR), pp. 115–124. ACM, New York (2017)

8. Mouhamadou, M.C.: Efficient extreme classification. Data Structures and Algorithms. [cs.DS]. Université Pierre et Marie Curie - Paris VI (2014)

9. Niculescu-Mizil, A., Abbasnejad, E.: Label filters for large scale multilabel classification. In: Proceedings of the 20th International Conference on Artificial Intelligence and Statistics (AISTATS), pp. 1448–1457 (2017)

10. Prabhu, Y., Varma, M.: FastXML: a fast, accurate and stable tree- classifier for extreme multi-label learning. In: Proceedings of the 20th ACM SIGKDD International Conference on Knowledge Discovery and Data Mining (KDD), pp. 263–272. ACM, New York (2014)

11. Prabhu, Y., Kag, A., Harsola, S., Agrawal, R., Varma, M.: Parabel: partitioned label trees for extreme classification with application to dynamic search advertising. In: Proceedings of the 2018 World Wide Web Conference (WWW), pp. 993–1002. International World Wide Web Conferences Steering Committee, Republic and Canton of Geneva (2018)

12. Yen, I.E.H., Huang, X., Dai, W., Ravikumar, P., Dhillon, I., Xing, E.: PPDsparse: a parallel primal-dual sparse method for extreme classification. In: Proceedings of the 23rd ACM SIGKDD International Conference on Knowledge Discovery and Data Mining (KDD), pp. 545–553. ACM, New York (2017)

13. Yen, I.E.H., Huang, X., Zhong, K., Ravikumar, P., Dhillon, I.S.: PD- Sparse: a primal and dual sparse approach to extreme multiclass and multilabel classification. In: Proceedings of the 33rd International Conference on Machine Learning (ICML), vol. 48, pp. 3069–3077. JMLR.org (2016)

14. Yu, H., Jain, P., Kar, P., Dhillon, I.S.: Large-scale multi-label learning with missing labels. In: Proceedings of the 31st International Conference on Machine Learning (ICML), vol. 32, pp. I-592–I-601. JMLR.org (2014)

15. Zhang, W., Yan, J., Wang, X., Zha, H.: Deep extreme multi-label learning. In: proceedings of the 2018 ACM on International Conference on Multimedia Retrieval (ICMR), pp. 100–107. ACM, New York (2018)

Statistical Analysis and Prediction of Parking Behavior

Ningxuan Feng[1], Feng Zhang[1(✉)], Jiazao Lin[2], Jidong Zhai[3], and Xiaoyong Du[1]

[1] Key Laboratory of Data Engineering and Knowledge Engineering (MOE), and School of Information, Renmin University of China, Beijing 100872, China
`fengzhang@ruc.edu.cn`
[2] Department of Information Management, Peking University, Beijing 100871, China
[3] Department of Computer Science and Technology, Tsinghua University, Beijing 100084, China

Abstract. In China, more and more families own cars, and parking is also undergoing a revolution from manual to automatic charging. In the process of parking revolution, understanding parking behavior and making an effective prediction is important for parking companies and municipal policymakers.

We obtain real parking data from a big parking company for parking behavior analysis and prediction. The dataset comes from a shopping mall in Ningbo, Zhejiang, and it consists of 136,973 records in 396 days. Specifically, we mainly explore the impact of weather factors on parking behavior. We study several models, and find that the random forest model can make the most accurate parking behavior prediction. Experiments show that the random forest model can reach 89% accuracy.

Keywords: Prediction model · Regression · Weather condition

1 Introduction

Currently, China has more than 217 million cars, and has a huge demand for parking lots [12]. It becomes very important to improve the utilization of parking space because the cars have faster growth. It also increases the demand for developing intelligent parking system, which can provide better parking management and higher profits for the owners of parking lots.

In the past few years, several parking-related types of research have been conducted to improve parking from different perspectives. For example, some studies [4,23,27,28,31] aim to provide parking information to drivers for free parking; Fang and others [7] proposed an algorithm to allocate cars to parking grid, aiming to improve the utilization of parking space.

The requirement of parking space is an important part of intelligent parking; the studies above considered the prediction of the requirement. However, few

© IFIP International Federation for Information Processing 2019
Published by Springer Nature Switzerland AG 2019
X. Tang et al. (Eds.): NPC 2019, LNCS 11783, pp. 93–104, 2019.
https://doi.org/10.1007/978-3-030-30709-7_8

of them involve weather conditions in parking prediction. In daily life, weather condition has a remarkable impact on our travel plan.

In this paper, we analyze the parking behaviors with weather considered, and then explore various models for parking prediction. In detail, we obtain real parking dataset from a big parking company for parking behavior analysis and prediction. The dataset comes from a shopping mall in Ningbo, Zhejiang, and it consists of 136,973 records in 396 days. We consider the influence of temperature, humidity, rainfall and wind speed. We use the Anova test [9] to analyze different categorical features, and test the correlation between all numerical features by pair plot. Moreover, we also separate workdays from holidays.

For the parking behavior prediction, we have explored linear regression [26], ridge regression [14], Lasso regression [10], decision tree [24], and random forest [15] to depict parking behaviors with weather considered. We find that the random forest is the most suitable model for parking behavior analysis and prediction. Experiments show that it achieves 94% accuracy; its root mean square error (RMSE) can be narrowed down to 0.1662, which is smaller than the other models.

2   Background

2.1   Parking Behavior

Parking behavior refers to the range of actions and mannerisms related to parking. In this paper, we mainly refer to the number of parking each day. In our life, traveling out with cars and demand for off-car activities lead to parking behavior. The purpose of parking can be business, shopping or accommodation. Parking behavior has increased significantly in recent years because of the rapid growth of the number of cars.

The parking behavior is changeable because it can be affected by many factors, especially weather. When it rains heavily, people would more likely to choose traveling out with cars if the activity is necessary. There are also other important determinants for travel plan related to parking behavior. For example, in holidays, the location of the parking lot also has a great influence on parking behavior.

2.2   Motivation

Prediction is meaningful in many fields, not only in computer architecture [16,32], but also in parking behavior [18,21,29]. Parking behavior plays an important role in the city's traffic management. Policymakers can optimize the traffic control strategy in real time based on parking behavior, such as changing the duration of some traffic lights. For parking lot managers, accurate prediction of parking behavior helps develop policies that can improve parking space utilization and get more benefits.

Predicting parking behavior makes a lot of sense. Several related works have been developed in recent years [1,18,21,29]. These studies proposed models to predict parking space availability and occupancy, which partially depicts parking behavior. However, none of them consider the influence of weather condition on parking behavior. This paper is the first to involve weather in parking behavior analysis and prediction.

2.3 Challenges

To conduct an extensive study of parking behavior, we face three major challenges.

Challenge 1: Irregular Data. The data we used to train the prediction model is disorganized. To eliminate the effect of impurity, we need to fully understand the data, and conduct specified data cleaning.

Challenge 2: Various Weather Factors. Weather condition is composed of many detailed factors, such as temperature, humidity, and wind speed. They all affect the prediction accuracy of parking space demand.

Challenge 3: Model Selection. Since there is no research before for weather-related parking behavior analysis and prediction, it is difficult to select the most appropriate model for training.

3 Solution Overview

3.1 Experimental Setup

In this paper, our parking dataset is composed of the parking records of 21-Wharf shopping mall parking lot in Ningbo City, Zhejiang Province, China. The dataset spans 13 months from March 1st, 2018 to March 31th, 2019. It consists of 136,973 parking records. For each parking records, we obtain parking information including the starting time, the ending time, and the selected parking space identity number. We regard the parking record with a duration less than five minutes as noise data. The weather dataset is weather-by-hour data for Ningbo. For each hour, we got precipitation, temperature, relative humidity, and wind speed. We also add some extra categorical features into the dataset that may potentially influence the analysis and prediction. These features include holiday, month, year, weathersit (decided by precipitation), weekday, and season.

3.2 Analysis and Prediction Framework of Parking Behavior

As stated in Sect. 2.3, we have three major challenges, irregular data, various weather factors, and model selection. For the first challenge, we visualize the data to assess the distribution of features, and then present a regularization to reduce the effect of impurity. For the second challenge, we implement a feature selection module to find out whether all the features are necessary for training, and eliminate the outliers. As to the last challenge, we explore five models for park behavior prediction.

The analysis and prediction framework consists of three modules, (1) data preprocessing module, used for data visualization and regularization (Sect. 4), (2) feature selection module, used to clean data and select major features (Sect. 5), and (3) parking space modeling, which explores related models.

4   Preprocessing Methodology

In order to perform an overall analysis of the relevance, we first perform visualization for both parking records and the related features. The features can be divided into two categories: numerical features and categorical features. The numerical features include temperature, wind speed, and humidity, which can be represented as numbers. The categorical features are features that belong to some categories, such as season, working day or holiday, and weather categories (sunny, windy, rainy, and so on).

4.1   Numerical Features

In this part, we analyze the numerical features and use temperature, wind speed, and humidity for illustration. We first normalize features using Eq. 1, and then check for Gaussian distribution [22]. According to our observation, the distribution of these features is in accordance with Gaussian distribution.

$$X_{norm} = \frac{X - X_{min}}{X_{max} - X_{min}} \tag{1}$$

We show the scatter plot of numerical features versus car parking count (denoted as cnt) in Fig. 1. Figure 1(a) exhibits the relation between **normalized temperature** and cnt. It shows that as the temperature increases, the cnt also increases, and the relation between **temperature** and cnt has a positive relationship, though there are some outliers.

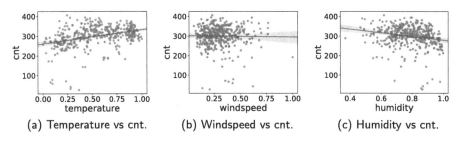

(a) Temperature vs cnt.   (b) Windspeed vs cnt.   (c) Humidity vs cnt.

Fig. 1. Linear regression model fit of numerical features to cnt. The line represents the regression trend.

Figure 1(b) shows a scatter plot of **normalized wind speed** versus cnt. We can see that when we compare the feature alone with cnt, the distribution is little scattered with concentration mainly on the lower side of the `normalized wind speed`.

The scatter plot of `humidity` versus `cnt` in Fig. 1(c) shows that as humidity increases, `cnt` decreases, which implies that people tend to avoid parking cars in 21-Wharf shopping mall when the humidity is high.

4.2 Categorical Features

In this part, we explore categorical features, including season, year, month, holiday, and weathersit. We show the relation of categorical features versus `cnt` in Fig. 2.

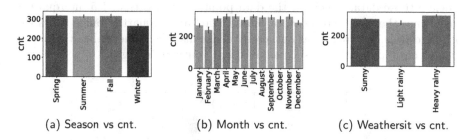

(a) Season vs cnt.   (b) Month vs cnt.   (c) Weathersit vs cnt.

Fig. 2. The relation between categorical features and `cnt`.

For the feature of the season, it has four categories: spring, summer, fall, and winter. Our dataset includes both March 2018 and March 2019, so we have about 120 days of spring, and 90 days for the other seasons. The season-related variation of car parking in Fig. 2(a) reveals that `cnt` in winter is much less than that in the other seasons. This phenomenon infers that people may not willing to travel out in winter.

The feature `year` has two values, 2018 and 2019. Our dataset has more days from 2018 than from 2019, because there are nine months in 2018 and four months in 2019 in our dataset. However, we find that the year 2019 has more car parking on average than the year 2018 does, which probably relates to the call of low-carbon traveling.

As to the feature of `month`, Fig. 2(b) shows that some months have fewer car parking, such as January, February, and December. It indicates that people tend not to drive out in these months, which is consistent with the phenomenon of `season`.

The number of `holidays` is less than that of working days in our dataset. We count the average of the parking times, `cnt`, for holidays and working days. Our analysis shows low `cnt` for working days than for holidays, which indicates that people travel out with cars more on holidays considering this parking lot.

For the influence from the categorical feature of weathersit, we consider three categories: sunny, light rainy, and heavy rainy, as shown in Fig. 2(c). Our dataset has more sunny days than rainy days. However, we count the average of `cnt`, and it shows that `cnt` are higher in heavy rain than in the others.

5   Feature Selection

In order to choose the right set of predictors, we need to perform feature selection before applying predictors to our model. Although more features imply more information on our dataset, they also lead to higher variance. In this section, we start with the outlier analysis.

5.1   Outlier Analysis

Outliers are the data points that differ greatly from other observations, which should be removed from our dataset. In our study, we use the method in [2] to delete those data. Specially, the data points with less than 1.5 interquartile range times the 25th percentile, or more than 1.5 interquartile range times the 75th percentile, are treated as outliers. We visualise the numerical features with (such as cnt) and without (such as fixed cnt) outliers in Fig. 3.

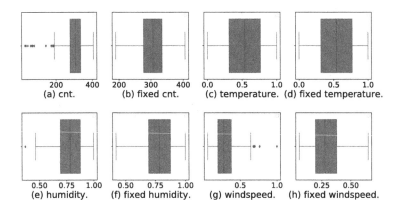

Fig. 3. Numerical features with and without outliers.

In addition, please note that the location of the parking lot also plays an important role in parking behavior. Because we only analyze one parking lot, we do not consider the location influence. We leave it to our future work.

5.2   Feature Analysis

We first show the pair plot for all numerical features in Fig. 4 to see the correlation between a pair of variables. Figure 4 shows that each pair of variables is uniformly distributed, no evident linear correlation between any pair of variables. In a word, each numerical feature is independent of the others.

As the target variable cnt is continuous (we turn it to continuous in the normalization of Sect. 4.1), we perform Anova (analysis of variance) [9] validation for checking the variation in the target variable explained by the categorical

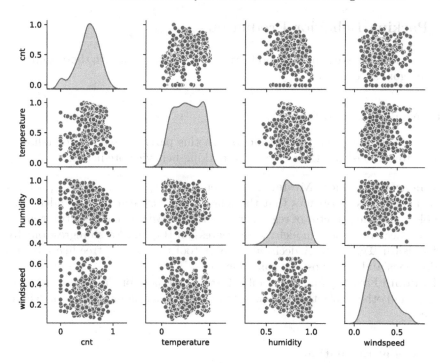

Fig. 4. Pair polt for all numerical features.

feature set. Considering 95% confidence interval, feature variables with p-value more than 0.05 shall be discarded.

We demonstrate the Anova for all categorical features in Table 1. The F-statistic represents the variation between sample means divided by the variation within the samples. It is the probability of the observed result the same as the one obtained in the experiment, assuming the null hypothesis [9] is true. Low P-values are indications of strong evidence against the null hypothesis. It can be seen from Table 1 that no feature has P-value more than 0.05.

Table 1. Anova results on categorical dataset.

Categorical feature	Season	Year	Month	Holiday	Weekday	Weathersit
F-statistic	211.46	893.92	1089.08	87.31	608.05	304.70
P-value	1.26e−42	5.70e−132	8.42e−151	9.26e−20	5.18e−100	5.90e−58

After the introduction of data preprocessing and feature selection, we have normalized the numerical features, eliminated the effects of the outlier and selected a workable set for our training. Next, we shall explore the parking behavior prediction with various models.

6 Parking Behavior Prediction

In this section, we are exploring models that can predict `cnt` with those numerical and categorical features.

6.1 Modeling Methods

Regression is widely used for prediction. In this part, we explore the following models to demonstrate their efficacy in parking behavior prediction.

- **Linear Regression Model** [26]. Given a set $\{y_i, x_{i1}, \ldots, x_{ip}\}_{i=1}^{n}$, a linear regression model assumes that the relationship between the dependent variable y and the p-vector of regressors x is linear. We also consider two generalized linear regression models: ridge regression [14] and Lasso regression [10].
- **Decision Tree** [24]. A decision support tool that uses a tree-like model of decisions and their possible consequences.
- **Random Forest** [15]. An ensemble learning method for classification, regression, and other tasks that are operated by constructing multitudes of decision trees.

6.2 Model Evaluation

In this part, we use linear regression, decision tree, and random forest models for parking behavior prediction, and use Eq. 2 to verify the model accuracy. The dataset covers 396 days. We randomly select 75% days (297 days) as training data, and 25% days (99 days) for validation.

$$accuracy = \frac{|cnt_{real} - cnt_{predicted}|}{cnt_{real}} \tag{2}$$

Linear Regression Model. We first perform an Ordinary Least Squares regression (OLS) model [25] shown in Table 2. The three features with the highest absolute value of coefficient are temperature, humidity, and wind speed. Their coefficients are positive, which means that when these three features are high, the parking lot has a higher utilization. In addition, the coefficient of temperature is 0.148, which is less than the coefficient of wind speed (0.166); this shows that wind speed has a higher impact on `cnt` than temperature does.

We show the output of the predictor using linear regression in Fig. 5(a). The accuracy of linear regression is 78%. In addition, ridge regression model [14] and Lasso regression model [10] are used to regularize the linear regression. We calculate the R-square and RMSE (Root Mean Squared Error) to test the predictors. For the ridge regression model, the best alpha is 0.1, the R-square is 0.3672, and the RMSE is 0.1714. We acquire similar results for the Lasso regression model with best alpha 0.001, R-square 0.3656, and RMSE 0.1719.

Table 2. OLS regression results.

Feature	coef	std err	t	P >|t|	[0.025	0.975]
Season	−0.0694	0.014	−5.121	0.000	−0.096	−0.043
Year	−0.0090	0.033	−0.268	0.789	−0.075	0.057
Month	0.0205	0.005	4.328	0.000	0.011	0.030
Holiday	0.0892	0.024	3.768	0.000	0.043	0.136
Weekday	0.0238	0.005	4.429	0.000	0.013	0.034
Weathersit	0.0045	0.031	0.145	0.885	−0.057	0.066
Temperature	0.1483	0.050	0.2962	0.003	0.050	0.0247
Humidity	0.3534	0.079	4.496	0.000	0.199	0.508
Wind speed	0.1664	0.077	2.167	0.031	0.015	0.318

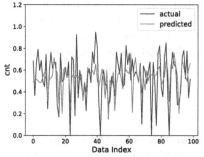

(a) Predicted result from linear model.

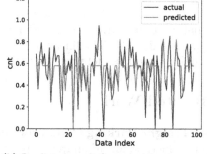

(b) Predicted result from decision tree model.

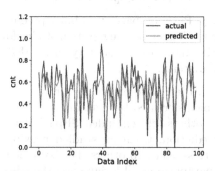

(c) Predicted result from random forest model.

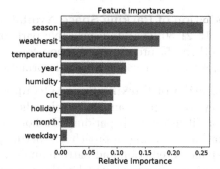

(d) Feature ranking from random forest model.

Fig. 5. Predicted results. Data index refers to the index of records in the test set.

Decision Tree. We also use a decision tree model for our predictor. The output of the predictor using the decision tree model is shown in Fig. 5(b). Its accuracy is 72%. The R-square of the predictor is 0.2747, while the RMSE is 0.1818. We can see that the predictor using a decision tree model has a worse result than the predictor using a linear regression model does.

Random Forest. We now explore a predictor using the random forest model. The maximum depth of the random tree regressor is set to eight, and the amount of estimators is set to 100. The output of the predictor using the random forest is shown in Fig. 5(c). Its accuracy is 89%. The R-square of the predictor is 0.3941, while the RMSE is 0.1662. It can be seen that the predictor using a random forest model is more suitable for the parking behavior prediction.

We then show the ranking of features using random forest model in Fig. 5(d). We can see that `season` is the most important features, and `weathersit`, which relates to precipitation, is also important to the model. Among the numerical features, the feature `temperature` has the most significant impact on the target variable `cnt`.

6.3   Results

As presented in Sect. 6.2, we have implemented five regression models (three linear regression models, a decision tree model, and a random forest model) for park behavior prediction. The decision tree model gives the worst result; its accuracy is only 72%. The linear regression model achieves an accuracy of 78%. The random forest model presents the best result; its accuracy is 89%.

7   Related Work

Urban Freight Parking Demand Prediction. Alho and others [3] proposed a prediction method for urban freight parking demand using ordinary least squares (OLS) linear regression and generalized linear models (GZLMs). This work helps parking lot managers to prediction the demand for parking space for freight cars.

Prediction of Parking Space Availability. Parking space availability prediction [18,21,30] is an indispensable part for intelligent parking system. Caicedo and others [5] proposed a method for predicting space availability in an IPR architecture for parking facility information systems.

Prediction of Parking Space Occupancy. Pierce and others [20] proposed a framework, SFpark, aiming to solve the problems created by charging too much or too little for curb parking. Simhon and others [29] extended SFpark with a machine learning approach for better prediction. Chen [6] studied parking occupancy prediction and pattern analysis. Hossinger and others [11] developed a real-time occupancy model of short-term parking zones. Florian and others [8] presented a model for predicting parking occupation.

Influencing factors of Parking Space Usage. There are many works about influencing factors of parking space usage, including pricing strategy, traffic condition, and parking lot locations. Pierce and others [19] provided an evaluation of pricing parking by demand. Ottosson and others [17] studied the sensitivity of on-street parking demand in response to price changes. Lam and others [13] proposed a bilevel programming model to determine the minimum supply of parking spaces.

8   Conclusion

In this paper, we have analyzed parking behavior with weather conditions considered. We exhibit our method about how to perform preprocessing and feature selection from data, and also explore different regression models for parking behavior prediction. Experiments show that the random forest model has the best results, which achieves 89% accuracy.

Acknowledgments. This work is partially supported by the National Key R&D Program of China (Grant No. 2017YFB1003103), National Natural Science Foundation of China (Grant No. 61722208, 61732014, 61802412). Feng Zhang is the corresponding author (fengzhang@ruc.edu.cn).

References

1. Abdullatif, A., Masulli, F., Rovetta, S.: Tracking time evolving data streams for short-term traffic forecasting. Data Sci. Eng. **2**(3), 210–223 (2017)
2. Aggarwal, C.C.: Outlier Analysis. Data Mining, pp. 237–263. Springer, Cham (2015). https://doi.org/10.1007/978-3-319-14142-8_8
3. Alho, A.R., Silva, J.D.A.E.: Freight-trip generation model: predicting urban freight weekly parking demand from retail establishment characteristics. Transp. Res. Rec. **2411**(1), 45–54 (2014)
4. Banti, K., Louta, M., Karetsos, G.: ParkCar: a smart roadside parking application exploiting the mobile crowdsensing paradigm. In: 2017 8th International Conference on Information, Intelligence, Systems & Applications (IISA), pp. 1–6. IEEE (2017)
5. Caicedo, F., Blazquez, C., Miranda, P.: Prediction of parking space availability in real time. Expert Syst. Appl. **39**(8), 7281–7290 (2012)
6. Chen, X.: Parking occupancy prediction and pattern analysis. Dept. Comput. Sci., Stanford Univ., Stanford, CA, USA, Technical Report CS229-2014 (2014)
7. Fang, J., Ma, A., Fan, H., Cai, M., Song, S.: Research on smart parking guidance and parking recommendation algorithm. In: 2017 8th IEEE International Conference on Software Engineering and Service Science (ICSESS), pp. 209–212. IEEE (2017)
8. Florian, M., Los, M.: Impact of the supply of parking spaces on parking lot choice. Transp. Res. Part B: Methodol. **14**(1–2), 155–163 (1980)
9. Girden, E.R.: ANOVA: Repeated Measures, No. 84. Sage, Thousand Oaks (1992)
10. Hans, C.: Bayesian lasso regression. Biometrika **96**(4), 835–845 (2009)
11. Hössinger, R., Widhalm, P., Ulm, M., Heimbuchner, K., Wolf, E., Apel, R., Uhlmann, T.: Development of a real-time model of the occupancy of short-term parking zones. Int. J. Intell. Transp. Syst. Res. **12**(2), 37–47 (2014)
12. Kong, D., Li, F., Zhang, B.: Design and implementation of intelligent management system for urban road parking. In: Journal of Physics: Conference Series, vol. 1087, p. 062061. IOP Publishing (2018)
13. Lam, W.H., Tam, M., Yang, H., Wong, S.: Balance of demand and supply of parking spaces. In: 14th International Symposium on Transportation and Traffic Theory Transportation Research Institute (1999)
14. Le Cessie, S., Van Houwelingen, J.C.: Ridge estimators in logistic regression. J. Roy. Stat. Soc.: Ser. C (Appl. Stat.) **41**(1), 191–201 (1992)

15. Liaw, A., Wiener, M., et al.: Classification and regression by randomForest. R. News **2**(3), 18–22 (2002)
16. Liu, L., Yang, S., Peng, L., Li, X.: Hierarchical hybrid memory management in OS for tiered memory systems. IEEE Trans. Parallel Distrib. Syst. (2019)
17. Ottosson, D.B., Chen, C., Wang, T., Lin, H.: The sensitivity of on-street parking demand in response to price changes: a case study in Seattle, WA. Transp. Policy **25**, 222–232 (2013)
18. Pflügler, C., Köhn, T., Schreieck, M., Wiesche, M., Krcmar, H.: Predicting the availability of parking spaces with publicly available data. Informatik **2016** (2016)
19. Pierce, G., Shoup, D.: Getting the prices right: an evaluation of pricing parking by demand in San Francisco. J. Am. Plann. Assoc. **79**(1), 67–81 (2013)
20. Pierce, G., Shoup, D.: SFpark: pricing parking by demand (2013)
21. Quinn, J.: System and method for predicting parking spot availability, February 28 2008. US Patent App. 11/849,493
22. Rasmussen, C.E.: Gaussian processes in machine learning. In: Bousquet, O., von Luxburg, U., Rätsch, G. (eds.) ML-2003. LNCS (LNAI), vol. 3176, pp. 63–71. Springer, Heidelberg (2004). https://doi.org/10.1007/978-3-540-28650-9_4
23. Roman, C., Liao, R., Ball, P., Ou, S., de Heaver, M.: Detecting on-street parking spaces in smart cities: performance evaluation of fixed and mobile sensing systems. IEEE Trans. Intell. Transp. Syst. **19**(7), 2234–2245 (2018)
24. Safavian, S.R., Landgrebe, D.: A survey of decision tree classifier methodology. IEEE Trans. Syst. Man Cybern. **21**(3), 660–674 (1991)
25. Seabold, S., Perktold, J.: Statsmodels: econometric and statistical modeling with python. In: 9th Python in Science Conference (2010)
26. Seber, G.A., Lee, A.J.: Linear Regression Analysis, vol. 329. Wiley, Hoboken (2012)
27. Shahzad, A., Choi, J.Y., Xiong, N., Kim, Y.G., Lee, M.: Centralized connectivity for multiwireless edge computing and cellular platform: a smart vehicle parking system. Wirel. Commun. Mob. Comput. **2018**, 1–23 (2018)
28. Shin, J.H., Kim, N., Jun, H.b., Kim, D.Y.: A dynamic information-based parking guidance for megacities considering both public and private parking. J. Adv. Transp. **2017**, 1–19 (2017)
29. Simhon, E., Liao, C., Starobinski, D.: Smart parking pricing: A machine learning approach. In: 2017 IEEE Conference on Computer Communications Workshops (INFOCOM WKSHPS), pp. 641–646. IEEE (2017)
30. Tayade, Y., Patil, M.: Advance prediction of parking space availability and other facilities for car parks in smart cities. Int. Res. J. Eng. Technol. **3**(5), 2225–2228 (2016)
31. Tilahun, S.L., Di Marzo Serugendo, G.: Cooperative multiagent system for parking availability prediction based on time varying dynamic markov chains. J. Adv. Transp. **2017**, 1–14 (2017)
32. Zhang, F., et al.: An adaptive breadth-first search algorithm on integrated architectures. J. Supercomput. **74**(11), 6135–6155 (2018)

Big Data+Cloud

ASTracer: An Efficient Tracing Tool for HDFS with Adaptive Sampling

Yang Song, Yunchun Li, Shuhan Wu, Hailong Yang$^{(\boxtimes)}$, and Wei Li

School of Computer Science and Engineering, Beihang University,
Beijing 100191, China
{yangsoon,lych,wushuhan,hailong.yang,liw}@buaa.edu.cn

Abstract. Existing distributed tracing tools such as HTrace use static probabilistic samplers to collect the function call trees for performance analysis, which may fail to capture important but less executed function call trees and thus miss the opportunities for performance optimization. To address the above problem, we propose *ASTracer*, a new distributed tracing tool with two adaptive samplers. The advantage of adaptive samplers is that they can adjust the sampling rate dynamically, which is able to capture comprehensive function call trees and in the meanwhile maintain the size of trace file acceptable. In addition, we propose an auto-tuning mechanism to search for the optimal parameter settings of the adaptive samplers in *ASTracer*. The experiment results demonstrate the adaptive samplers are more effective in tracing the function call trees compared to probabilistic sampler. Moreover, we provide several case studies to demonstrate the usage of *ASTracer* in identifying potential performance bottlenecks.

Keywords: HDFS · Distributed tracing tool · Adaptive sampling

1 Introduction

With the rapid development of the computing technologies, cloud computing has been widely adopted in large scale applications. Understanding the behavior of distributed systems and tracing the performance bottlenecks is becoming more complicated in the scenario of cloud computing. This is because services are deployed on different nodes, which is particularly difficult to locate abnormal behaviors within the massive volume of log files. Therefore, the distributed tracing tools are proposed to solve the above problems, which can be used to trace function calls in distributed systems to help users understand the system behaviors and analyze performance bottleneck. Currently, distributed tracing tools are widely used inside the large Internet service providers.

Moreover, popular big data analyzing frameworks such as Spark and Hadoop universally use distributed file systems such as HDFS [5] to store the large amount of data. Targeting HDFS, Htrace [1] is a distributed tracing tool for

© IFIP International Federation for Information Processing 2019
Published by Springer Nature Switzerland AG 2019
X. Tang et al. (Eds.): NPC 2019, LNCS 11783, pp. 107–119, 2019.
https://doi.org/10.1007/978-3-030-30709-7_9

guiding the performance analysis and optimization of HDFS. Although tracing every function call within HDFS seems ideal for performance analysis, the huge volume of trace data generated would make the data analysis infeasible. Therefore, Htrace relies on probabilistic samplers to collect a subset of all possible traces. The sampler used in Htrace determines the way how the function calls are collected based on probability.

The drawback with Htrace probabilistic sampler is that it determines to sample a call tree at the root node based on probability, therefore it decides either to sample the entire call tree or nothing. In some cases, such design of probabilistic sampler leads to low sampling rate, and thus fails to provide enough information of the function calls for the developers, especially the information of the abnormal functions. For instance, Table 1 shows the execution statistics of several functions for *nweight* in Hibench [8]. Some functions (e.g., *DFSOutputStream#writeChunk*) are executed for a large number of times, but take a quite short time to execute. Whereas, some functions (e.g., *FileSystem#createFileSystem*) are executed for only a few times, but take a long time to execute, which are more likely to be the performance bottlenecks. However, when using probabilistic samplers in Htrace, the low sampling rate is more likely to ignore these functions. At the same time, some function calls may be called more frequently than others, which may generate very large the trace file that buries the abnormal behaviors with tremendous less useful information. For instance, Table 2 shows the number of calls of several function for *kmeans* in Hibench. The function *DFSInputStream#byteArrayRead* has been executed for a large number of times, which greatly increases the size of the trace file.

Table 1. The execution statistics of several functions in *nweight*.

Function name	Number of calls	$Time_{mean}(ms)$	$Time_{std}(ms)$
DFSOutputStream#writeChunk	3489	0.042	0.350
DFSOutputStream#write	400	1.520	1.882
BlockSender#sendPacket (transferTo)	177	31.717	75.610
BlockSender#doSendBlock	48	121.145	156.751
DFSOutputStream#close	40	267.175	229.670
FileSystem#createFileSystem	20	1244.350	641.715

Table 2. The number of calls for several functions in *kmeans*.

Function name	Number of calls
DFSInputStream#byteArrayRead	1644123
DFSOutputStream#writeChunk	4963
BlockReaderRemote2#readNextPacket	251
ClientNamenodeProtocol#getFileInfo	219
DFSInputStream#fetchBlockAt	131

To solve the above problems, we propose a new tracing tool *ASTracer* for HDFS. The *ASTracer* extends *HTrace* with two adaptive samplers, which records the number of function calls at the root node of the call tree in the sampler, and generates sampling decisions for different root nodes based on the recorded information. For instance, *ASTracer* limits the sampling rate of the call tree that is executed frequently, and ensures that the call trees that are executed less frequently have at least the minimum number of samples. Because the sampling decision is made for each call tree, it guarantees to capture the execution information of more functions. Moreover, *ASTracer* reduces the number of samples from the frequently executed call trees, which is effective to compress the size of the trace file. In addition, we propose several metrics from various aspects such as efficiency, storage and sampling quality to evaluate the effectiveness of the proposed samplers. Compared to the probabilistic samplers, *ASTracer* is able to capture more function call relationships while maintaining a small size of trace file.

Specifically, the main contributions of this paper are as follows:

- We propose *ASTracer*, a new distributed tracing tool with two adaptive samplers for increasing the coverage of function call sampling, as well as maintaining the size of the trace file acceptable.
- We design an auto-tuning mechanism to search for the optimal parameter settings within *ASTracer*, which eliminates the overhead of human effort and time cost of exhaustive search.
- We present several important metrics from various apsects, including efficiency, storage and sampling quality to evaluate the effectiveness of the proposed samplers in *ASTracer*.
- We provide a case study by applying *ASTracer* to analyze representative workloads, which identifies potential performance bottlenecks and gives guidance for performance optimization.

The rest of this paper is organized as follows: Sect. 2 introduces the background of distributed tracing tools as well as the motivation of this paper. Section 5 presents the related work on the samplers of distributed tracing systems. We present the design and implementation of our *ASTracer* with two adaptive samplers, as well as the automatic tuning method for the sampler parameters in Sect. 3. We evaluate the effectiveness of *ASTracer* in Sect. 4, and conclude this paper in Sect. 6.

2   Background and Motivation

2.1   HDFS

HDFS is a distributed file system proposed in Hadoop, but it is also used in other distributed computing frameworks such as Spark. HDFS is highly fault-tolerant and suitable for deployment on commodity clusters. It provides functionalities such as error checking and automatic data recovery. The HDFS cluster adopts the master-slave model, which consists of a *NameNode* and several *DataNodes*. The *NameNode* is responsible for managing the namespace, storing metadata, etc., whereas the *DataNode* performs operations such as creating, deleting, and copying the blocks under the scheduling of the *NameNode* in order to meet the requests from the *Client*.

2.2   Distributed Tracing Tool

To cope with the complicated tracing demand in the distributed systems, Google proposes Dapper [14] that builds the tracing tool based on call tree and span. Another typical tracing tool is Xtrace [7], which is able to provide a comprehensive view of the system service behaviors. However, it is incapable to handle distributed systems at very large scale. Currently, the widely used distributed tracing tools include Zipken [2], Jaeger [3] and Htrace [1]. Among them, Htrace is a tracing tool specially designed for HDFS. The design of Htrace is based on the following concepts: *(1)* a *Span* object represents a function being traced. *(2)* *TraceScopes* manages the life time of *Span* objects, and the *Tracers* are responsible for creating a *TraceScope*. *Tracer* determine whether to sample a function call by calling *Sampler*. *(3)* *Spanreciver* is a collector, which is responsible for receiving *Span* objects sent from *Tracer* and serializing trace data. In this paper, we leverage the *LocalFileSpanReciver* to periodically write sampling data to trace files.

2.3   Motivation

Certain call trees in HDFS application may be executed frequently. Sampling such call trees is not only unnecessary, but also consumes significant computation and storage resources. In addition, generating large trace files could severely degrade the performance of the running application. Moreover, the huge volume of the trace data is also difficult to analyze. However, there are few research works focusing on the design of adaptive samplers in distributed tracing tools, especially in the field of big data application. The samplers in Dapper [14] all adopt a global sampling rate. Zipkin [2] supports more samplers such as counting sampler and boundary sampler, however it fails to consider the execution behaviors of different call trees. Jaeger [3] also misses the dynamic sampling functions in its current implementation [4]. Htrace [1] only provides probabilistic sampling and equidistant sampling that are infeasible to change during the tracing.

It is clear that there is still much work to do for improving the effectiveness of samplers used in distributed tracing tool. For instance, how to improve the coverage of call trees during sampling, and in the meanwhile reduce the size of the trace file. With detailed function call trees sampled, especially when abnormal behaviors happened during the execution, the developers can effectively identify the performance bottlenecks and optimize accordingly. All the above needs motivate this paper.

3   The Design and Implementation of ASTracer

3.1   The Design Overview

The design overview of *ASTracer* is shown in Fig. 1. First, the HDFS application is instrumented. When the application is executed, *ASTracer* decides whether to sample certain call trees. The samplers in *ASTracer* make sampling decisions for the root nodes of each call trees, which can be approximated as sampling the call tree. In *ASTracer*, we use the *record* table to record how many times each call tree has been called. The sampler determines whether to sample a call tree based on the number of occurrences of the call tree.

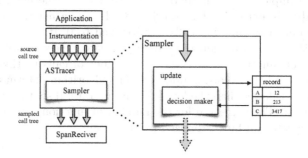

Fig. 1. The design overview of *ASTracer*.

To solve the problem of missing call tree with global sampling rate, *ASTracer* adjusts the sampling rate dynamically according to the number of occurrences of the call trees. The workloads that contain a large number of iterations, the number of occurrences of different call trees could differ by even 5 to 6 orders of magnitude. Sampling such workloads requires dynamically adjusting the sample rate in order to capture enough call trees without generating too large trace files.

The sampler works in the following way within *ASTracer*. The sampler is consulted for sampling decision when the root node of a function call tree is traced by the *Tracer* in *ASTracer*. The sampler updates the record and then generates a sampling decision based on the record.

3.2   Bump Sampler

Bump sampler uses the bump function to generate the sampling decision with probability. The advantage for using the bump function is that the sampling probability changes significantly when the input variable exceeds a certain threshold. With this property, we can guarantee that each function has a high probability of being sampled before a specified threshold. However, after exceeding the threshold, the sampling rate drops dramatically.

The bump function used in the bump sampler is shown in Eq. 1, where x represents the number of times a function is being called. The property of the bump function is that when the number of occurrence of a function is small, the sampling probability is almost 1. However, when a function is being called more often, the sampling rate starts to decrease rapidly. In order to avoid the non-sampling problem with the functions that are being called for a large number of times in the later, we set a minimum sampling rate. Moreover, in order to prevent the frequently executed functions being sampled too less, we create a new thread when instantiating the sampler, and reset the number of function calls in the record table to be 0 every second.

$$f(x) = 1 - e^{-\frac{\lambda^2}{x^2}} \tag{1}$$

The bump sampler works as follows. It first checks whether the record of the function already exists in the record table. If not, a new entry is created, in which the number of function calls is initialized to 0. If there is a record of the function, the bump function is used to generate a new sample rate based on the number of function calls. Then, the number of calls to this function is increased by one and the record is updated. The algorithm determines whether the sampling rate is lower than the lowest sampling rate. If so, the sampling rate is set to the lowest. The threshold for the number of function calls as well as the lowest sampling rate can be customized by the users.

3.3 Token Bucket Sampler

Token bucket sampler is based on the idea of token buckets [9]. The design of token bucket sampler is to maintain a bucket with a certain number of tokens. The number of tokens in the bucket only vary within the range of 0 and bucket capacity. Each time a function is called, the tokens in the bucket are decremented by one. The tokens are replenished to the bucket at a certain rate.

In our token bucket sampler, we set a bucket for the root node of each call tree during workload execution. When the sampler is consulted, it first looks up the bucket to see if there are any tokens left, updates the tokens according to the policy of the token bucket, and decides whether to sample. Instead of using the static sampling rate, it decides whether to sample based on the remaining tokens in the bucket. The advantage of this sampler is that frequently occurring call trees are suppressed, and the call trees that occur less frequently are almost always taken. In particular, when a function call occurs in a burst for a short time period, the sampler can effectively compress the number of samples taken.

The token bucket sampler works as follows. It first checks whether there is an entry for the function in the table. If not, a new entry is created and initialized. Based on whether there is at least one token for the function remained in the token bucket, the token is updated according to the time elapsed from the last execution, however without exceeding the bucket capacity. The bucket capacity as well as the rate for replenishing tokens can be customized by the users.

3.4 Auto Tuning the Sampler Parameters

Since the optimal parameter settings for the samplers vary across different applications as well as distributed systems, it is more effective to use an auto-tuning mechanism to search for the optimal parameter settings for the samplers in *ASTracer*. Therefore, we propose an auto-tuning mechanism using the simulated annealing algorithm [10].

The objective function $f(x)$ as shown in Eq. 2. For bump sampler, $x = (\lambda, threshold)$, whereas for token bucket sampler, $x = (bucket\_size, increase\_step)$. The $entropy(x)$ represents the information entropy of the sampling result. The larger the entropy is, the more information the trace collects. The $dist(x)$ measures the similarity between the sampled results and the full instrumented results, which uses the Euclidean distance. The smaller the Euclidean distance is, the higher the similarity is.

$$f(x) = \frac{dist(x)}{entropy(x)} \tag{2}$$

The constraints to the objective function $f(x)$ is shown in Eq. 3, where $S_{p0.1}$ indicates the trace size sampled using 0.1 probability, and S represents the trace size sampled by the adaptive sampler after the parameter auto-tuning. That is, while ensuring a small size of compressed trace file, it will not lose too much information.

$$0.1 \cdot S_{p0.1} \leq S \leq S_{p0.1} \tag{3}$$

The parameter auto-tuning using the simulated annealing algorithm works as follows. First, it generates a random initial solution x and calculates its objective function $f(x)$. A new solution x' is then proposed by adding a perturbation, and then a new objective function $f(x')$ is calculated. If the constraint is not met, a new solution x' is re-proposed. In order to choose a better solution, let: $\delta f = f(x) - f(x')$, if $\delta f \leq 0$, replace x with x'. However, in order to prevent the algorithm trapping in a local optimal solution, it is necessary to accept a sub-optimal solution with certain probability. The simulating annealing algorithm accepts x' with probability $p = e^{-\frac{\delta f}{T}}$, where T is the current temperature to control the acceptance probability of a sub-optimal solution. The above process iterates until the upper limit is reached. Then the temperature T is decreased and the number of iterations is reset. The above procedure is repeated until the condition is met.

The optimal parameter settings of the samplers after auto-tuning using the simulated annealing algorithm are shown in Sect. 4.1.

4 Evaluation

4.1 Experimental Setup

Our experiments are conducted on a cluster with five nodes, which includes one master node, three slave nodes, and one client node running HDFS v2.8.3.

Each node is equipped with 2 Intel Xeon E5-5620 processors and 16 GB DDR3 memory. The operating system on each node is 64 bit CentOS v6.5. We collect trace file from the *Client* and *Namenode* for result analysis. Representative workloads are selected in Table 3 to demonstrate the robustness of *ASTracer*. To the best of our knowledge, there is no public tracing tool available on HDFS except for Htrace. Therefore, we compare with the static samplers in Htrace with the sampling rate set to 0.1 and 0.01, which is commonly used in literature [14]. The parameter settings for the samplers in *ASTracer* are also shown in Table 3.

Table 3. The parameter settings in *ASTracer*.

	Probability sampler	Bump sampler		Token bucket sampler	
	Sampling rate	λ	Threshold	Bucket size	Replenish rate
dfsioe_read	0.1	128	0.022	1047	21
dfsioe_write	0.1	215	0.020	2595	9
terasort	0.1	552	0.014	3728	12
wordcount	0.1	1034	0.013	5002	14
kmeans	0.1	3490	0.010	21035	19
pagerank	0.1	102	0.021	1083	116

4.2 Evaluation Metrics

To better evaluate the samplers in *ASTracer*, we propose the five metrics including execution time (ET), trace file compression ratio (TFCR), sampling coverage (SC), sample similarity (SS) and information entropy (IEn), to measure the effectiveness of the samplers from different aspects. We provide a brief description about SS and IEn in the following subsections.

Sample Similarity represents the similarity to the trace results with call trees all sampled. The calculation of SS is as follows: for a sampler B, assume that it samples function m and function n. Then we use the feature vector $\boldsymbol{F_B} = ((mean_m, std_m), (mean_n, std_n))$ to represent the sampling characteristics of the sampler, and $\boldsymbol{F_A}$ represents the feature vector with call trees all sampled. After that, we calculate the Euclidean distance between $\boldsymbol{F_A}$ and $\boldsymbol{F_B}$ as shown in Eq. 4, where n represents the number of all functions. A closer Euclidean distance means higher similarity.

$$SS = dist(\boldsymbol{F_A}, \boldsymbol{F_B})$$
$$= \sqrt{\sum_{k=1}^{n} \left[(A_k.mean - B_k.mean)^2 + (A_k.std - B_k.std)^2 \right]} \quad (4)$$

Information Entropy can be used to describe the information uncertainty in a system [13]. The higher uncertainty means higher information entropy. IEn is calculated using Eq. 5. The three properties of information entropy are monotonicity, non-negativeness and additivity. According to monotonicity, the more likely a sample occurs, the less information it carries. In other words, the samples with low probability to occur are more valuable to us. Whereas the non-negativity and additivity ensure that we should focus on high-value samples.

$$H(X) = -\sum_{x \in \mathcal{X}} p(x) \log p(x) \tag{5}$$

The calculation information entropy is as follows: for a sample result, it calculates the execution time $(count_x)$ for each function as well as the total number of calls for all functions $(C = \sum_{x \in \mathcal{X}} count_x,)$. Then, it calculates the frequency of each function $p(x) = count_x/C$ and applies $p(x)$ to Eq. 5.

4.3   Sampler Evaluation

To reduce the impact of system noise, we run each workload for 10 times under each evaluating metric and report the mean of the results. The results are shown in Fig. 2.

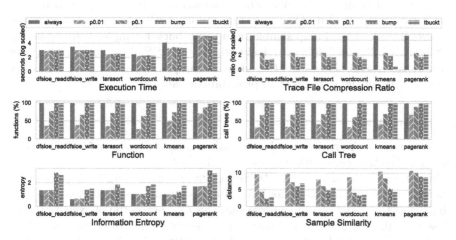

Fig. 2. The evaluation results with different samplers under different metrics. The legend *always*, *p0.01*, *p0.1*, *bump* and *tbuckt* mean the methods of sampling all functions, with probability 0.01, with probability 0.1, bump sampler and token bucket sampler respectively. The execution time, trace file compression ratio, function and call tree of all sampling methods are normalized to *always*.

In terms of execution time, because our sampler uses *ConcurrentHashMap* to store information for parallel accesses, it has less impact on the performance of the workload. Compared to the workload execution time without the sampler, the average sampling latency with bump sampler and token bucket sampler is 6.99% and 5.49% respectively across different workloads, whereas the average sampling latency with probability sampler (rate = 0.1) is 7.92%.

In terms of trace file compression ratio, compared to collecting all samples, the trace file size generated by *ASTracer* is compressed to about 5%, probability sampler (rate = 0.1) is approximately 10%. ASTracer significantly reduces the number of samples of functions that performs too many times, so the size of the trace is also reduced considerably. Note that reducing the number of samples seldom leads to insufficient information about such functions. In addition, we can adjust the sampler parameters to achieve the optimal results.

In terms of sampling coverage, our samplers can capture more functions and call trees, whereas probabilistic samplers fail to capture more functions as the sampling rate decreases. Compared to collecting all samples, both bump sampler and token bucket sampler can achieve close to 100% coverage across different workloads, whereas the average coverage of the probabilistic sampler is 72%.

In terms of information entropy, our adaptive samplers reduce the sampling rate of some high-probability functions and improves the sampling rate of some low-probability functions, therefore it can obtain more information. The average information entropy of the bump sampler and token bucket sampler across different workloads is 2.03 and 2.02, whereas the probabilistic sampler (rate = 0.1) only achieves 1.21.

In terms of sample similarity, when comparing to itself, the SS is 0 when all call trees are sampled. Therefore, the sampler with SS close to 0 is better. The average SS of the bump sampler and token bucket sampler across different workloads is 5.20 and 5.40 respectively, whereas the SS of probabilistic sampler (rate = 0.1) is 6.80. Therefore, the adaptive samplers preserve more statistical characteristics of the samples than probabilistic sampler.

4.4   Case Study

In this section, we provide several case studies using *AStracer* to identify several abnormal function calls with workloads in Hibench.

In general, all calls to the *DFSInputStream#byteArrayRead* function in Hibench are considered as abnormal. After analyzing the execution time distribution of this function, we observe that the 75% percentile of execution time is less than 0.1 ms, however the maximum execution time is as long as 100 ms. This indicates that when the workloads read data, the size of data block is extremely unbalanced.

The machine learning algorithm such as *kmeans* requires multiple iterations, and thus calls the *DFSInputStream#byteArrayRead* function frequently. Table 4 shows the sampled function information of *kmeans*. We can see that the *DFSInputStream#byteArrayRead* function is called more often and the execution time is unbalanced. Therefore, the problem of data skew has a significant impact on the performance of such workload.

Pagerank is an algorithm for measuring the importance of a particular web page. In particular, *pagerank* is a computation intensive workload. A lot of work

Table 4. The sampled functions of *kmeans* and *pagerank*.

Workload	Function	Number of calls	$Time_{mean}$ (ms)	$Time_{std}$ (ms)	$Time_{median}$ (ms)	$Time_{max}$ (ms)
kmeans	DFSInputStream#byteArrayRead	22635	0.041	1.042	0.1	120
	ClientNamenodeProtocol#getFileInfo	219	1.01	4.712	1	70
	ClientNamenodeProtocol#addBlock	22	12.10	9.310	12.5	52
pagerank	DFSInputStream#byteArrayRead	1713	0.149	0.804	0	24
	DFSOutputStream#write	1726	0.162	0.417	0	5
	ClientNamenodeProtocol#create	247	8.846	12.022	7	106

is used to build directed graphs through link relationships. The number of function calls to *DFSInputStream#byteArrayRead* and *DFSInputStream#write* is much fewer than other workloads as shown in Table 4. In addition, the execution time is quite even across function calls. Therefore, the I/O operations are unlikely to become a performance bottleneck.

We also observe that the execution time of the *Client* (e.g., *ClientNamenodeProtocol#create* and *ClientNamenodeProtocol#getFileInfo*) varies significantly as shown in Table 4, which obtains the metadata from *NameNode* via RPC. Although such function is only called for a few times, its execution time is usually long and thus could become the potential performance bottleneck. Whereas, Htrace fails to capture the above information and thus loses the opportunity for performance optimization.

5   Related Work

In the design and optimization of samplers for distributed systems, the Dapper experience from Google [14] emphasizes the dynamic adjustment of the sampling strategy for different workloads, which reduces the sampling rate under high load conditions, and increases the sampling rate under low load conditions to ensure that the coverage of the trace. In addition, Liu et al. [11] use Htrace to analyze the performance of HDFS, and propose a compressed tree algorithm to reduce the size of the trace file, however their algorithm can only be used for offline compression.

Jaeger [3] is a distributed tracing system developed by Uber. It is used to monitor the health of the system. Its implementation is based on Dapper. Jaeger is mainly composed of *jaeger-client*, *jaeger-agent* and *jager-collector*. The *jaegar-agent* is responsible for forwarding the recorded data to the *jaegar-collector*. And it can dynamically adjust the sampling frequency.

Adaptive features are widely studied in performance analysis and tracing systems [6,12,15]. The main idea of these works is based on the runtime information, dynamically adjusting the pre-set parameters to achieve a certain purpose. However, there is little research work on sampling. Therefore, this paper attempts to introduce adaptive sampling into the tracing system in order to achieve better sampling results.

Different from existing works, this paper proposes adaptive samplers by extending the tracing system Htace. Each time the sampler is called, the number of calls to the function (the root node of the call tree) is recorded. According to this record, the sampler can adjust its sampling rate according to the dynamic strategies.

6 Conclusion

In this paper, we propose a new distributed tracing tool *ASTracer* with two adaptive samplers that adjusts the sampling rate dynamically to improve the effectiveness of function tracing from various aspects. The experiment results show that our proposed samplers are better than the probabilistic sampler under various evaluating metrics. Moreover, we provide several case studies to apply *ASTracer* in identifying the performance bottlenecks with representative workloads.

Acknowledgement. This work is supported by National Key Research and Development Program of China (Grant No. 2016YFB1000304) and National Natural Science Foundation of China (Grant No. 61502019). Hailong Yang is the corresponding author.

References

1. https://github.com/apache/incubator-retired-htrace/
2. https://zipkin.io/
3. https://www.jaegertracing.io/
4. https://github.com/jaegertracing/jaeger/issues/365/
5. Borthakur, D.: The hadoop distributed file system: architecture and design. Hadoop Project Website **11**(2007), 21 (2007)
6. Ehlers, J., van Hoorn, A., Waller, J., Hasselbring, W.: Self-adaptive software system monitoring for performance anomaly localization. In: Proceedings of the 8th ACM International Conference on Autonomic Computing, pp. 197–200. ACM (2011)
7. Fonseca, R., Porter, G., Katz, R.H., Shenker, S., Stoica, I.: X-trace: a pervasive network tracing framework. In: Proceedings of the 4th USENIX Conference on Networked Systems Design & Implementation, p. 20. USENIX Association (2007)
8. Huang, S., Huang, J., Dai, J., Xie, T., Huang, B.: The HiBench benchmark suite: characterization of the MapReduce-based data analysis. In: 2010 IEEE 26th International Conference on Data Engineering Workshops, ICDEW 2010, pp. 41–51. IEEE (2010)
9. Humayun, F., Babar, M.I.K., Zafar, M.H., Zuhairi, M.F., et al.: Performance analysis of a token bucket shaper for MPEG4 video and real audio signal. In: 2013 IEEE International Conference on Smart Instrumentation, Measurement and Applications (ICSIMA), pp. 1–4. IEEE (2013)
10. Kirkpatrick, S., Gelatt, C.D., Vecchi, M.P.: Optimization by simulated annealing. Science **220**(4598), 671–680 (1983)
11. Liu, Y., Li, Y., Zhou, H., Zhang, J., Yang, H., Li, W.: A fine-grained performance bottleneck analysis method for HDFS. In: Zhang, F., Zhai, J., Snir, M., Jin, H., Kasahara, H., Valero, M. (eds.) NPC 2018. LNCS, vol. 11276, pp. 159–163. Springer, Cham (2018). https://doi.org/10.1007/978-3-030-05677-3_17

12. Mos, A., Murphy, J.: COMPAS: adaptive performance monitoring of component-based systems. In: Proceedings of 2nd ICSE Workshop on Remote Analysis and Measurement of Software Systems. Citeseer (2004)
13. Shannon, C.E.: A mathematical theory of communication. Bell Syst. Tech. J. **27**(3), 379–423 (1948)
14. Sigelman, B.H., et al.: Dapper, a large-scale distributed systems tracing infrastructure (2010)
15. Wert, A., Schulz, H., Heger, C.: AIM: adaptable instrumentation and monitoring for automated software performance analysis. In: Proceedings of the 10th International Workshop on Automation of Software Test, pp. 38–42. IEEE Press (2015)

BGElasor: Elastic-Scaling Framework for Distributed Streaming Processing with Deep Neural Network

Weimin Mu[1,2], Zongze Jin[1,2(✉)], Junwei Wang[1,2], Weilin Zhu[2],
and Weiping Wang[2]

[1] Institute of Information Engineering, Chinese Academy of Sciences, Beijing, China
{muweimin,jinzongze,wangjunwei}@iie.ac.cn
[2] School of Cyber Security, University of Chinese Academy of Sciences,
Beijing, China
{zhuweilin,wangweiping}@iie.ac.cn

Abstract. In face of constant fluctuations and sudden bursts of data stream, elasticity of distributed stream processing system has become increasingly important. The proactive policy offers a powerful means to realize the effective elastic scaling. The existing methods lack the latent features of data stream, it leads the poor prediction. Furthermore, the poor prediction results in the high cost of adaptation and the instability. To address these issues, we propose the framework named BGElasor, which is a proactive and low-cost elastic-scaling framework based on the accurate prediction using deep neural networks. It can capture the potentially-complicated pattern to enhance the accuracy of prediction, reduce the cost of adaptation and avoid adaptation bumps. The experimental results show that BGElasor not only improves the prediction accuracy with three kinds of typical loads, but also ensure the end-to-end latency on QoS with low cost.

Keywords: Data stream processing · Load prediction ·
Deep neural network · Gated recurrent units · Elasticity

1 Introduction

With the rapid development of the Internet and the rise of the Internet of Things (IoT), various software and sensors continuously generate massive amounts of continuous data streams. Distributed stream processing systems (DSPSs) [1–5] offer a powerful means to carry out data stream processing applications (DSPAs). The Quality of Service (QoS) is important for the DSPAs, which is commonly measured through end-to-end latency and throughput. For example, when an intrusion occurs, determining and warning operations should be made in a certain time window. Nevertheless, data streams have the characteristics of load

© IFIP International Federation for Information Processing 2019
Published by Springer Nature Switzerland AG 2019
X. Tang et al. (Eds.): NPC 2019, LNCS 11783, pp. 120–131, 2019.
https://doi.org/10.1007/978-3-030-30709-7_10

varying and sudden burst. In this case, to ensure the QoS requirements of the DSPAs, the elasticity of the DSPSs has become more and more important.

Many researches have focused on improving the elasticity of the DSPSs. They can be divided into two classes, one is based on the reactive policy and the other one is based on the proactive policy. Although the reactive policy [6–8] is widely used by many DSPSs, it results in some severe issues, such as the QoS degradation and the frequent scaling actions. The elastic scaling happens when the performance of DSPAs does not match the work load.

To address these problems of the reative policy, many researchers propose the proactive policy. With the proactive policy, the scaling actions are executed in advance based on the prediction result of some performance metrics. The existing predicting methods are mainly based on time-series models and some machine learning methods. Traditional time series models, such as MA, ARIMA [9], and Holt-Winters [10] have been widely used. Meanwhile, some methods based on machine learning are often used to predict. Zacheilas et al. [11] provides an adaptive algorithm based on the prediction of the load and latency in upcoming time windows using Gaussian Processes. Repantis et al. [12] proposes a hot-spot prediction technique based on the linear regression for the purpose of alleviating application hot-spots in the DSPA. Hidalgo et al. [13] proposes a method to adjust the parallelism of the operators using Markov chain model. However, these methods lack the latent features of data stream, it leads the poor prediction. The reason is that they can not capture the nonlinear characteristics of drastic fluctuating data stream [14,15] well. Furthermore, the poor prediction results in the high cost of the elastic scaling and the instability.

In order to statisfy the QoS of the DSPAs well, in this paper, we propose a proactive and low-cost elastic-scaling framework based on the accurate prediction using deep neural networks. In summary, our paper makes following contributions.

- We propose the framework named BGElasor, which is a proactive and low-cost elastic-scaling framework based on the accurate prediction using deep neural networks. It can capture the potentially-complicated pattern to improve the accuracy of prediction, reduce the cost and frequency of the elastic scaling.
- As far as we know, our work is the first to use the bidirection gated recurrent units neural networks (BiGRU) to catch the features of the fluctuations of the data stream and build the prediction module (Predictor) to predict the input rate of operators in the DSPSs.
- Besides, we propose a cost-based elastic-scaling algorithm, named CBA and build the elastic scaling module (ElasticityController). It invokes Predictor for multiple time windows ahead of the current state and finds the right point to increase (scale-out) or decrease (scale-in) the parallelism degree of each operator with low cost.
- Finally, our BGElasor runs on DataDock, which is our data stream processing system. The experimental results show that BGElasor not only improve the accuracy of prediction on three kinds of typical loads, but also ensure the end-to-end latency on the QoS.

The rest of our paper is organized as follows. In Sect. 2, we introduce the background, including model of the data stream processing, our distributed stream processing system, DataDock. Section 3 mainly describes the design of BGElasor. We show experimental results of our framework in Sect. 4. Finally, Sect. 5 concludes our paper.

2   Background

In this section, we first present the model of the data stream processing. Then we present our data stream processing system, DataDock.

2.1   DSP Model

A DSPA over data stream is usually organized as a directed acyclic graph $G = (O, S)$, where O is the operator set and S is the stream set. An operator $o \in O$ represents a sort of computation logic, such as filter, join, aggregate or user-define function. **Src** and **Sink** are two special operators in G, which are responsible for spouting source streams and collecting final results, respectively. A stream $s \in S$ is a directed arc (o_p, o_c), $o_p, o_c \in O$, where o_p and o_p are the producer and consumer respectively of s. When a DSPA is submitted to the underlying cluster to execute, its logic operator graph will be transformed into the execution graph, in which each operator o is parallelized into multiple instances $I_o = \{o^1, ..., o^\alpha\}$, where $\alpha \in \mathbb{N}^+$ is the parallelism.

2.2   DataDock

DataDock is our distributed data stream processing system implemented in Java. It is mainly aimed at satisfying the heterogeneous data preprocessing requirements. In order to ensure that the system executes efficiently, DataDock only retains core functions of a DSPS, such as the DSPA definition, the job scheduling.

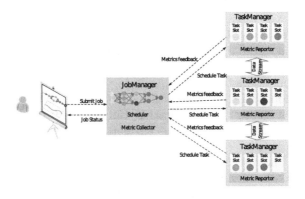

Fig. 1. Architecture of DataDock.

As shown in Fig. 1. DataDock offer users an user interface (UI) to define the DSPA. The JobManager receives a DSPA and turns it into an execution graph. Then the execution graph is scheduled to execute in a set of TaskManagers. The TaskManager runs on a node and is responsible for local resource allocation and instance management. During the execution, the TaskManager continuously collects the performance metrics of each operator instance and reports them to the JobManager. The MetricCollector is in charge of gathering these metrics and storing into MetricDatabase. At the runtime, DataDock allows the instances to quickly discard input data to make sure the latest data receives priority processing. For the fault tolerance, DataDock adopts the similar fault-tolerant policy as Storm to ensure that each event is processed at least once.

3   Framework

3.1   Overview

We show our framework BGElasor as shown in Fig. 2, it contains two important modules, the Predictor and the ElasticityController. We use the Predictor to predict the input rate of all operators. Then we refer the results of the prediction and use the ElasticityController to adjust the parallelism of operators.

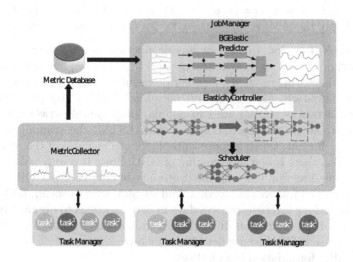

Fig. 2. The architecture of BGElasor

At the runtime, Predictor runs periodically and reconstructs the input rate model for each operator using BiGRU. Then we use the ElasticityController to update the execution graph based on predictions given by Predictor. At last, we use the JobManager to schedule according to the lastest execution graph.

3.2   Predictor

In this section, we build the Predictor to predict the input rate of each opera-tor based on BiGRU. The Predictor contains three parts, the model input, the prediction networks and the model output.

Model Input. At the runtime, our model is trained offline. The Predictor peri-odically reads input rate metrics over the past b time period of each operator from MetricDatabase and normalizes them to the range $[0, 1]$ with the Min-Max scaler.

For the prediction of input rates sequences, formally, we use t to denote the current time and use o to denote the operator whose input rates will be predicted. We use b and f to denote the length of historical time window and the length of future time window respectively. $v(o, t)$ denotes the input rates of o at time t, so, we use $(v(o, t-b), v(o, t-b+1), ..., v(o, t))$ to denote the input rates sequence of o over the past b time period, that is, the input of our predictor. Correspondingly, $(v(o, t+1), (v(o, t+2), ..., (v(o, t+f))$ denotes the input rates sequence of o in the future f time period, which is the output of the Predictor.

For example, assuming the current time is t, the model input is $X_t = (v(o, t-b), v(o, t-b+1), ..., v(o, t))$, the label at t is $y_t = (v(o, t+1), (v(o, t+2), ..., (v(o, t+f))$, and each element ranges between $[0, 1]$. Similarly, the model input at time $t+1$ is $X_{t+1} = (v(o, t\text{-}b+1), v(o, t-b+2), ..., v(o, t+1))$ and correspondingly the label is $y_{t+1} = (v(o, t+2), (v(o, t+3), ..., (v(o, t+f+1))$.

Prediction Networks. In face of drastic fluctuating characteristics of data stream, methods based on deep neural networks have shown better performance, compared with existing methods, for their powerful nonlinear generalization abil-ities. Long short-term memory neural networks (LSTM) and gated recurrent units neural networks (GRU) are two popular deep neural networks to predict the trends of the time-series. Compared with LSTM, the training process of GRU is more efficient, which is more suitable for the scenario of the data stream processing. But there is one shortcoming of GRU. Since it is only able to process the data in one direction ignoring the continuity of data changes, GRU can only capture the partial features of metrics. And its bidirection version, BiGRU, uses two separate hidden layers to process data in two directions to obtain more infor-mation in the time dimension during the training stage. To achieve the higher accuracy, we use BiGRU to predict the input rate.

The BiGRU formulation is as follows:

$$F_t = \sigma(W_F x_t + U_F F_{t-1} + b_F) \tag{1}$$
$$B_t = \sigma(W_B x_t + U_B B_{t+1} + b_B) \tag{2}$$
$$y_t = \sigma(V_F F_t + V_B B_t + b_o) \tag{3}$$

W_F and U_F denote the input-to-forward layer weight matrices. W_B and U_B are the weight matrices of the output-to-backward weight matrices layer.

b_F, b_B and b_o denote biases of forward, backward and output layer, respectively. σ denotes the nonlinear activation function, such as Sigmoid function and Rectified Linear Unit.

Model Output. We first use BiGRU hidden layers to construct the BiGRU network. As mentioned in above sections, BiGRU can capture both forward and backward dependencies to make full use of the input data and learn the complex and comprehensive features. Then, we add a dense layer to transform high-dimensional data into the low-dimensional data to make predictions. During the training process, we use the mean square error (MSE) as the loss function which is computed using the following equation:

$$MSE = \frac{1}{n} \sum_{i=1}^{n} (x_i - \hat{x}_i)^2 \tag{4}$$

3.3   ElasticityController

In this section, we build the ElasticityController. The ElasticityController can ensure the end-to-end latency with the minimum elastic-scaling cost. It contains two parts, the Cost Model and Cost-Balance-Algorithm (CBA).

Cost Model. We build a cost model to evaluate the total cost of all elastic-scaling actions for an operator from the current epoch C to the future epoch F. The cost is defined as:

$$W_o(\mathbf{n}) = p_o \sum_{t_C}^{t_F} n_{o \cdot t} + \sum_{t_C}^{t_{F-1}} (p_o^u C_{o \cdot t}^u(n) + p_o^d C_{o \cdot t}^d(n) + p_o^r C_{o \cdot t}^r(n)) \tag{5}$$

$$C_{o \cdot t}^u(n) = \max(0, n_{o \cdot (t+1)} - n_{o \cdot t}) \tag{6}$$

$$C_{o \cdot t}^d(n) = |\min(0, n_{o \cdot (t+1)} - n_{o \cdot t})| \tag{7}$$

$$C_{o \cdot t}^r(n) = \begin{cases} 0 & n_{o \cdot (t+1)} = n_{o \cdot t} \\ 1 & n_{o \cdot (t+1)} \neq n_{o \cdot t} \end{cases} \tag{8}$$

$W_o(\mathbf{n})$ is the total cost. p_o is the cost of system resources used by the single instance for each operator o. $n_{o \cdot t}$ is the instance number of operator o at time t. p_o^u is the startup-cost of a single o instance. p_o^d is the shutdown-cost of a single o instance. $C_{o \cdot t}^u(n)$ is the startup times of instances of o. $C_{o \cdot t}^d(n)$ is the stop times of instances of o. p_o^r is the cost of each re-routing. $C_{o \cdot t}^r(n)$ is o's re-routing times. To satisfy the end-to-end latency, we ensure that the processing capability of each operator is not less than the data input rate. In other word, $p_o \geq in_o$ at any time.

The processing capability of o is expressed as $p_o = \lambda_o n_o$, where λ_o is the capability of one instance of o. We know that $W_{o \cdot base}(\mathbf{n}) = \sum_{t_C}^{t_F} p_o n_{base}$, $\Delta W(\mathbf{n}) = W_o(\mathbf{n}) - W_{o \cdot base}(\mathbf{n})$ and $\Delta n_o = n_o - n_{base}$. The Cost Mode is defined as:

$$\min \quad \Delta W_o(\mathbf{n}) = p_o \sum_{t_C}^{t_F} \Delta n_{o \cdot t} + \sum_{t_C}^{t_F - 1} (p_o^u C_{o \cdot t}^u(n) + p_o^d C_{o \cdot t}^d(n) + p_o^r C_{o \cdot t}^r(n)) \tag{9}$$

$$\text{s.t.} \quad \Delta n_o \geq 0$$

Cost-Balance-Algorithm. As mentioned above, we get the cost expression ΔW_o for operator o. The aim of the optimization should be $min(\Delta W_o)$ with the constraint: $\Delta n_o \geq 0 (\forall o \in O, \forall t \in T)$. In order to address this issue, we propose the Cost-Balance-Algorithm (CBA). CBA improves the basic simple proactive elasticity algorithm by taking the cost of instance operations (e.g. re-routing and startup/shutdown) into account. CBA balances these three parts of the cost to guarantee lower system cost. CBA is divided into 3 steps. Firstly, we only consider computing capability to find the optimal scheduling timetable. Secondly, we refer the cost of re-routing to optimize timetable. At last, we refer startup and stop cost to optimize timetable. In CBA, the *act* denotes the result of the previous step and the input for the next step.

4   Experiments

4.1   Settings and Datasets

Settings. Our evaluations run on a cluster consisting of ten machines. These machines are all comprised of two eighteen-core Intel Xeon E5-2697 2.30 GHz CPUs, 256 GB memory, and 500 GB disks. One of the machines is used as Job-Manager and MetricDataBase, seven machines are used as TaskManagers, and the remaining machines with NVIDIA TESLA P4 GPUs are used to train and evaluate our prediction models.

Datasets. We collect the input rates records from our online DataDock system in 60 days as the dataset and you can download our dataset at https://github.com/alexmu/DSP-R-BGElasor. The dataset contains three type loads, which are stable load, periodic load and fluctuating load, as shown in Fig. 3. We divide them into three sets: a training set (from the beginning to the 40th day), a validation set (from the 40th day to the 50th day) and the test set (the 50th day to the last day). The training set is used to train prediction models, the validation set is used to optimize hyper-parameters and prevent overfitting, and the test set is used to evaluate the effectiveness of the prediction models.

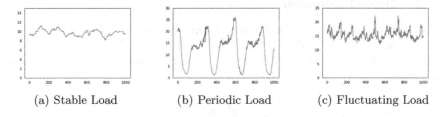

(a) Stable Load   (b) Periodic Load   (c) Fluctuating Load

Fig. 3. Three types of data stream load.

Algorithm 1. CBA

Step 1

1: **for** cur : seq : prediction.non-reducings **do**
2:    **if** cur < act[cur-1].max **then**
3:        act[cur-1].up(roundUp((cur- act[cur-1].max)/p))
4:    **end if**
5: **end for**
6: **for** cur : seq : prediction.reducings **do**
7:    **if** cur < act[cur].min **then**
8:        act[cur].down(takeDown((act[cur].max - cur)/p))
9:    **end if**
10: **end for**

Step 2

1: **for** cur : seq.reverse : prediction.non-reducings **do**
2:    **if** act[cur].hasAct **then**
3:        cost0 = p^r, cost1 = act[cur].upNum * t
4:        act.mv(cur → cur-1) **when** cost0 > cost1
5:    **end if**
6: **end for**
7: **for** cur : seq : prediction.reducings **do**
8:    **if** act[cur].hasAct **then**
9:        cost0 = act[cur].downNum * t, cost1 = p^r
10:        act.mv(cur → cur+1) **when** cost0 > cost1
11:    **end if**
12: **end for**

Step 3

1: act.setHasChanged()
2: **while** act.hasChanged **do**
3:    **for** (down, up, ΔT) $\in$ act.adjacency **do**
4:        min_num = min(down.num , up.num)
5:        cost0 = (p^d +p^u) * min_num + p^r
6:        cost1 = p * min_num * ΔT
7:        act[down, up].cancelAct(min_num) **when** cost0 > cost1
8:    **end for**
9: **end while**
10: **return** act

4.2   Predictor Evaluation

In this section, we compare the prediction performance of different algorithms, including ARIMA, SVR, LSTM, GRU, BiLSTM and BiGRU. We use the Root Mean Square Errors (RMSE) and Mean Absolute Errors (MAE) as the evaluation metrics.

$$RMSE = \sqrt{\frac{1}{n}\sum\nolimits_{i=1}^{n}(x_i - \hat{x}_i)^2} \text{ and } MAE = \frac{1}{n}\sum\nolimits_{i=1}^{n}|x_i - \hat{x}_i|$$

where x_i is the observed input rate, and $\hat{x}_i$ is the predicted input rate. All the compared models in this section are trained and tested multiple times to eliminate outliers, and the results of them presented are averaged to reduce random errors.

The experimental results are shown in Table 1. In our experiment, to balance the results of results, we repeat 10 times and get the results. For the stable load, the models all get low RMSE and MAE results. For periodic load, deep learning models significantly outperform other traditional models, such as ARIMA and SVR, because neural networks can learn more latent features from historical data. For the fluctuating load, the traditional linear function can not deal with the situation better, because the data load changes drastically. But the neural networks, which leverage nonlinear representations, can extract the hidden features from the time series data and detect the load change more accurately. And in our experiments, our model achieves the best performance.

Table 1. The RMSE and MAE of each models

Model	Load Type					
	Stable load		Periodic load		Fluctuate load	
	RMSE	MAE	RMSE	MAE	RMSE	MAE
ARIMA	0.207	0.156	1.252	0.888	1.001	0.713
SVR	0.163	0.127	0.789	0.589	0.720	0.504
LSTM	0.141	0.113	0.541	0.397	0.639	0.433
GRU	0.124	0.098	0.531	0.381	0.637	0.434
BiLSTM	0.119	0.094	0.529	0.385	0.624	0.426
BiGRU	**0.101**	**0.081**	**0.490**	**0.341**	**0.581**	**0.401**

4.3   ElasticityController Evaluation

We evaluate the ElasticityController from three aspects, the total cost, the adaptation frequency and the latency guarantee. We compare CBA with 2 other algorithms, the Standard Reactive Elasticity Algorithm (SREA) and the Simple Proactive Elasticity Algorithm (SPEA). SREA considers only the current load and adjusts the instance number reactively. SPEA considers the load of the current epoch and the next epoch.

Total Cost. In this part, we evaluate the total cost of the three algorithms. Firstly, we get the instance number of operators o at the same epoch with three algorithms. Then we use the data prediction result to generate the scheduling process. At last, we calculate the cost of each time and get the total cost of the experiment.

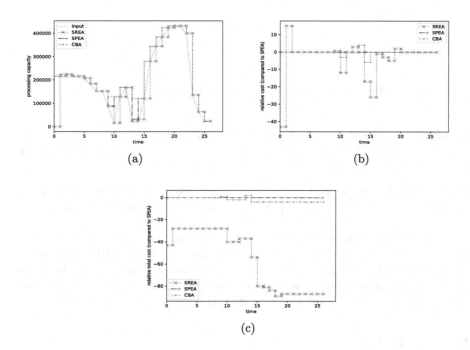

Fig. 4. Scheduling result

In Fig. 4, we set $p_o = 8000$, $p_o^u = 5000$, $p_o^d = 5000$ and $p_o^r = 100$. The Fig. 4(a) represents the processing capacity. The Fig. 4(b) represents the relative cost of each algorithm. The value is compared with the baseline of SPEA to get the relative cost. The Fig. 4(c) represents the total relative cost. The baseline is the same as Fig. 4(d).

When we consider both epoch 0 and 1, we find that the SPEA and CBA cost more than SREA because SPEA and CBA take the prediction into account. CBA considers the cost of the instance startup or shutdown and sometimes does not stop and start instances. Thus CBA costs less than SPEA.

Adaptation Frequency. In this part, we evaluate the adaptation frequency of the three algorithms. We collect the numbers of the instance number change at each epoch. The result is shown in Fig. 5(a).

SREA considers only the current load, so it starts or stops instances later than the others from epoch 1 to epoch 10. However, after the epoch 10, SREA starts and stops more instances because of the sudden input rate fluctuation. Compared

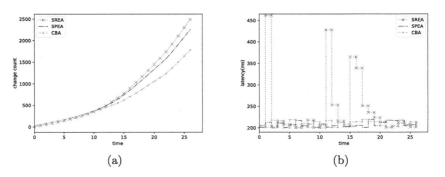

Fig. 5. Adaptation frequency and end-to-end latency

with SPEA, CBA takes the cost of instance startup and shutdown into account, so it is not necessary to deal with the instance startup and shutdown during the input rate fluctuation. So CBA changes less than SPEA.

End-to-end Latency Guarantee. In this part, we focus on the latency guarantee. We use the DataDock to compare three algorithms and record the end-to-end latency of each algorithm. The result is shown in Fig. 5(b).

CBA and SPEA take the prediction into account and start instances before input rate rises, so they deal with the data fast and stably. SREA only starts instances when the input rate is beyond the processing capacity. It results in the processing waiting everytime when input rate rises beyond processing capacity. So SREA is unstable. In our experiments, CBA outperforms the others.

5   Conclusion

In this paper, we propose a framework, BGElasor, contains two important modules, the Predictor and the ElasticityController. The Predictor based on BiGRU to get the precise prediction result of the input rate of each operator. Then we refer the results of prediction and use the ElasticityController to adjust the parallelism of operators. Experiments on the real load demonstrate our framework is better than the state-of-the-art methods, which not only improves the prediction accuracy with three kinds of the data loads, but also ensure the end-to-end latency on the QoS with the low cost.

References

1. Arasu, A., et al.: STREAM: the stanford stream data manager. IEEE Data Eng. Bull. **26**(1), 19–26 (2003)
2. Abadi, D.J., et al.: The design of the borealis stream processing engine. In: CIDR 2005, Second Biennial Conference on Innovative Data Systems Research, pp. 277–289 (2005)

3. Neumeyer, L., Robbins, B., Nair, A., Kesari, A.: S4: distributed stream computing platform. In: ICDMW 2010, The 10th IEEE International Conference on Data Mining Workshops, pp. 170–177 (2010)
4. "Storm." http://storm.apache.org/
5. Carbone, P., Katsifodimos, A., Ewen, S., Markl, V., Haridi, S., Tzoumas, K.: Apache flink^{TM}: stream and batch processing in a single engine. IEEE Data Eng. Bull. **38**(4), 28–38 (2015)
6. Fernandez, R.C., Migliavacca, M., Kalyvianaki, E., Pietzuch, P.R.: Integrating scale out and fault tolerance in stream processing using operator state management. In: Proceedings of the ACM SIGMOD International Conference on Management of Data, SIGMOD 2013, pp. 725–736 (2013)
7. Gulisano, V., Jiménez-Peris, R., Patiño-Martínez, M., Soriente, C., Valduriez, P.: Streamcloud: an elastic and scalable data streaming system. IEEE Trans. Parallel Distrib. Syst. **23**(12), 2351–2365 (2012)
8. Gedik, B., Schneider, S., Hirzel, M., Wu, K.: Elastic scaling for data stream processing. IEEE Trans. Parallel Distrib. Syst. **25**(6), 1447–1463 (2014)
9. Herbst, N.R., Huber, N., Kounev, S., Amrehn, E.: Self-adaptive workload classification and forecasting for proactive resource provisioning. In: ACM/SPEC International Conference on Performance Engineering, ICPE 2013, pp. 187–198 (2013)
10. Balkesen, C., Tatbul, N., Özsu, M.T.: Adaptive input admission and management for parallel stream processing. In: The 7th ACM International Conference on Distributed Event-Based Systems, DEBS 2013, pp. 15–26 (2013)
11. Zacheilas, N., Kalogeraki, V., Zygouras, N., Panagiotou, N., Gunopulos, D.: Elastic complex event processing exploiting prediction. In: 2015 IEEE International Conference on Big Data, Big Data 2015, pp. 213–222 (2015)
12. Repantis, T., Kalogeraki, V.: Hot-spot prediction and alleviation in distributed stream processing applications. In: The 38th Annual IEEE/IFIP International Conference on Dependable Systems and Networks, DSN 2008, pp. 346–355 (2008)
13. Hidalgo, N., Wladdimiro, D., Rosas, E.: Self-adaptive processing graph with operator fission for elastic stream processing. J. Syst. Softw. **127**, 205–216 (2017)
14. Xing, Y., Hwang, J., Çetintemel, U., Zdonik, S.B.: Providing resiliency to load variations in distributed stream processing. In: Proceedings of the 32nd International Conference on Very Large Data Bases, pp. 775–786 (2006)
15. Xing, Y., Zdonik, S.B., Hwang, J.: Dynamic load distribution in the borealis stream processor. In: Proceedings of the 21st International Conference on Data Engineering, ICDE 2005, pp. 791–802 (2005)

High Performance DDoS Attack Detection System Based on Distribution Statistics

Xia Xie[✉], Jinpeng Li, Xiaoyang Hu, Hai Jin, Hanhua Chen, Xiaojing Ma, and Hong Huang

National Engineering Research Center for Big Data Technology and System,
Services Computing Technology and System Lab,
Cluster and Grid Computing Lab, School of Computer Science and Technology,
Huazhong University of Science and Technology, Wuhan 430074, China
shelicy@hust.edu.cn

Abstract. Nowadays, web servers often face the threat of distributed denial of service attacks and their intrusion prevention systems cannot detect those attacks effectively. Many existing intrusion prevention systems detect attacks by the state of per-flow and current processing speed cannot fulfill the requirements of real-time detection due to the high speed traffic. In this paper, we propose a powerful system TreeSketchShield which can improve sketch data structure and detect attacks quickly. First, we discuss a novel structure TreeSketch to obtain statistics of network flow, which utilizes the stepped structure of binary tree to map the distribution and reduces the complexity of the statistic calculation. Second, we present a two-level detection scheme that could make a compromise between the detection speed and detection accuracy. Experimental results show that our method can process more than 100,000 records per second. The false alarm rate can achieve 2% to 25% performance improvement.

Keywords: DDoS attack · Intrusion prevention system ·
Sketch data structure · Real-time

1 Introduction

People have enjoyed numerous high-quality services when the Internet technologies develop rapidly. Indeed, *Distributed Denial of Service* (DDoS) attacks have been mentioned more and more frequently since it adds huge burden to Internet services. In a DDoS attack, legitimate users' access to information or network resource are discarded, because the server cannot afford such numerous requests generated by a huge amount of compromised computers. Theses compromised computers can be controlled by attackers and ordered to perform some malicious tasks unintentionally. With the advancement of modern technology and the complexity of networking environment, this tendency is becoming more serious than before.

© IFIP International Federation for Information Processing 2019
Published by Springer Nature Switzerland AG 2019
X. Tang et al. (Eds.): NPC 2019, LNCS 11783, pp. 132–142, 2019.
https://doi.org/10.1007/978-3-030-30709-7_11

According to the Kaspersky lab research, more than 1/3 organizations in the world have suffered DDoS attacks in 2017, compared to just 17% in 2016. To detect DDoS attacks, many intrusion detection and prevention systems have been proposed. However, these systems usually make a compromise between scalability and accuracy. For example, fine grain traffic monitoring can increase the detection accuracy, but cannot scale well. There are many other schemes to detect anomalies based on the statistics of the traffic state, such as entropy-computing [1], deep learning [2]. These methods have high accuracy in detection but the calculation is too heavy to be applied. Since the Internet traffics increase rapidly every year, it is foreseeable that monitoring numerous network traffics in real-time is becoming more and more difficult in anomaly detection. Though dimensionality reduction may be an effective method to process such huge amount of data, it requires complicated computation, and thus can not be applied on a large scale in real-time detection.

Recently, a sort of methods based on sketch have been raised to deal with anomaly detection. Sketch is an effective data structure, which is used to store a summary of a large data set for space efficiency. However, these methods are either very computationally intensive or requiring a large amount of storage capacity, which limits their application in intrusion prevention systems.

In this paper, we address a new system *TreeSketchShield*, which can process the statistics of network traffic and can support a two-level detection to defend against DDoS attacks. The remaining parts are listed as follows: in Sect. 2, we introduce the related works about how to prevent DDoS attacks; in Sect. 3, a brief overview of the detection is discussed; in Sect. 4, we introduce the architecture and implementation of our system. We evaluate the whole system in Sect. 5 and draw a conclusion in Sect. 6.

2   Related Work

The purpose of DDoS attacks is to make the Internet servers or network resource unavailables to its normal users. This is usually implemented by masquerading the normal flash crowd requests. Flash crowd means special situations many different users access a website at the same time, causing sudden huge access pressure on the website or database that could make the website inaccessible [3]. Based on these characteristics, most signature-based intrusion detection and prevention systems are difficult to effectively identify DDoS attacks. To distinguish the DDoS traffic and the normal flash crowds, researchers proposed a series of schemes. Xie et al. [4] observed that these attackers launched application-level DDoS attacks in the flow pattern similar to normal traffics. By using an access matrix, they could identify the spatial-temporal characteristics about the normal traffics. Besides that, a hidden semi-markov model was used to present the dynamics access pattern for detecting these DDoS attacks. Chonka et al. [5] proposed a model based on chaos theory to distinguish a normal traffic flow from the attack traffic flow. A novel system based on neural network was raised to detect anomalous traffic. In order to detect abnormal traffics, Rahmani et al. [6] utilized joint entropy to record the characteristics of the traffic flows. They used connection coherence to identify the links of packets and the quantity of connections for both normal and abnormal traffic flow. Their result showed that the aggregated traffic with a DDoS attack was nearly doubled compared with normal traffics.

However, these schemes only considered about how to find the DDoS traffic in a flash crowds and ignored the issue that the detection speed could not meet the requirement of the increasing network traffic sometime. When a DDoS attack occurs suddenly and violently, in order to minimize the loss and mitigate the impact on normal users' access, a security system is in need to quickly discover and defend against those attacks. The sketch data structure can reduce the dimension of multidimensional data streams. With efficiently estimating the initial signal [7,8], high-speed network links are very effective in detecting DDoS attacks with the sketch data structure when people deal with huge network traffic, especially under flooding attacks. Currently, there are many methods for detecting huge network traffic anomalies based on sketch [9–11].

Since sketches do not preserve the detailed information about the malicious hosts, we cannot use them to mitigate DDoS attacks. To solve this problem, Schweller et al. [12] proposed a reverse hashing scheme, which could be used to identify the keys of malicious flows from reversible sketches. Liu et al. [13] proposed an online DDoS detection scheme which adopted the sketch structure to cope with problems raised by DDoS attacks and used the distinction of IP addresses to pinpoint victims. Wang et al. [14] proposed an efficient system SkyShield to combine sketch data structure and special distance to detect DDoS attacks. Moreover, abnormal sketches are used to help identify malicious hosts with a DDoS attack.

These methods are based on the sketch data structure to make statistics, and they often use the distribution distance of the traffic attribution to detect the DDoS attacks. Even though these sketch-based schemes can well mitigate DDoS attacks, the high frequency of statistics cause not only high calculation cost but also the upping false alarm rate. Another problem needed to be solved is to meet the requirement of reducing computational complexity and decrease the time interval of detection.

Because the internet traffic flow grows rapidly, it is difficult to meet the challenge that DDoS detection technologies should handle requests as much as possible within affordable response time. Moreover, to achieve real-time detection, the time of calculating statistics needs to be set very short and this could result in the increasing of false alarm rate due to the insufficient statistic of network traffic.

3   Background

In this section, we give a brief overview of the detection based on the sketch data structure.

The process of the detection based on the sketch data structure is depicted in Fig. 1. First, the intput streaming passes through the filter layer. In the filter layer, a bloom filter is used to filter incoming requests. Malicious requests identified by the blacklist will be filtered and recorded. This can be explained by the fact that a large number of requests is needed to initiate a valid DDoS attack. Therefore, sketch-based detection can identify malicious hosts using the volume of the malicious requests. Second, the detection employs the divergence between $Sketch1$ and $Sketch2$ to signal an abnormal situation that are raised by a large number of requests from malicious hosts. In this step, sketch-based detection conducts the detection cyclically with a fixed time interval ΔT. During a detection cycle, any access received by this system will be aggregated into $Sketch1$, and their IP addresses are set as the input keys. Another $Sketch2$ is used to store the results of $Sketch1$ in the last normal mode. Finally, at the end of each detection cycle, the divergence $d(Sketch1, Sketch2)$ is calculated and compared to a threshold θ_t. If $d(Sketch1, Sketch2)$ exceeds θ_t, the system is considered to be suffering a DDoS attack and an alarm needs to be raised.

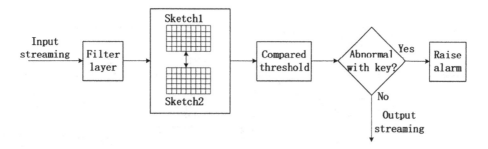

Fig. 1. Overview of sketch-based detection

Sketch-based schemes are based on the statistic of network traffic to detect DDoS attacks. However, the statistical time interval of each detection cycle cannot be set too short. On the one hand, the statistic of distribution will become incomplete due to the insufficiency of the time length, which leads high false alarm rate. The high frequency of statistics generate high calculation costs and cause long time delay. For above reasons, the speed of sketch-based detection cannot meet the requirement of real-time monitoring.

Besides sketch-based detection, other methods are also discussed. For example, the bloom filter can be used to add the requests from malicious hosts into the blacklist, which is implemented by its special data structure. Sketch is another data structure that can efficiently compute raw signals by reducing high dimensional data streams to low dimensions. We can deal with the statistics of traffic and then detect DDoS attacks by it. When DDoS attack occurs, the distribution of bucket values in a sketch is unstable compared with the normal situation due to the steadiness of the normal network traffic. Therefore, we can use the divergence between $Sketch1$ and $Sketch2$ to detect DDoS attacks efficiently.

4   System Design and Implementation

In this section, we introduce the structure of *TreeSketchShield* and describe the difference with sketch-based detection, then we explain how to implement it.

4.1   Process of TreeSketchShield

TreeSketchShield is depicted in Fig. 2. Similar to the sketch-based detection, the intput streaming first goes through the filter layer. But in the detection phase, the detection process is divided into the coarse grain detection and fine grain detection. The first one is used to quickly detect whether there is an abnormal traffic, while the last one determines whether the abnormal traffic is a real DDoS attack traffic. The principle of this scheme is that short statistical interval in coarse grain detection can reduce the computation time and the long statistical interval of fine grain detection can decrease the rate of false alarm.

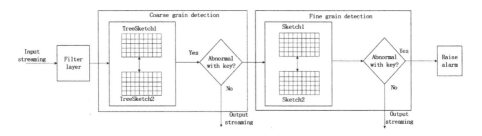

Fig. 2. Process of TreeSketchShield

In the step of coarse grain detection, a novel structure *TreeSketch* is employed, and the statistical time interval of detection cycle is set very short. All requests received and their corresponding IP addresses are saved as key-value pairs, which will be aggregated into $TreeSketch1$. Then $TreeSketch2$ is ready to store the results generated in $TreeSketch1$ during the last normal mode. As soon as the detection cycle is end, the divergence $d(TreeSketch1, TreeSketch2)$ will be calculated and compared to a threshold θ_t. If $d(TreeSketch1, TreeSketch2)$ exceeds θ_t, the fine grain detection will start. In the step of fine grain detection, the time interval of detection cycle is set longer and the fine grain detection uses the sketch data structure instead of the *TreeSketch* to deal with the statistic. If the distance exceeds the value of threshold, the alarm will be raised. After the detection, *TreeSketchShield* uses the volume of all buckets in fine grain detection to identify malicious hosts. Compared to the sketch-based detection, a special two-level detection and a novel structure are adopted to decrease the false alarm rate and improve the detection speed in $TreeSketchShield$.

4.2   TreeSketch

The *TreeSketch* data structure is shown in Fig. 3. *TreeSketch* is a special data structure with H rows of size K, while the source data stream is composed with key-value pairs. For every row in the *TreeSketch*, there is a binary tree and their leaf nodes are associated with different hash functions. When a key-value pair comes, the value will be added into the leaf nodes corresponding to the key. Besides, values in child nodes will be added up and stored in their parent nodes.

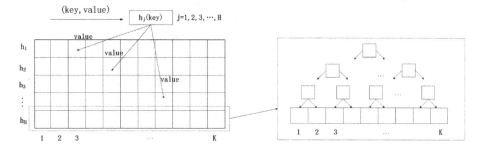

Fig. 3. TreeSketch data structure

In the sketch data structure, the attribute change is mapped to the sketch which represents the compact summaries of a data stream. The attribute is signed by each row of the sketch. It will result in a high computational cost. To avoid this problem, we use the trapezoidal data structure to map the attribute which is based on the *TreeSketch*. Different from the divergence calculation between sketches, when computing the divergence of TreeSketches, we do not need to calculate each button in every row of the *TreeSketch* in order. Denoting the attribution of i-th row in the *TreeSketch* as a vector $\langle n_{i1}, n_{i2}, ..., n_{iK} \rangle$, each row of *TreeSketch* is a binary tree and n_{ij} presents the first node in j-th layer of the binary tree. Let $N_i = \sum_{j=1}^{K} n_{ij}$ represents all the requests received. Denote $P_i = \langle p_{i1}, p_{i2}, ..., p_{iK} \rangle$ for the corresponding row, where $p_{ij} = n_{ij}/N_i$ means the probability that an incoming request is mapped into the j-th bucket of the i-th vector. Denote $Q_i = \langle q_{i1}, q_{i2}, ..., q_{iK} \rangle$ are the probability similar to P_i, then the distance $d(P_i, Q_i)$ can be calculated. The metric is that when the server cannot afford these numerous requests, the clients will receive little responses from the server, which could bring about the traffic attribute change. As shown in Fig. 3, each layer of an row in *TreeSketch* contains the next estimation of distribution. If there is any abnormity in the leaf layer, it will cause the upper node to be abnormal and the abnormality will continuously present to the root node.

4.3   Cycle Synchronization of Detection

In the detection of *TreeSketchShield*, the difference of statistical time between coarse grain detection and fine grain detection leads to the out-of-synchronization problem at both detection points. As shown in Fig. 4, when a cycle of coarse grain detection is finished and an alarm is raised, a cycle of fine grain detection is still going on. To solve this problem, we employ a sliding time window scheme to keep the pace between two detections. Figure 5 shows that in the fine grain detection, the interval time of each cycle is unchanged but the sliding distance of each cycle is set to a short time. This scheme makes the detection combine the sufficient statistic and the speedy computation, which decreases the false alarm rate and reaches the real-time detection at the same time.

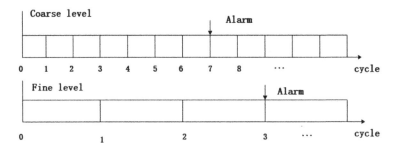

Fig. 4. Cycle of the two-level detection with different interval time

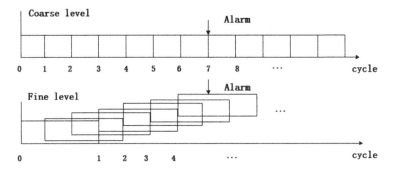

Fig. 5. Cycle of the two-level detection with sliding time window

5   Evaluation

In this section, we design a comparative experiment to evaluate the performance of *TreeSketchShield* about the detection speed and detection accuracy between *TreeSketchShield* and sketch-based detection. The experiments are tested on a local machine equipped with 2.5 GHz I5-7300HQ CPU, 24 GB RAM.

Table 1. Summary of datasets

Dataset	Requests	Hosts	Attacks
Dataset0611	4,315,622	5,400	40
Dataset0630	5,332,901	5,880	40
Dataset0710	5,834,324	5,563	45

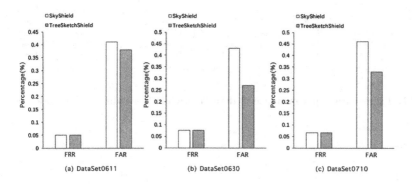

Fig. 6. The accuracy performance of detection while the sliding time window is 5 s

5.1 Datasets

At first, we introduce these datasets used in this experiment. These datasets are generated from WordCup98 [15], which records all requests from April 30, 1998 to July 26, 1998. The website received nearly 1.3 billion requests during this time. Each dataset in this experiment is composed of access logs of two days in the WordCup98 dataset. To assess the performance of the *TreeSketchShield*, the data from 1998/06/11 to 1998/06/12 is denoted as Dataset0611, the data from 1998/06/30 to 1998/07/01 as Dataset0630, and the data from 1998/07/10 to 1998/07/11 as Dataset0710. Table 1 presents a brief summary of the datasets.

5.2 Performance

We employ the *False Rejection Rate* (FRR) and the *False Acceptance Rate* (FAR) to evaluate the performance of DDoS attacks detection. FAR measures the proportion of normal requests that are mistakenly identified as DDoS attacks in all requests, and FRR is the probability of wrongly identifying normal requests as DDoS attacks in all attacks. We make a comparison between the performance of sketch-based detection and *TreeSketchShield* on three datasets, with the parameters set as: $\alpha = 0.3, \beta = 0.4, \lambda = 3, k = 16384, H = 8$.

Figure 6 shows the FRR and the FAR on the three datasets with the sliding time window set as 5 s. In Dataset0611, *TreeSketchShield* is equal to SkyShield in terms of FRR, which is 5%, but the FAR has decreased from 40% to 38%. In Dataset0630, *TreeSketchShield*'s FRR is 7.5%, the same as SkyShield's, and the

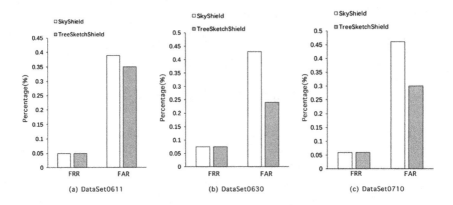

Fig. 7. The accuracy performance of detection while the sliding time window is 10 s

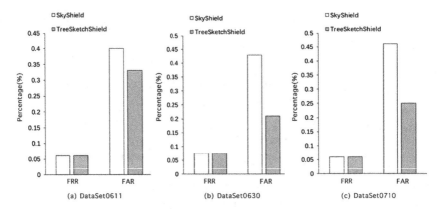

Fig. 8. The accuracy performance of detection while the sliding time window is 15 s

FAR is much lower than that of SkyShield, from 44% to 27%. In Dataset0710, the FRR of both methods is 6.7%, but the *TreeSketchShield*'s FAR dropped from 46% to 33%. This is because the *TreeSketch* data structure decreases the sensitivity of attribute detection. The result shows that *TreeSketchShield* is better than sketch-based detection in the accuracy of DDoS attack detection.

Figures 7 and 8 shows the results when the sliding time window is set as 10 s and 15 s respectively. According to the experimental results of these three datasets, the *TreeSketchShield* system reduces the FAR by 5%−20% on the basis of ensuring the FRR.

We also employ the number of requests per second to evaluate the throughput of detection. Figure 9 shows the performance of sketch-based detection and *TreeSketchShield* on three datasets. We can see the throughput of *TreeSketchShield* is more than 100,000 records per second. Moreover, compared with sketch-based detection, the average throughput of *TreeSketchShield* is 8 times faster in

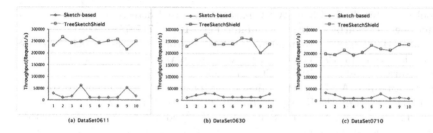

Fig. 9. The throughput performance of detection

Dataset0611, 10 times faster in Dataset0630, and 10 times faster in Dataset0710 respectively. The result shows that *TreeSketchShield* is better than sketch-based detection in processing speed.

6   Conclusion and Future Work

In this paper, we address a new defense system *TreeSketchShield*, which can detect DDoS attacks quickly. First, a novel structure *TreeSketch* is addressed, which can greatly reduce the complexity of statistical calculations of network flow. Then a two-level detection scheme is designed to decrease the false alarm rate. *TreeSketchShield* still needs to be improved: first, although the FAR is decreased, it is still unacceptable. The reason is that the statistic of *TreeSketchShield* adopts the partial substitution strategy, which reduces the sensitivity of detection; Second, *TreeSketchShield* increases the memory consumption when obtaining statistics of net flow. In the future, we will continue to improve the structure to eliminate the rate of missing DDoS alarm and decrease the memory consumption of detection.

Acknowledgements. This work is supported in part by the National Key Research and Development Program of China under grant No. 2016QY02D0302, the Fundamental Research Funds for the Central Universities (HUST No. 3020210111).

References

1. Osanaiye, O., Choo, K.K.R., Dlodlo, M.: Distributed denial of service (DDoS) resilience in cloud: review and conceptual cloud DDoS mitigation framework. J. Netw. Comput. Appl. **67**, 147–165 (2016)
2. Zargar, S.T., Joshi, J., Tipper, D.: A survey of defense mechanisms against distributed denial of service (DDoS) flooding attacks. IEEE Commun. Surv. Tutor. **15**(4), 2046–2069 (2013)
3. Yu, S., Zhou, W., Jia, W., Guo, S., Xiang, Y., Tang, F.: Discriminating DDoS attacks from flash crowds using flow correlation coefficient. IEEE Trans. Parallel Distrib. Syst. **23**(6), 1073–1080 (2012)
4. Xie, Y., Yu, S.: Monitoring the application-layer DDoS attacks for popular websites. IEEE/ACM Trans. Netw. **17**(1), 15–25 (2009)

5. Chonka, A., Singh, J., Zhou, W.: Chaos theory based detection against network mimicking DDoS attacks. IEEE Commun. Lett. **13**(9), 717–719 (2009)
6. Rahmani, H., Sahli, N., Kammoun, F.: Joint entropy analysis model for DDoS attack detection. In: Proceedings of the 5th International Conference on Information Assurance and Security, pp. 267–271 (2009)
7. Ben, U., Bremler, A., Levy, H.: Vulnerability of network mechanisms to sophisticated DDoS attacks. IEEE Trans. Comput. **62**(5), 1031–1043 (2013)
8. Tang, J., Cheng, Y., Hao, Y., Song, W.: SIP flooding attack detection with a multi-dimensional sketch design. IEEE Trans. Dependable Secur. Comput. **11**(6), 582–595 (2014)
9. Liu, Y., Chen, W., Guan, Y.: A fast sketch for aggregate queries over high-speed network traffic. In: Proceedings of the IEEE International Conference on Computer Communications, pp. 2741–2745 (2012)
10. Gangam, S., Sharma, P., Fahmy, S.: Pegasus: precision hunting for icebergs and anomalies in network flows. In: Proceedings of the IEEE International Conference on Computer Communications, pp. 1420–1428 (2013)
11. Wang, P., Guan, X., Zhao, J., Tao, J., Qin, T.: A new sketch method for measuring host connection degree distribution. IEEE Trans. Inf. Forensics Secur. **9**(6), 948–960 (2014)
12. Schweller, R., et al.: Reverse hashing for high-speed network monitoring: algorithms, evaluation, and applications. In: Proceedings of the IEEE International Conference on Computer Communications, pp. 1–12 (2006)
13. Liu, H., Sun, Y., Kim, M.: Fine-grained DDoS detection scheme based on bidirectional count sketch. In: Proceedings of the 20th International Conference on Computer Communications and Networks, pp. 1–6 (2011)
14. Wang, C., Miu, T.N., Luo, X., Wang, J.: SkyShield: a sketch-based defense system against application layer DDoS attacks. IEEE Trans. Inf. Forensics Secur. **13**(3), 559–573 (2018)
15. Worldcup98 (2016). http://ita.ee.lbl.gov/html/contrib/WorldCup.html

DDP-B: A Distributed Dynamic Parallel Framework for Meta-genomics Binary Similarity

Mengxian Chi$^{(\boxtimes)}$, Xu Jin$^{(\boxtimes)}$, Feng Li, and Hong An$^{(\boxtimes)}$

University of Science and Technology of China, Hefei, China
{mxchi10,jinxu,fli186}@mail.ustc.edu.cn, han@ustc.edu.cn

Abstract. Great efforts have been made on meta-genomics in the field of new species exploration in the past decades. With the development of next-generation sequencing technology, meta-genomics datasets have been produced as large as dozens of hundreds of gigabytes or even several terabytes, which brings a severe challenge to data analysis. Besides, conventional meta-genomics comparing algorithms may not take full advantage of powerful computing capacity from parallel computing techniques due to lack of parallelism. In this paper, we propose DDP-B, a distributed dynamic parallel framework for meta-genomics binary similarity analysis, to overcome these limitations. In this framework, we introduce a binary distance algorithm for meta-genomics similarity measurement and develop different levels of parallel granularity of the algorithm utilizing MPI, OpenMP, and SIMD techniques. Moreover, we establish a dynamic scheduling method to deliver asynchronous parallel computing tasks and design a distributed cluster to deploy the dynamic parallel system, which completes 2.97K pairs of meta-genomics vectors comparison per second and achieves an 134.79x speedup versus the baseline in the optimal condition. Our framework shows stable scalability when assigned larger workloads.

Keywords: Meta-genomics · Big data · Parallel computing · Binary distance · Dynamic scheduling · Distributed scalability

1 Introduction

Great efforts have been made on exploring new species in the last several decades since the Woese significant work [23]. Meta-genomics [25], which involves the total DNA sequences extracted directly from the natural environment (e.g. ocean, soil, and the human body) samples, occasionally preserves the molecular signatures of potential unexplored or undiscovered microorganisms [24].

The work is supported by the National Key Research and Development Program of China (Grants No. 2016YFB1000403).

© IFIP International Federation for Information Processing 2019
Published by Springer Nature Switzerland AG 2019
X. Tang et al. (Eds.): NPC 2019, LNCS 11783, pp. 143–155, 2019.
https://doi.org/10.1007/978-3-030-30709-7_12

Many novel techniques such as cultivation-independent shotgun genomics and next-generation sequencing [16] have been widely applied in this domain and dramatically aggrandized the scale of sequencing data as well as the speed of genome sequencing. As a result, meta-genomics datasets could be as large as dozens of hundreds of gigabytes or even several terabytes, which makes it a typical big data problem.

On the other hand, the rise of large scale computing clusters brings huge opportunities for meta-genomics research [2]. High-performance clusters support parallel computing and scalable architectures at multiple levels, which delivers extraordinary powerful performance theoretically. However, most of the conventional meta-genomics similarity algorithms barely take the most of high-performance parallel computing because there is a lack of excavation and utilization of their parallelism and scalability, which turns out the major limitations to deploy the algorithms on high-performance clusters. Moreover, the vast data scale brings a severe challenge to data storage and transmission in the field of high-performance computing.

Recently, binary distance measurements have been comprehensively applied in the field of biology [11], ethnology [6], and taxonomy [21]. Furthermore, genome sequence compressing methods (such as Hash map [18]) have contributed significantly to reducing the scale of meta-genomics datasets and enabling efficient search of massive sequences collections. Genome sequence data could be converted from character strings into binary vectors by the Hash map. As a result, we can measure the genome similarity through binary distance methods [5].

In this paper, we introduce a binary distance coefficient based comparing algorithm to measure meta-genomics similarity. The binary vectors generated from meta-genomics sequences is still too long (even more than 10^8 bits) to calculate the binary distance coefficient straightforward. To develop the algorithm efficiently, we divide a whole binary vector into 64-bit sub-sequences and process the calculation with Intel intrinsic instructions [15], which is easy to be paralleled as an atomic operation. Besides, we develop the binary similarity algorithm with hierarchical parallelism taking advantage of multiple parallel techniques such as SIMD [13], OpenMP [3], and MPI [10]. The hybrid parallel optimization delivers an 87.9x speedup compared with the original baseline.

Moreover, the data loading procedure is difficult to accelerate because of the memory read/write speed limitation, which constrains further optimization of the algorithm. And with the growth of data size, how to balance the workloads among large scale distributed clusters becomes a huge challenge [20]. To overcome these challenges, we design a dynamic scheduling system based on a master-slave structure and deploy it on a 9-node cluster. The scheduler distributes parallel computing tasks to the unoccupied worker nodes dynamically so that the communication and computation are decoupled and the data loading time is overlapped with computing time. As a result, we achieve a 7.5x speedup with 8 worker nodes under basic workload, which is close to linear acceleration. Meanwhile, we design a grouping strategy to organize worker nodes into extensive worker groups based on the workloads. Every worker group will be reorganized automatically if the workload exceeds its capacity. We achieve an extra speedup benefit under 4 times of basic workload (15.55x versus 9.27x), which exhibits our framework having stable scalability under larger workloads.

With all the above contributions, we propose DDP-B, a distributed dynamic framework taking advantage of multiple parallel levels for a binary similarity algorithm. We deliver 2.97K pairs of meta-genomics similarity comparison per second and achieve an 134.8x speedup overall in the optimal condition using this framework.

The rest of the paper is organized as follows. Section 2 introduces related work as well as the background of our research. Section 3 explains our methodologies in detail. Section 4 presents the implementation and experimental results. Section 5 delivers a conclusion of whole work and discusses future research.

2   Background

In this section, we introduce some related techniques concerning genome comparing algorithms.

2.1   Genome Sequences Alignment

The next-generation sequencing techniques usually gather massively short genome reads and then align them into longer reads. Experimental evidence shows that a whole chromosome sequence usually covers millions to billions of base pairs [7,22]. Though it is difficult to compare the chromosomes from end to end because we can hardly find the exact beginning of them, there is a requirement to investigate the similarity between whole long genome reads. The main reason is that quite a few similar basic genome functional units may be carried by the chromosomes of many different organisms. So it is not easy to figure out whether the short reads belong to different species.

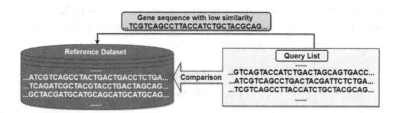

Fig. 1. Meta-genomics compares the new collected query genome list with the reference datasets, and extracts the genome sequences with low similarity to investigate whether there is a possibility of unknown species existing.

As meta-genomics sequences are captured from the environment randomly, it is almost impossible to extract every single microorganism's information through biological techniques. However, computer-aided analysis technology provides a feasible method for the study of meta-genomics. Taking advantage of the redundant and overlapping information generated by genome sequencing techniques, genome fragments (also mentioned as *reads* by biologists) can be assembled into longer reads (also called *contigs*) and finally spliced into a chromosome sequence [14]. Consequently, meta-genomics research would be transformed into sequence alignment tasks [19]. We can speculate on the possibility of the existence of unknown organisms based on the results of sequence alignment. Potential undiscovered species information will be dug out through genome sequence compare processing if there exists a quite different sequence compared with every known reference sequence. Figure 1 shows the meta-genomics comparing processing.

2.2 K-mer, Hash Map, and Binary Distance

The conventional comparing algorithms designed to quantitatively evaluate the similarity among meta-genomics based on computing system usually regard genome sequences as character strings (i.e. representing DNA's four bases with 'A', 'G', 'C', and 'T' and assembling them into strings in order.) and compare these strings to measure the similarities among genomes. Therefore, numerous algorithms of string similarity comparison have been applied on meta-genomics, which can be roughly divided into two categories: exact matching (such as Boyer Moore Algorithm and Shift Or Algorithm [4]) and approximate matching (such as Edit Distance and Haiming Distance [17]).

Except for the naive sequence comparing algorithm, there are also many other distinguished methods obtaining remarkable achievements [4,17]. Among them, K-mer similarity [1] is widely applied in bioinformatics which generates k-length sub-sequences of a long read step by step, therefore, we could just compare the much shorter K-mers. Although an L-length read still produces $L-k+1$ K-mers, we could focus on the distinct K-mers, so that the scale of the datasets will be reduced appreciably.

Besides, K-mers of a whole genome sequence can be mapped into a binary vector using Hash map algorithms. Figure 2 shows the detail of K-mers Hashing conversion. The vector's i-th position will be set to 1 only if the Hash value of the K-mer equals to i, which is expressed as

$$Vec[i] = \begin{cases} 1, & Hash(K\text{-}mer) = i \\ 0, & others \end{cases} \tag{1}$$

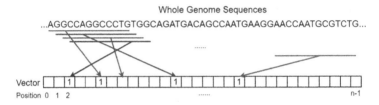

Whole Genome Sequences

...AGGCCAGGCCCTGTGGCAGATGACAGCCAATGAAGGAACCAATGCGTCTG...

Fig. 2. Convert K-mers into binary vector. The vector's i-th position will be set to 1 only if the Hash value of the K-mer equals to i.

Therefore, the meta-genomics similarity problem is transformed into a binary distance problem. Choosing the appropriate Hash function is not within the scope of this paper. We focus the research on how to take advantage of the parallelism and the scalability of the binary distance algorithm.

3   Methodologies

3.1   Binary Distance Coefficient

The binary distance method can be regarded as an approximate matching algorithm and applied on the domain of meta-genomics with two major advantages: (1) the inaccuracy results are almost surely generated during the genome sequencing procedure limited by the transcription properties of genetic information, so that the approximate methods can provide sufficient effectiveness; (2) data size and computational complexity are likely to be obviously reduced so that the approximate algorithms usually perform more efficiently. In the field of binary distance research, the Jaccard coefficient [12] is one of the most famous measurements and the Forbes coefficient [8] is proposed for clustering ecologically related species especially, so that the Forbes coefficient could reveal the genome similarity quantitatively. In this paper, we define a modified Forbes-II coefficient to measure the similarity score between the genome binary vectors.

$$S_{M-F_{II}} = \frac{na - (a + b)(a + c)}{(a + b)^2 + (a + c)^2 - (a + b)(a + c)} \tag{2}$$

where the definitions of n, a, b, c, and d are referred to Table 1. The coefficient $S_{M-F_{II}}$ is only related to the parameters $n, a, (a + b), (a + c)$. In other words, the binary distance between two genome vectors depends on the length of two vectors, the number of bit set to 1 (abbreviated as bit-1) inside the bit-wise logic AND result from two vectors, and the number of bit-1 inside each vector respectively.

Table 1. Vec 1, 2 are two n-length binary vectors, a is the number of attributes where the values of Vec 1 and Vec 2 are both 1, b is the number of attributes where the value of Vec 1 and Vec 2 is (0,1), c is the number of attributes where the value of Vec 1 and Vec 2 is (1,0), and d is the number of attributes where both Vec 1 and Vec 2 have 0.

&		Vec 1		
		1	0	Sum
Vec 2	1	a	b	a+b
	0	c	d	c+d
	Sum	a+c	b+d	n = a+b+c+d

Algorithm 1. Binary Coefficient Calculating Parallel Hierarchy

---
Phase 0:
Receive $Query_{[p_i:p_j]}[0:n]$, Allocate $QueryBit_{[p_i:p_j]} \leftarrow \{0\}$
#pragma omp parallel
for m in $[p_i:p_j]$ **do**
  #pragma simd
  $QueryBit_{[m]} \mathrel{+}= popcnt(Query_{[m]}[0:n])$
end for
Phase 1:   Send Ready Signal
Phase 2:
Receive $RefVec_{[q_i]}[0:n]$, Allocate $RefBit_{[q_i]} \leftarrow 0$, $AndBit_{[q_i]}[p_i:p_j] \leftarrow \{0\}$
Allocate $S_{M-F_{II}[q_i][p_i:p_j]} \leftarrow \{0\}$
#pragma simd
$RefBit_{[q_i]} \mathrel{+}= popcnt(RefVec_{[q_i]}[0:n])$
#pragma omp parallel
for m in $[p_i:p_j]$ **do**
  #pragma simd
  $AndBit_{[q_i]}[m] \mathrel{+}= popcnt(Query_{[m]}[0:n]\ \&\ RefVec_{[q_i]}[0:n])$
end for
Calculate $S_{M-F_{II}[q_i][p_i:p_j]}$
Phase 3: Send $S_{M-F_{II}[q_i][p_i:p_j]}$, Ready Signal

3.2   Parallel Hierarchy Design

Therefore, how to count the number of bit-1 inside a binary vector becomes the first challenge. Here we evaluate three counting methods: *left shift*, *look-up table*, and *popcnt*. The left shift method shifts the vector to the left continuously and counts the number of bit-1 according to the sign bit. The look-up table means preparing a table consisting of the number of bit-1 in advance and looking for the exact number based on the binary vector's decimal form. Both of them are inefficient because they are both at a computation complexity of $O(n)$ for an n-bit vector. Besides, the look-up table method requires an extra $O(2^n)$ memory consumption.

Intrinsic Instruction Optimization. Fortunately, Intel intrinsic instructions provides an operator to count the number of bit-1 inside a 64-bit unsigned integer (named *popcnt*) bound with an assembly instruction. So that the theoretical computation complexity of this operation is $O(1)$ for a 64-bit vector. We have measured all the above methods and find that the performance of the popcnt is at least 2x faster than the other two methods. Moreover, popcnt is easy to parallelize as an atomic operation. Therefore, we use the popcnt operator to calculate the number of bit-1 inside vectors.

Data-Level Parallel. Since the length of binary genome vectors usually exceeds the capacity of the popcnt operator (64-bit), we need to separate a whole vector into several 64-bit sub-vectors. Here we utilize the SIMD (Single Instruction Multiple Data) vectoring techniques to deal with two 64-bit binary sub-vectors simultaneously as the Intel AVX (Advanced Vector Extensions) supplies 128-bit registers. This method could deliver a 2x speedup theoretically.

Thread-Level Parallel. When comparing long-winded binary vectors, we allocate the popcnt and bit-wise AND operations to multiple threads uniformly as the two operations are both regular and aligned. Here we apply compiler directives of OpenMP (Open Multi-processing) to provide multi-threading parallel with a portable, scalable model. Meanwhile, we fork and synchronize the threads dynamically to balance the workloads among threads.

Overall, we accumulate each parameter and organize the binary distance coefficient computation through different levels of parallel granularity. Algorithm 1 shows the detail of the parallel hierarchy to calculate the distance coefficient.

3.3 Distributed Dynamic Schedule Design

We construct the dynamic scheduling system based on two major components in this system: Master Node and Slave Nodes. The communication interface between the master node and the slave nodes is established by MPI (Message Passing Interface). Figure 3 shows the distributed scalable dynamic programming architecture and exhibits the detail of control flow. And we will describe the behavior of the master node detailedly in the following content.

Master Node. It is the core module of our distributed dynamic system. The master node is responsible for managing all worker groups, fetching binary vectors from reference and query lists, broadcasting vectors to targeted worker

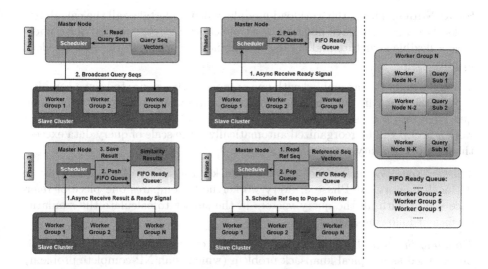

Fig. 3. Distributed scalable dynamic programming. The left part shows the data flow path of dynamic scheduling. The right part shows the worker group structure and FIFO ready queue.

groups, and gather all comparing results from them. We design two kernels for the master node: *Ready Queue* and *Scheduler*.

Ready Queue is a two-way first-in-first-out (FIFO) queue, preserving the standby state information of worker groups. A new ready worker group will be pushed back to the end of the ready queue, while an assigning worker group will be popped up from the top.

Scheduler is the controller unit and decides all behaviors of the master node. The control flow can be divided into four asynchronous phases as below. The scheduler will scan through these phases until all the comparing tasks distributed and all calculating results gathered.

– Phase 0: Scheduler fetches the query vectors while the query list is not empty and then broadcasts them to all worker groups. Every worker group will keep a complete copy of the query list.
– Phase 1: Every worker group sends a ready signal to scheduler asynchronously after receiving all the query vectors. Scheduler pushes the worker group ID into the end of the FIFO ready queue once receives the response signal from the worker.
– Phase 2: Scheduler pops a worker group ID from the ready queue and distributes one reference vector to the popped worker group while the reference list is not empty.
– Phase 3: After finishing the comparing task, the worker group sends the similarity results as well as a ready signal back to the scheduler. The scheduler will save the results with a tag and push the ready worker into the queue again.

Slave Nodes. This is the workload module which undertakes all the calculating tasks. We construct slave nodes as a set of scalable *worker groups*. Slave nodes have a flexible structure and can be recombined according to the real workload in practice.

Worker Group. It is assembled with one or more worker nodes to keep a complete copy of the query list with all memory consumption. And we separate the whole query list into k subsets if there are k worker nodes in one worker group. Every worker group will be reorganized automatically if the scale of query data exceeds this group's capacity.

Worker Node. We make each worker node keep a subset of the query list in its local memory. At Phase 3, every worker node compares one piece of reference vector with all the query vectors in the subset to calculate the similarity coefficient following Algorithm 1.

Grouping Strategy. Distributed heterogeneous clusters grouping strategy could be regarded as a typical knapsack problem (which is an NP-complete problem), so we will leave it for further research. For simplicity of implementation, in this paper we adopt a *naive grouping strategy* to organize worker nodes, i.e. we will add worker node one by one from scratch into a worker group until the worker group's capacity is enough to hold a whole copy of query list.

4   Experiment Results

4.1   Implementation

In order to deploy our framework, we adopt a 9-nodes cluster to build a dynamic system with 1 master node and 8 worker nodes. There are two CPU sockets on each node, 8 physical cores with hyper-threading enabled in each CPU. Every worker group has one worker node in default. The master node is connected with every worker node by Infiniband. Other hardware and software information is provided in Table 2.

Table 2. Hardware and software information

Item	Description
CPU	Intel(R) Xeon(R) E5-2660 @ 2.2 GHz * 2
Memory	DDR3 1333 MHz 96 GB
Hard disk	SAS HDD 300 GB
Network	Intel Ethernet Adapter I350 with 1 Gb/s
Connection	Mellanox QDR Infiniband 40 Gb/s
Operating system	CentOS 7.2-1511 Linux 3.10.0
MPI	Intel MPI 2017.0.098

The binary vectors are aligned at 3×10^8-bit length, which is supplied by DOE Joint Genome Institute [9]. As mentioned in Sect. 3.2, we separate each vector into 4.69M 64-bit sub-vectors. So one $S_{M-F_{II}}$ coefficient computation requires about 4.69M bit-wise ANDs, 14.06M popcnts, and 18.75M accumulations (37.5M operations altogether). We set both query list and reference list to 203 vectors so that the traverse comparison requires 1.55T operations. Every experiment is repeated three times and adopted the average results to avoid environmental instability. The serial computing baseline completes 22.07 coefficients calculating per second.

4.2   Performance Analysis

We summarize the runtime performance of the distributed dynamic programming system in a parallel hierarchy.

Multi Threads. Figure 4 shows the multi-threading performance optimized on a single node. We observe that the speedup curve has a sub-linear growth trend when threads number goes from one to eight. And then, it will convergence and even suffer performance damage after that. As mentioned above, every CPU has eight physical cores. Hence CPU may be overloaded when forked into more than eight threads which increases the overhead of thread switching. The multithreading delivers 4.2x speedup with 16 threads compared with baseline.

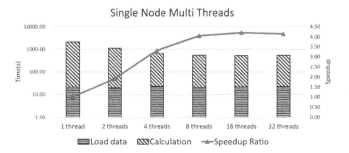

Fig. 4. Single Node Performance. The bar chart uses exponential coordinates for clarity.

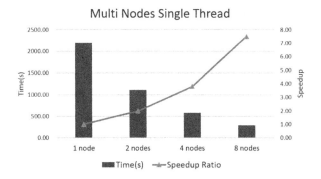

Fig. 5. Multi Nodes Single Thread Performance. The master node is not counted.

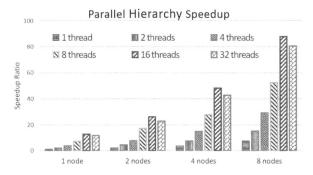

Fig. 6. Hybrid Parallel Performance. The bar chart shows the speedup ratio compared with single node single thread baseline.

Multi Nodes. Moreover, we evaluate the data loading time and coefficient computing time consumption separately. Although the data loading procedure is not time-consuming, it is difficult to parallelize this part because of the memory read/write speed limitation. However, we can efficiently overlap this procedure with computing in multi-nodes architecture. Figure 5 shows the benefits from multi-nodes single thread parallel. The speedup curve demonstrates a favorable accelerating trend and this method delivers 7.5x speedup compared with baseline as a result.

Hybrid Parallel. Then we nest MPI and OpenMP programming techniques to take the most of this distributed dynamic architecture. Figure 6 exhibits the comprehensive results with a various number of nodes as well as threads. The optimized performance trends similarly with those mentioned above two orthogonal strategies. System performance increases with the number of nodes as well as threads but will degenerate when too many threads assigned on one worker node. The optimal performance is achieved on 8 nodes 16 threads, which gives an 87.9x speedup compared with the baseline.

Table 3. SIMD performance

8 nodes 16 threads	With/SIMD	With SIMD	Speedup
Time	21.25 s	13.85 s	1.5x

Table 4. Scalability under larger workloads

Query vecs	Reference vecs	1 node 16 threads with SIMD	8 nodes 16 threads with SIMD	Speedup
203	203	128.42 s	13.85 s	9.27x
203	812	506.54 s	32.58 s	15.55x

SIMD. We investigate the SIMD optimization as well as the scalability of our system. Table 3 shows that an extra 1.5x speedup obtained by the SIMD on multi-nodes multi threads condition approaching the theoretical 2x peek.

Scalability. Furthermore, we enlarge the reference data size to examine the scalability of the system. As Table 4 shows, our system achieves a 15.55x speedup under a larger workload while the original speedup is 9.27x when we keep 16 threads on every worker node and scale the number of worker nodes from 1 to 8 which exhibits super-linear scalability.

In a summary, we accomplish 2.97K binary coefficients calculating per second which gives an 134.8x speedup compared with baseline as well as 111.6 GOPS (Giga Operations per Second) at the condition of 8 worker nodes, 16 threads per worker node with SIMD applied.

5 Conclusion

In the filed of computer-aid meta-genomics research, how to design similarity measurement algorithms with high efficiency remains an enormous challenge.

In this paper, we propose PPD-B, a distributed dynamic parallel framework based on a binary similarity coefficient to support the meta-genomics analysis. Our framework modifies the Forbes coefficient to quantitatively evaluate the similarity among Hashed meta-genomics binary vectors and utilizes a hierarchical parallel architecture to optimize the computing process of coefficients computation. The experimental results show that the framework operates efficiently and achieves an 134.8x speedup compared with the baseline. And we design a scalable distributed dynamic programming system scheduling the whole system to decouple the communication and computation, which proven stable scalability on large workloads. Our work can be a novel standard instance implicating for designing efficient meta-genomics algorithms on distributed parallel clusters.

In the future, we plan to further research on several respects: assembling heterogeneous machines into our dynamic architecture with balanced workloads, which can be regarded as a typical knapsack problem; applying our system on multi-core accelerators or processors such as GPUs and Sunway to accelerate our algorithm; investigating more efficient binary similarity measurements and related Hash algorithms; and comparing our framework with prospective meta-genomics similarity analysis systems in terms of performance and effectiveness.

References

1. Bernard, G., Greenfield, P., Ragan, M.A., Chan, C.X.: k-mer similarity, networks of microbial genomes, and taxonomic rank. mSystems **3**(6), e00257–18 (2018)
2. Buyya, R., et al.: High Performance Cluster Computing: Architectures and Systems (Volume 1), vol. 1, p. 999. Prentice Hall, Upper Saddle River (1999)
3. Chapman, B., Jost, G., Van Der Pas, R.: Using OpenMP: Portable Shared Memory Parallel Programming, vol. 10. MIT Press, Cambridge (2008)
4. Charras, C., Lecroq, T.: Handbook of Exact String Matching Algorithms. Citeseer (2004)
5. Choi, S.S., Cha, S.H., Tappert, C.C.: A survey of binary similarity and distance measures. J. Syst. Cybern. Inform. **8**(1), 43–48 (2010)
6. Driver, H.E., Kroeber, A.L.: Quantitative Expression of Cultural Relationships, vol. 31. University of California Press, Berkeley (1932)
7. Fleischmann, R.D., et al.: Whole-genome random sequencing and assembly of haemophilus influenzae RD. Science **269**(5223), 496–512 (1995)
8. Forbes, S.A.: On the local distribution of certain Illinois fishes: an essay in statistical ecology, vol. 7. Illinois State Laboratory of Natural History (1907)
9. Grigoriev, I.V., et al.: The genome portal of the department of energy joint genome institute. Nucleic Acids Res. **40**(D1), D26–D32 (2011)
10. Gropp, W., Lusk, E., Doss, N., Skjellum, A.: A high-performance, portable implementation of the MPI message passing interface standard. Parallel Comput. **22**(6), 789–828 (1996)
11. Hubalek, Z.: Coefficients of association and similarity, based on binary (presence-absence) data: an evaluation. Biol. Rev. **57**(4), 669–689 (1982)
12. Jaccard, P.: Étude comparative de la distribution florale dans une portion des alpes et des jura. Bull. Soc. Vaudoise Sci. Nat. **37**, 547–579 (1901)
13. Jeong, H., Kim, S., Lee, W., Myung, S.H.: Performance of SSE and AVX instruction sets. arXiv preprint arXiv:1211.0820 (2012)

14. Li, D., Liu, C.M., Luo, R., Sadakane, K., Lam, T.W.: Megahit: an ultra-fast single-node solution for large and complex metagenomics assembly via succinct de bruijn graph. Bioinformatics **31**(10), 1674–1676 (2015)

15. Lomont, C.: Introduction to Intel advanced vector extensions. Intel White Paper, pp. 1–21 (2011)

16. Metzker, M.L.: Sequencing technologies-the next generation. Nat. Rev. Genet. **11**(1), 31 (2010)

17. Navarro, G.: A guided tour to approximate string matching. ACM Comput. Surv. (CSUR) **33**(1), 31–88 (2001)

18. Ondov, B.D., et al.: Mash: fast genome and metagenome distance estimation using minhash. Genome Biol. **17**(1), 132 (2016)

19. Rognes, T., Flouri, T., Nichols, B., Quince, C., Mahé, F.: Vsearch: a versatile open source tool for metagenomics. PeerJ **4**, e2584 (2016)

20. Schroeder, B., Gibson, G.: A large-scale study of failures in high-performance computing systems. IEEE Trans. Dependable Secur. Comput. **7**(4), 337–350 (2009)

21. Sneath, P.H.A.: The principles and practice of numerical classification. Numer. Taxon. **573**, 263–268 (1973)

22. Wilming, L.G., Gilbert, J.G., Howe, K., Trevanion, S., Hubbard, T., Harrow, J.L.: The vertebrate genome annotation (vega) database. Nucleic Acids Res. **36**(suppl_1), D753–D760 (2007)

23. Woese, C.R., Fox, G.E.: Phylogenetic structure of the prokaryotic domain: the primary kingdoms. Proc. Natl. Acad. Sci. **74**(11), 5088–5090 (1977)

24. Woyke, T., Rubin, E.M.: Searching for new branches on the tree of life. Science **346**(6210), 698–699 (2014)

25. Wrighton, K.C., et al.: Fermentation, hydrogen, and sulfur metabolism in multiple uncultivated bacterial phyla. Science **337**(6102), 1661–1665 (2012)

Optimal Resource Allocation Through Joint VM Selection and Placement in Private Clouds

Hongkun Chen[1], Feilong Tang[1(✉)], Linghe Kong[1], Wenchao Xu[2],
Xingjun Zhang[3], and Yanqin Yang[2]

[1] Department of Computer Science and Engineering, Shanghai Jiao Tong University,
Shanghai, China
tang-fl@cs.sjtu.edu.cn
[2] Department of Computer Science and Technology, East China Normal University,
Shanghai, China
[3] School of Computer Science and Technology, Xi'an Jiaotong University,
Xian, China

Abstract. It is the goal of private cloud platforms to optimize the resource allocation process and minimize the expense to process tasks. Essentially, resource allocation in clouds involves two phases: virtual machine selection (VMS) and virtual machine placement (VMP), and they can be jointly considered. However, existing solutions separate VMS and VMP, therefore, they can only get local optimal resource utilization. In this paper, we explore how to optimize the resource allocation globally through considering VMS and VMP jointly. Firstly, we formulate the joint virtual machine selection and placement (JVMSP) problem, and prove its NP hardness. Then, we propose the *Resource-Decoupling* algorithm that converts the JVMSP problem into two independent sub-problems: *Max-Capability* and *Min-Cost*. We prove that the optimal solutions of the two sub-problems guarantees the optimal solution of the JVMSP problem. Furthermore, we design the efficient *Max-Balanced-Utility* and *Extent-Greedy* heuristic algorithms to solve *Max-Capability* and *Min-Cost*, respectively. We evaluate our proposed algorithms on datasets with different distributions of resources, and the results demonstrate that our algorithms significantly improve the resource utilization efficiency compared with traditional solutions and existing algorithms.

Keywords: Resource allocation · VM selection · VM placement ·
Resource utilization efficiency · Private clouds

1 Introduction

With the rise of cloud services, it is becoming increasingly common for enterprises to build their own cloud platforms. Typically, there are two phases in the resource allocation process of modern cloud platforms [2], being *virtual machine selection*

© IFIP International Federation for Information Processing 2019
Published by Springer Nature Switzerland AG 2019
X. Tang et al. (Eds.): NPC 2019, LNCS 11783, pp. 156–168, 2019.
https://doi.org/10.1007/978-3-030-30709-7_13

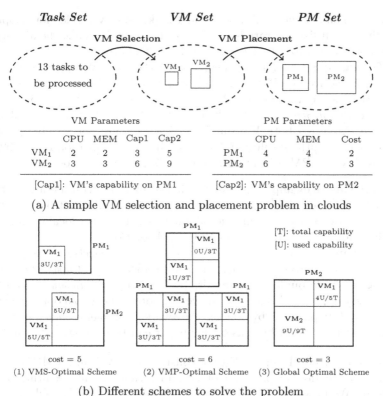

	CPU	MEM	Cap1	Cap2
VM$_1$	2	2	3	5
VM$_2$	3	3	6	9

	CPU	MEM	Cost
PM$_1$	4	4	2
PM$_2$	6	5	3

[Cap1]: VM's capability on PM1   [Cap2]: VM's capability on PM2

(a) A simple VM selection and placement problem in clouds

cost = 5   cost = 6   cost = 3

(1) VMS-Optimal Scheme   (2) VMP-Optimal Scheme   (3) Global Optimal Scheme

(b) Different schemes to solve the problem

Fig. 1. VM selection and placement in clouds

(VMS) and *virtual machine placement (VMP)*, respectively. The VMS phase aims to select proper VMs to process the tasks, and the VMP phase places the selected VMs on proper PMs.

Although such a division of roles provides a clear organization of cloud resources and is widely used in existing public clouds, it is actually not very suitable for private clouds. For a public cloud, VMS scheme is decided by users or their brokers and VMP by the platform, therefore, they have to be separated. However, for a private cloud, where the platform has the opportunity to decide the VMS scheme, such a functional division only results in inefficient resource utilization. We use the following example to demonstrate our point of view.

Suppose there is a simple private cloud platform, where there are two different types of VMs and PMs, with their parameters of CPU, memory, task processing capability and cost shown in Fig. 1(a). In particular, different PMs have different costs, when the same VM is placed on different PMs, it has different task processing capabilities due to the different hardware configurations of PMs. Now, there are 13 tasks is to be processed, and we need to figure out a VM selection scheme and a VM placement scheme so that all the tasks can be processed with a minimum cost. We compare the possible schemes in Fig. 1(b).

- *VMS-Optimal Scheme.* This scheme first makes sure VM selection is optimal, and the VM utilization rate is $(3 + 5 + 5)/13 = 100\%$. While the average PM utilization rate is $(6/10 + 6/9)/2 = 63.33\%$. The total cost is $2 + 3 = 5$.
- *VMP-Optimal Scheme.* This scheme first makes sure VM placement is optimal, and the average PM utilization rate is 100%. While the resulting VM utilization rate is $13/18 = 72.22\%$. The total cost is $2 + 2 + 2 = 6$.
- *Global-Optimal Scheme.* This scheme solves the problem from a global perspective. Although both VMS and VMP scheme are not locally optimal, it gives a global optimal solution. The resulting VM utilization rate is 92.86%, and the average PM utilization rate is 91.67%. The total cost is 3.

From the above example, we observe that a separated consideration of VMS and VMP may lead to significant resource wastage in either of the two stages. Even if these two stages can individually achieve their own local optimal solutions, they can not guarantee a global optimal solution.

In this paper, we convert the original joint VM selection and placement (JVMSP) problem into two independent sub-problems *Max-Capability* and *Min-Cost*, making it decoupled as a result, by our proposed *Resource-Decoupling* algorithm. By applying this algorithm, we can obtain the global optimal solution of entire JVMSP problem by solving the two sub-problems independently. In summary, the main contributions of this paper can be summarized as follows.

1. We propose a novel approach that considering VMS and VMP jointly for resource allocation in private clouds, and formulate the resulting *JVMSP* problem. We prove that the JVMSP problem is a NP-hard problem.
2. We propose the *Resource-Decoupling* algorithm which can obtain the global optimal solution of the *JVMSP* problem. It decouples the JVMSP problem into two independent sub-problems *Max-Capability* for maximizing task processing capabilities of PMs and *Min-Cost* for minimizing the cost.
3. We propose the efficient *Max-Balanced-Utility* algorithm by considering both variance and utility to solve the *Max-Capability* sub-problem, and the efficient *Extent-Greedy* algorithm to solve the *Min-Cost* sub-problem.

2   Related Work

The VMS and the VMP mechanisms are studied separately in previous researches. We briefly review these related studies as follows.

Virtual Machine Placement. The related algorithms proposed to solve the VMP problem can be categorized by their mathematical ideas. Among them, solving VMP problem by bin-packing algorithms [1] is the most straightforward way. Besides, linear programming and stochastic integer programming strategy [4] are other common methods. Finally, a large part of the research works use the heuristic strategy [5], from the simple best-fit strategy and greedy-based method to the genetic algorithm and PSO-based algorithm. As the comparison algorithm used in this paper, the authors applied the classic PageRank algorithm in VMP problem in [10]. They compared many state-of-the-art heuristics and showed the proposed *PageRankVM* brings very good performance.

Virtual Machine Selection. The VMS problem involves many aspects in resource management of clouds. Usually, it is regarded as a sub-problem of the whole dynamic VM consolidation process, where it is used to select VMs for migration [11]. Besides, VMS strategy is also used by the cloud brokers to select proper VMs among multiple cloud resource providers [8]. VMS problem also exists in pay-per-use related deployments, where proper resources are to be selected for specific applications and are charged to application providers [3]. In this paper, VMS helps to decide a set of VMs with different types and quantities, so that the tasks can be processed with minimum VM resource wastage.

3   Problem Statement

3.1   Problem Formulation

We firstly list the notations used in problem formulation in Table 1. Particularly, we use the PM's market price as the cost in this paper, which aims to help the private cloud owners process the tasks with minimum economic expenses.

Table 1. Notations for problem formulation

Inputs	Explanations
T	Amount of total tasks, $T \geq 0$
V	Total VM types, $V \in N^+$ and $v \in \{1, 2, \cdots, V\}$
P	Total PM types, $P \in N^+$ and $p \in \{1, 2, \cdots, P\}$
D	Resource dimensions, $D \in N^+$ and $d \in \{1, 2, \cdots, D\}$
$s = \{s_v^d\}$	VM scales, s_v^d is the v_{th}-type VM's resource value on dimension d
$t = \{t_v^p\}$	VM capabilities, t_v^p is the v_{th}-type VM's capability on p_{th}-type PM
$S = \{S_p^d\}$	PM scales, S_p^d is the p_{th}-type PM's resource value on dimension d
$C = \{C_p\}$	PM costs, C_p is the p_{th}-type PM's usage cost
$K = \{K_p\}$	Maximum quantities, K_p is the p_{th}-type PM's maximum quantity
Outputs	Explanations
$N = \{n_v\}$	VM selection scheme, n_v is the v_{th}-type VM number
$M = \{m_p\}$	PM selection scheme, m_p is the p_{th}-type PM number
$G = \{g_i^v\}$	Placement scheme, g_i^v is the v_{th}-type VM number on the i_{th} PM, use $\hat{i}$ to represent the i_{th} PM's type, where $i \in \{1, 2, \cdots, |M|\}$

The JVMSP problem is formed through jointly considering VMS and VMP problem. In order to analyze their relationship from the mathematical point of view, we first formulate VMS and VMP problems, and then jointly consider them to formulate the JVMSP problem.

VMS Formulation. The VMS problem aims to select proper VMs N to process all the tasks T with minimum cost, as shown in Eq. (1). The optimization goal shows the total cost of all the selected VMs, where $f(s_v)$ is the cost of VM with scale v. The constraint shows the selected VMs' capabilities are enough for all the tasks.

$$\min_{\{N\}} \sum_{v=1}^{V} n_v \cdot f(s_v) \quad s.t. \quad \sum_{v=1}^{V} n_v \cdot t_v \geq T \tag{1}$$

VMP Formulation. The VMP problem aims to find a PM scheme M and the mappings G so that all the VMs N can be placed on PMs with minimum cost, as shown in Eq. (2). We optimize the total cost of all the PMs, where $f(S_p)$ is the cost to use a p_{th}-type PM. The constraints shows all the VMs are needed to be placed, and for every PM, the placed VMs can not exceed its resource capacity.

$$\min_{\{M,G\}} \sum_{p=1}^{P} m_p \cdot f(S_p) \quad s.t. \quad \sum_{i=1}^{|M|} g_i^v = n_v \quad and \quad \sum_{v=1}^{V} g_i^v \cdot s_v^d \leq S_i^d \tag{2}$$

JVMSP Formulation. The optimization goal for JVMSP problem is formulated in Eq. (3), where we aim to decide proper VMs N and proper PMs M as well as the placement method G so that all the tasks can be processed and the total cost is minimum.

$$\min_{\{N,M,G\}} \sum_{p=1}^{P} m_p \cdot C_p \tag{3}$$

The constraints are shown in Eq. (4). The first constraint shows the selected VMs' capabilities are enough for all the tasks. The second one shows the selected PM's quantity of each type is limited by its maximum available number. The third one makes sure all of the selected VMs are placed on PMs. The last one shows every PM's resources should be enough for all the VMs placed on it.

$$\sum_{i=1}^{|M|} \sum_{v=1}^{V} g_i^v \cdot t_v^{\hat{i}} \geq T \quad m_p \leq K_p \quad \sum_{i=1}^{|M|} g_i^v = n_v \quad \sum_{v=1}^{V} g_i^v \cdot s_v^d \leq S_{\hat{i}}^d \tag{4}$$

3.2 Complexity Analysis

It is easy to know that VMP problem is NP-hard as Eq. (2) is equivalent to a multidimensional bin-packing problem, and VMS problem is also NP-hard as Eq. (1) is equivalent to a dual problem of a bin-packing problem. We now show the complexity of JVMSP problem by proving the theorem below.

Theorem 1. *JVMSP is NP-hard, and it is harder than either VMP or VMS.*

Proof. The theorem can be proved by a reduction from both VMP problem and VMS problem to JVMSP problem.

Firstly, we show that VMS can be reduced to JVMSP. Suppose we apply a placement scheme for PMs (i.e., fix G), and use T_p to denote the capability for each type of PM. Then, the last constraint in Eq. (4) is satisfied and the first constraint becomes $\sum_p T_p \cdot m_p \geq T$. Moreover, variable N can be canceled by removing the third constraint. That finally becomes a VMS problem. Therefore, VMS problem is nothing but a special case of the associated JVMSP problem where variable G is set to be a constant.

Secondly, we show that VMP can be reduced to JVMSP. Similarly, suppose we apply a selection scheme for VMs (i.e., fix N) so that all the tasks can be processed, which essentially makes M a function of G. Then, the problem becomes determining M and G to optimize Eq. (3) under Eq. (4) without the first constraint (N is fixed to meet this constraint), which is obviously a VMP problem. Therefore, VMP problem is also a special case of the associated JVMSP problem where variable M is set to be a function of variable G.

4   Joint VM Selection and Placement

4.1   JVMSP Problem Conversion

Traditional solutions greedily split resource allocation into VMS and VMP phases. In this way, even if VMS and VMP can individually achieve their own optimal solutions, it does not guarantee a global optimal solution.

We propose the *Resource-Decoupling* algorithm to derive the global optimal solution, as shown in Algorithm 1. Specifically, the *Resource-Decoupling* algorithm converts the JVMSP problem into two sub-problems *Max-Capability* and *Min-Cost*. *Max-Capability* aims to determine an optimal placement scheme for a given PM so that it has the maximum task processing capability. And *Min-Cost* aims to select the well-placed PMs to process tasks so that the cost is minimum. Line 2 in Algorithm 1 shows we obtain the placement scheme $[\hat{n}_p^1, \cdots, \hat{n}_p^v]$ and the maximum capability $\hat{T}_p$ for a PM of type p by solving the *Max-Capability* problem. Line 4 shows we obtain the PM selection scheme $[\hat{m}_1, \cdots, \hat{m}_p]$ to process all the tasks with a minimum cost by solving the *Min-Cost* problem.

The *Resource-Decoupling* algorithm decouples the JVMSP problem into two independent sub-problems. Now, we prove that the optimal solutions of the two sub-problems guarantees the optimal solution of the JVMSP problem.

Theorem 2. *Resource-Decoupling will give an optimal solution for JVMSP problem if the Max-Capability and Min-Cost sub-problems' solutions are optimal.*

Algorithm 1. Resource-Decoupling

Input: $T, V, P, D, s, t, S, C, K$
Output: $N = \{n_v\}, M = \{m_p\}, G = \{g_i^v\}$
1 **for** $p = 1, 2, \cdots, P$ **do**
2 $\quad [\hat{n}_p^1, \cdots, \hat{n}_p^v], \hat{T}_p \leftarrow$ **Max-Capability**$(S_p, \{s_v^d\}, \{t_v^p\})$
3 **end**
4 $[\hat{m}_1, \cdots, \hat{m}_p] \leftarrow$ **Min-Cost**$(\{\hat{T}_p\}, \{K_p\}, T, \{C_p\})$
5 Construct M by $\{\hat{m}_p\}$, G by $\{\hat{n}_p^v\}$, N by G
6 **return** N, M, G

Proof. We use $\{\hat{T}_p\}$ and $\{\hat{m}_p\}$ to denote the optimal solutions of the *Max-Capability* and *Min-Cost* sub-problems, and $\hat{C} = \sum_p \hat{m}_p C_p$ to denote the corresponding cost. Now, we only need to show that $\hat{C}$ is minimum among all other $C' = \sum_p m_p' C_p$ which satisfies $\sum_p m_p' T_p' \geq T$, where m_p' and T_p' are the constant results obtained from any other strategies.

Let's first consider the optimization problem shown in Eq. (5). Now x_p becomes a variable to be optimized. We use m_p^s to denote the optimal solution for x_p, and C^s to denote the corresponding optimal cost. Then, it's obvious that $C^s \leq C'$, because m_p^s is the optimal case among all other m_p'.

$$min. \quad \sum_{p=1}^{P} x_p \cdot C_p \qquad s.t. \quad \sum_{p=1}^{P} x_p \cdot T_p' \geq T \qquad (5)$$

Let's now consider another optimization problem shown in Eq. (6). Now y_p is also a variable to be optimized, and we can see the optimal solution for y_p is just $\hat{m}_p$ (by the definition of $\hat{m}_p$), and the corresponding optimal cost is $\hat{C}$. As $\hat{T}_p \geq T_p'$ for any PM type p (by the definition of $\hat{T}_p$), it's not hard to see that the optimal cost of Eq. (6) is smaller than or equal to the optimal cost of Eq. (5), i.e., $\hat{C} \leq C^s$. Therefore, we have $\hat{C} \leq C^s \leq C'$.

$$min. \quad \sum_{p=1}^{P} y_p \cdot C_p \qquad s.t. \quad \sum_{p=1}^{P} y_p \cdot \hat{T}_p \geq T \qquad (6)$$

4.2 Algorithm for Max-Capability Sub-problem

The *Max-Capability* problem takes a PM with scale S_p, a VM candidate set with different scales $\{s_v^d\}$ and the task processing capabilities $\{t_v^p\}$ as inputs. It needs to choose proper VMs to place on the PM, where we use $[\hat{n}_p^1, \cdots, \hat{n}_p^v]$ to denote the selected VM quantities of different types, and $\hat{T}_p$ to denote the corresponding task processing capability of this PM after the placement.

The *Max-Capability* is essentially a bin-packing problem and can be optimally solved by the dynamic programming method. However, the time complexity is very high. In this paper, we propose a heuristic algorithm, where two main factors the balance of PM and the utility of VM are considered. The balance of PM aims to avoid using resource excessively in one dimension. Specifically, for each type of VM, we define $V(v) = Var(R^d + s_v^d/S^d)$ to measure the equilibrium effect it brings to the PM, where S^d and R^d is the PM's total and remaining resource, and s_v^d is the v_{th}-type VM's resource on dimension d. The utility of VM aims to choose the VM which brings more task processing capability while consuming less PM resource. We define $U(v) = t_v / \sqrt[d]{\prod_{d=1}^{D} s_v^d}$ to measure the utility of a given VM, where t_v is the v_{th}-type VM's capability.

Algorithm 2. Max-Balanced-Utility

Input: s, t, S_p
Output: $\hat{n}_p = [\hat{n}_p^1, \cdots, \hat{n}_p^v]$, $\hat{T}_p$
1 **for** *repeat R times* **do**
2   **while** *PM is not fully placed* **do**
3   **for** *each type of VM* **do**
4 | calculate $V(v)$ and $U(v)$
5   **end**
6   Sort the different types of VMs by $V(v)$ and drop the tail
7   Generate probabilities for the remaining VMs by $U(v)$
8   Select a VM by their probabilities, deploy the VM on PM
9   **end**
10   Update $\hat{T}_p$ and $\hat{n}_p$ if the placement scheme has a larger T_p
11 **end**
12 **return** $\hat{n}_p$, $\hat{T}_p$

The *Max-Balanced-Utility* algorithm is described in Algorithm 2, which works in the following steps. Firstly, for a given PM, it calculates $V(v)$ and $U(v)$ for each type of VM (lines 4–5). Secondly, it sorts the VMs' types by descending $V(v)$, retain the best VM types in a certain proportion, and assign a possibility for each type of VM according to their $U(v)$ (lines 7–8). Thirdly, it randomly chooses a VM according to the probabilities and repeat the above process until the PM is fully placed. Finally, for the same PM, it repeats the placing strategy R times to get the best scheme.

4.3   Algorithm for Min-Cost Sub-problem

The *Min-Cost* problem aims to determine a PM selection scheme from the well-placed PMs to process all the tasks with minimum cost. Specifically, it takes the capabilities $\{\hat{T}_p\}$, costs $\{C_p\}$, and maximum PM numbers $\{K_p\}$ as inputs. The goal is to determine the quantities $\{m_p\}$ for each type of PM.

We propose the *Extent-Greedy* algorithm to solve the *Min-Cost* problem, as shown in Algorithm 3. It firstly divides the task processing capability of each PM by its cost to get its extent $\{ext_p\}$ and then uses the extent to decide the

---

Algorithm 3. Extent-Greedy

Input: $\{\hat{T}_p\}$, $\{C_p\}$, $\{k_p\}$, T
Output: $\{\hat{m}_p\}$, cost
1 Calculate the extent for each type of PM by $ext_p = \hat{T}_p/C_p$.
2 Sort PM types by their extents $\{ext_p\}$.
3 **for** *the sorted PM types* **do**
4 | **for** *available pm number of this type* **do**
5 | | Select the PM, update $\{\hat{m}_p\}$, cost and remaining tasks
6 | | **if** *there no are tasks remained* **then**
7 | | | Re-select the last PM, update $\{\hat{m}_p\}$, cost
8 | | | **return** $\{\hat{m}_p\}$, cost
9 | | **end**
10 | **end**
11 **end**

---

selection order of the PMs (line 1–2) until the selected PMs are enough for tasks. Note that for the selection of the last PM, on the premise that its resources are enough for the remaining tasks, we select the one with the minimum cost rather than the one with the maximum extent to avoid resource wastage.

Actually, the *Extent-Greedy* algorithm will give a solution for M in Eq. (7) if *Max-Capability* is optimally solved, where $[s\_1, s\_2, \cdots, s\_P]$ is the sorted PM types in Algorithm 3, K is the maximum number and T is the capability for a certain type of PM. In Eq. (7), s_a is a boundary of the sorted PM types, ahead of which all the PMs are selected with the maximum available number (m_{s_x}), behind of which only one PM with the least enough capability is selected (m_{s_b}). Use C to denote the cost derived from Eq. (7), we have the Theorem 3.

$$M = \begin{cases} m_{s\_x} = K_{s\_x}, & x \ for \ any \ 1 \leq x < a \\ m_{s\_a} = \lfloor (T - \sum_{i=1}^{a-1} K_{s\_i} T_{s\_i})/T_{s\_a} \rfloor \\ m_{s\_b} = 1, & a \leq b \leq P \end{cases} \tag{7}$$

Theorem 3. C^o *is a lower bound of the global minimum cost for JVMSP.*

$$C^o = C - C_{s\_b} + \frac{C_{s\_a}}{T_{s\_a}}(T - \sum_{i=1}^{a} m_{s\_i} T_{s\_i}) \tag{8}$$

Proof. Firstly, it is easy to see that $C^o = C$ when $T = \sum_{i=1}^{a} m_{s\_i} T_{s\_i}$. Because the PMs are selected by the sorted extent order and no extra capabilities are wasted. Secondly, we consider the general case where $T > \sum_{i=1}^{a} m_{s\_i} T_{s\_i}$. We know that $C - C_{s\_b}$ is the global minimum cost for $\sum_{i=1}^{a} m_{s\_i} T_{s\_i}$ tasks as stated above. For the remaining tasks, the extent of the most efficient available PMs is C_{s_a}/T_{s_a}. Therefore, at least another $(T - \sum_{i=1}^{a} m_{s\_i} T_{s\_i})C_{s\_a}/T_{s\_a}$ cost is needed to process the remaining tasks, which finally explains the lower bound of the minimum cost show in the above theorem.

In the evaluation part, we will evaluate algorithms by comparing their costs with the lower bound of global minimum cost C^o.

5 Performance Evaluations

5.1 Datasets

The data in our datasets is made up of two parts, VM scales and PM scales, respectively. For VM scales, we combine the VM sizes in the trace-based dataset Google-Cluster [12] and the VM sizes in public cloud Amazon EC2 [7]. For PMs, we configure a number of servers with different specifications, and use their marked prices [6] as their costs. The PMs and VMs are with different ratio types (general type, high-performance type, large-memory type, large-storage type), and for each type there are different sizes of resources. We then divide the PMs and VMs into 9 different PM sets and 9 different VM sets so that they can form different datasets. Their types and quantities are recorded in Table 2.

Table 2. PM and VM sets used in experiments

Set index	1	2	3	4	5	6	7	8	9
PM type	18	36	54	72	90	108	126	144	162
PM quantity	1423	2825	4215	5592	6923	8245	9592	10853	12211
VM type	56	112	168	224	280	336	392	448	512
VM quantity	6129	10947	16317	21381	26511	32295	38526	43383	48741

5.2 Compared Algorithms

To evaluate the performance of our proposed framework and algorithms, we construct four different schemes to solve the JVMSP problem in Table 3, where

- Scheme1 uses our proposed framework and algorithms. It is our proposed scheme for efficient resource allocation in private clouds.
- Scheme2 uses the PageRankVM as the placement strategy. We use it as a comparison scheme to evaluate the different placement strategies.
- Scheme3 uses the Function-Separated framework. We use it as a comparison scheme to evaluate the different resource allocation frameworks.
- Scheme4 uses the Function-Separated framework, and it adopts the First-Fit-Decreasing strategy for placement. It is treated as the baseline.

Table 3. Comparison schemes to solve JVMSP problem

Scheme	Framework	Selection strategy	Placement strategy
1	Resource-Decoupling	Extent-Greedy	Max-Balanced-Utility
2	Resource-Decoupling	Extent-Greedy	PageRankVM [10]
3	Function-Separated [2]	Min-Waste	PageRankVM [10]
4	Function-Separated [2]	Min-Waste	First-Fit-Decreasing [9]

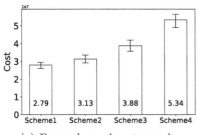

(a) Four schemes' costs results

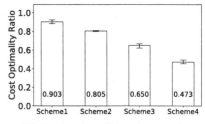

(b) Cost optimality ratio results

Fig. 2. Four compared schemes' cost results to solve the JVMSP problem

5.3   Results and Analysis

We perform the different schemes in Table 3 to solve the JVMSP problem by applying them on the dataset formed by PM set 5 and VM set 5 in Table 2. We perform 50 groups of simulations on the dataset by setting each simulation a different available PMs quantities and tasks inputs. We define the *Cost Optimality Ratio* as C^o/C_{alg} to measure the proximity of the algorithm-derived cost to the global optimal cost, where C^o is the lower bound of minimum cost shown in Theorem 3 and C_{alg} is the algorithm's cost. The results are shown in Fig. 2

By comparing scheme2 and scheme3 in Fig. 2(b), we can see that the *Feedback-Decoupling* framework improves the performance of JVMSP's solution up to 15.5% in average compared with the traditional resource management method Besides, by comparing scheme1 and scheme2, we can see that our proposed *Max-Balanced-Utility* placement strategy improves the solution another 10% when compared with *PageRankVM*. Overall, our proposed scheme improves the resource allocation efficiency 43% compared with the baseline.

We further evaluate the adaptability of our proposed framework and algorithms by applying the schemes on different datasets, and their results are shown in Fig. 3. The results show that scheme1 and scheme2 remain more stable and higher performance in all of the cases compared with scheme3 and scheme4, which illustrates that our proposed framework and algorithms outperform the traditional separated resource allocation methods in different datasets. In particular, with the number of PM and VM types increasing, our algorithms derive better allocation schemes, while the traditional methods tend to have more uncertainty on their performance. Finally, increasing PM types reduces the cost more

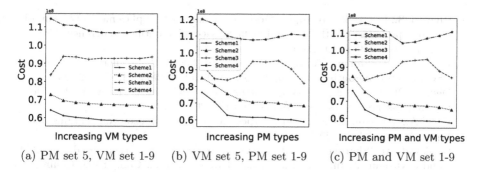

(a) PM set 5, VM set 1-9   (b) VM set 5, PM set 1-9   (c) PM and VM set 1-9

Fig. 3. Four compared schemes' adaptability on different distribution of datasets

significantly compared with increasing VM types for our proposed framework, and increasing both PM and VM types is to some extent equivalent to combining the effects of increasing them separately.

6   Conclusion and Future Work

This paper presents a global perspective to optimize the resource allocation in private cloud platforms, where we combine the original separated VMS and VMP processes, and formulate the joint VM selection and placement (JVMSP) problem. We analyze the hardness of the JVMSP problem and convert it into two sub-problems. A theoretical proof is provided to show the relationship between the JVMSP problem's optimal solution and its two sub-problems'. Besides, for each sub-problem, we provide a heuristic algorithm. Future work can be done to define other forms of cost function, so that this work can be applied to deal with other optimization goals more than economic expense in private clouds.

Acknowledgments. This work was supported in part by the National Natural Science Foundation of China projects under Grants 61832013 and 61672351, and in part by the Huawei Technologies Co., Ltd. project under Grant YBN2018125107.

References

1. Babu, K.R., Samuel, P.: Virtual machine placement for improved quality in IAAS cloud. In: 2014 Fourth International Conference on Advances in Computing and Communications, pp. 190–194. IEEE (2014)
2. Beloglazov, A., Abawajy, J., Buyya, R.: Energy-aware resource allocation heuristics for efficient management of data centers for cloud computing. Future Gener. Comput. Syst. **28**(5), 755–768 (2012)
3. Blaisse, A.P., Wagner, Z.A., Wu, J.: Selection of virtual machines based on classification of MapReduce jobs. In: 2015 IEEE 35th International Conference on Distributed Computing Systems Workshops (ICDCSW), pp. 82–86. IEEE (2015)

4. Chaisiri, S., Lee, B.S., Niyato, D.: Optimal virtual machine placement across multiple cloud providers. In: IEEE Asia-Pacific Services Computing Conference, APSCC 2009, pp. 103–110. IEEE (2009)
5. Dashti, S.E., Rahmani, A.M.: Dynamic VMS placement for energy efficiency by PSO in cloud computing. J. Exp. Theor. Artif. Intell. **28**(1–2), 97–112 (2016)
6. Dell: Dell PowerEdge Servers. https://www.dell.com/en-us/work/shop/dell-poweredge-servers/sc/servers. Accessed 4 Feb 2019
7. EC2, A.: Amazon EC2 instance types. https://aws.amazon.com/ec2/instance-types. Accessed 4 Feb 2019
8. Gahlawat, M., Sharma, P.: VM selection framework for market based federated cloud environment. In: 2015 International Conference on Computing, Communication and Automation, pp. 695–698. IEEE (2015)
9. Johnson, D.S.: Near-optimal bin packing algorithms (1973)
10. Li, Z., Shen, H., Miles, C.: PageRankVM: a PageRank based algorithm with anti-collocation constraints for virtual machine placement in cloud datacenters. In: 2018 IEEE 38th International Conference on Distributed Computing Systems (ICDCS), pp. 634–644. IEEE (2018)
11. Melhem, S.B., Agarwal, A., Goel, N., Zaman, M.: Minimizing biased VM selection in live VM migration. In: 2017 3rd International Conference of Cloud Computing Technologies and Applications, pp. 1–7. IEEE (2017)
12. Reiss, C., Tumanov, A., Ganger, G.R., Katz, R.H., Kozuch, M.A.: Heterogeneity and dynamicity of clouds at scale: Google trace analysis. In: Proceedings of the Third ACM Symposium on Cloud Computing (SOCC), p. 7. ACM (2012)

A Parallel Multi-keyword Top-k Search Scheme over Encrypted Cloud Data

Maohu Yang[1], Hua Dai[1,2](✉), Jingjing Bao[1], Xun Yi[3], and Geng Yang[1,2]

[1] Nanjing University of Posts and Telecommunications, Nanjing 210023, China
yangmh1234@163.com, {daihua,yangg}@njupt.edu.cn, jing874444051@163.com
[2] Jiangsu Security and Intelligent Processing Lab of Big Data, Nanjing 210023, China
[3] Royal Melbourne Institute of Technology University, Melbourne 3001, Australia
xun.yi@rmit.edu.au

Abstract. With searchable encryptions in the cloud computing, users can outsource their sensitive data in ciphertext to the cloud that provides efficient and privacy-preserving multi-keyword top-k searches. However, most existing top-k search schemes over encrypted cloud data are the centralize schemes which are limited in large scale data environment. To support scalable searches, we propose a parallel multi-keyword top-k search scheme over encrypted cloud data. In this scheme, the fragment-based encrypted inverted index is designed, which is indistinguishable and can be used for parallel searching. On the basis of such indexes, a Map-Reduce-based distributed computing framework is adopted to implement the parallel multi-keyword top-k search algorithms. Security analysis and experiment evaluation show that the proposed scheme is privacy-preserving, efficient and scalable.

Keywords: Cloud computing · Inverted index ·
Multi-keywords top-k search · Parallel computing ·
Searchable encryption

1 Introduction

With the rapid development of computer technology and internet application, data in many areas are growing exponentially, thus the demand for large and scalable storage and computation is becoming urgent. More and more enterprises and individuals outsource their storage and computation to the cloud for using data anytime and anywhere and saving costs of hardware and software [1].

Supported by the National Natural Science Foundation of China under the grant Nos. 61872197, 61572263, 61672297 and 61872193; the Postdoctoral Science Foundation of China under the Grand No. 2019M651919; the Natural Research Foundation of NJUPT under the grand No. NY217119; the Natural Science Foundation of Anhui Province under the grant No. 1608085MF127; the University Natural Science Foundation of Anhui Province under the grant No. KJ2017A419.

X. Tang et al. (Eds.): NPC 2019, LNCS 11783, pp. 169–181, 2019.
https://doi.org/10.1007/978-3-030-30709-7_14

However, while enjoying the benefits of cloud computing, users have to face the risk that sensitive outsourced data could be leaked or abused because cloud service providers can access the data without authorization. Therefore, data owners usually encrypt data before outsourcing [2]. Although encryption preserves the security of data, it also affects the data availability. In this scenario, searchable encryptions (SE) [3–18] that guarantee the security and availability of data have been proposed.

At present, most solutions are based on the vector space model (VSM) and TF-IDF model which extract keywords of documents into "points" in multi-dimensional space and describe the relevance scores between documents and search keywords. The top-k documents are determined by comparing the relevance scores. However, if a scheme calculates the relevance scores between every document and the search keywords, it will cost a large amount of time and computing resources. To improve search efficiency, researchers have provided a variety of schemes. Song et al. [3] proposed the first SE scheme where users need to traverse entire documents while searching, and the search time is proportional to the amount of data set. Goh et al. [4] proposed a search scheme based on the Bloom Filter. Curtmola et al. [5] proposed an efficient SE scheme based on the inverted index, but using this scheme could expose the privacy of keywords.

Cao et al. [6] proposed a new structure to adapt to multi-keyword search, but it's search time increases exponentially when document size grows. Xia et al. [8] proposed a secure and dynamic multi-keyword ranked search scheme which can reduce the large inner product calculation by pruning function. Jiang et al. [9] proposed a secure ciphertext search scheme based on the inverted index, which avoids calculating the relevance scores of irrelevant documents. Chen et al. [10] proposed a method based on data mining, which can achieve linear time complexity with the exponential growth of the document set. Our previous work [11] also proposed a hierarchical agglomerative clustering tree index scheme, which can perform an effective and verifiable ranked search.

However, the above existing schemes need to load the complete indexes into memory at one time for performing search. Because the index size is proportional to the number of documents, when the scale of documents grows to a certain level, the memory will be overflow. Moreover, the indexes of those schemes should be kept in integrity and cannot be segmented, which also limits their scalability. To conquer such limitation, we propose a parallel privacy-preserving top-k search (PPTS) scheme, which can meet the requirements of large scale data. First, we propose a fragment-based encrypted inverted index model. In data preprocessing and outsourcing phase, the indistinguishable fragment-based encrypted inverted indexes are constructed and outsourced to the cloud together with the encrypted documents. In the search phase, the Map-Reduce-based distributed computing framework is adopted and the parallel multi-keyword top-k search algorithms are proposed. After that, we analyze the security of PPTS and perform experiments to evaluate its efficiency.

The contributions of this paper are: (1) We present the fragment-based encrypted inverted index model which is indistinguishable through adding random paddings. (2) By adopting the Map-Reduce-based distributed computing framework, the parallel multi-keyword top-k search algorithms are proposed. (3) We analyze the security and evaluate the search performance. The result shows that the proposed scheme can realize parallel search while preserving data privacy.

2   Models and Problem Formulation

2.1   Notations and Preliminaries

- D: The document set, $D = \{d_1, d_2, ..., d_n\}$. $\widetilde{D}$ is the encrypted form.
- n: The number of documents in D.
- W: The dictionary, namely, the set of keywords, denoted as $W = \{w_1, w_2, ..., w_m\}$.
- m: The number of keywords in W.
- Q: The query consisting of a set of the search keywords, $Q = \{w_1, w_2, ..., w_q\}$.
- V_{d_i}: The m-dimensional document vector of d_i. $\widetilde{V}_{d_i}$ is the encrypted form.
- V: The document vector set, $V = \{V_{d_1}, V_{d_2}, ..., V_{d_n}\}$. $\widetilde{V}$ is the encrypted form.
- V_q: The m-dimensional query vector for Q. $\widetilde{V}_q$ is the encrypted form.
- TD: The trapdoor for the search request.
- RS: The result of the search.
- $P_{i,j}$: The posting corresponding to the document d_j containing keyword w_i, $P_{i,j} = <id(d_j), V_{d_j}>$. $\widetilde{P}_{i,j}$ is the encrypted form.
- PL_i: The posting list of keyword w_i, $PL_i = \{P_{i,1}, P_{i,2}, ..., P_{i,\delta}\}$. $\widetilde{PL}_i$ is the encrypted form.
- δ: The number of postings in PL_i.
- ε: The fragmentation parameter.
- F: The fragmented documents of D according to ε, $F = \{F_1, F_2, ..., F_t\}$.
- t: The number of fragments of F.
- β: The number of posting list in F_i.

Vector Space Model (VSM) and TF-IDF Model. The VSM and TF-IDF are widely used in multi-keyword privacy-preserving top-k search [6–13]. The term frequency (TF) refers to the number of times a given keyword or term appears in documents, while the inverse document frequency (IDF) is equal to the total number of documents in the set divided by the number of documents containing a given keyword. VSM is used to convert a given document d_i and search keywords Q into vectors V_{d_i} and V_q. The calculation of those vectors can be referred to [6–13].

Secure Inner Product Operation. This scheme uses the secure inner product operation to calculate the inner product of two encrypted vectors without knowing the plaintext value. The basic idea of this is as follows. Assuming that p and q are two n-dimensional vectors and M is a random $n \times n$-dimensional invertible matrix. M is treated as the secure key. The encrypted form of p and q are denoted as $\widetilde{p}$ and $\widetilde{q}$ respectively, where $\widetilde{p} = pM^{-1}$ and $\widetilde{q} = qM^T$. Then we have $\tilde{p} \cdot \tilde{q} = (pM^{-1}) \cdot (qM^T) = pM^{-1}(qM^T)^T = pM^{-1}Mq = p \cdot q$, i.e. $\tilde{p} \cdot \tilde{q} = p \cdot q$. Therefore, we have that the inner product of two encrypted vectors equals the inner product of the corresponding two plaintext vectors.

Inverted Index. Inverted index can be used to quickly find those documents containing a given keyword by mapping to improves search efficiency. It consists of dictionary and posting list. The dictionary is a collection of all keywords that appeared in the D. Each index item in inverted index records a keyword and a pointer to the posting list, which is the entry of posting. The posting list records a list of all documents that contain a specified keyword. Each record in the posting list is a posting that describes the information of the document.

2.2 System Model

The system model is shown in Fig. 1, which is the same as [6–11, 14–16]. It includes three different entities. Data owners (DO) are responsible for constructing fragment-based encrypted inverted indexes ($\tilde{I}$), and outsourcing the encrypted indexes and documents to the cloud server (CS). CS provide the search service in parallel according to the search request submitted by data users (DU). DU construct a search trapdoor based on its needs and send it to CS, then wait for CS to return the search results.

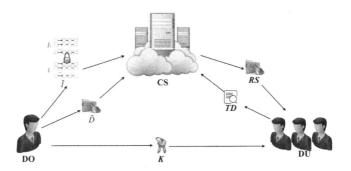

Fig. 1. The system model

2.3 Problem Description

We adopt the "Honest-but-Curious" threat model. In this model, CS honestly and correctly executes instructions in the designated protocol. However, CS can analyze stored data and try to snoop on sensitive information.

The search result of PPTS is represented as RS. V_q is the query vector of Q. V_{d_i} and V_{d_j} respectively represent the document vector of d_i and d_j. Then, RS meets the requirement:

$$|RS| = k \wedge \forall d_i, d_j(d_i \in RS \wedge d_j \in (D - RS)) \rightarrow V_{d_i} \cdot V_q > V_{d_j} \cdot V_q.$$

The PPTS should satisfy three goals. First, the contents are directly seen by CS only include encrypted documents, indexes, and trapdoors, that is, the confidentiality of documents, indexes and trapdoors cannot be leaked. Second, PPTS can handle the search requirements of large document sets in parallel with Map-Reduce parallel search framework. Third, PPTS should fully guarantee the accuracy of search, that is, to improve the efficiency without reducing the accuracy.

2.4   Search Framework

To clearly describe the scheme proposed in this paper, we define a framework for the PPTS scheme. As shown in Fig. 2, the search model is composed of five modules: *GenKey, Setup, BuildIndex, GenTrapdoor*, and *Search*.

- *Genkey*: DO generate the key for encryption, and share it with DU.
- *Setup*: DO preprocess the document set D, generate a document vector for each document, and encrypt the D.
- *BuildIndex*: DO fragment the D and then construct an indistinguishable inverted index to provide the CS to perform the search service.
- *GenTrapdoor*: DU generate a trapdoor based on the search keywords.
- *Search*: CS perform the top-k search service in parallel according to the TD and $\tilde{I}$, and return the RS that satisfy the condition to the DU.

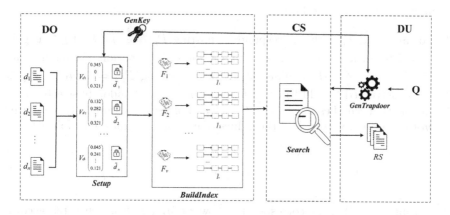

Fig. 2. PPTS search framework

3   Parallel Privacy-Preserving Top-k Search Scheme

3.1   Fragment-Based Encrypted Inverted Indexes Model

Definition 1. *Document Fragmentation.* The document set D is divided into equal lengths according to the parameter ε. The generated fragmented documents are denoted as $F = \{F_1, F_2, ..., F_t\}$, satisfying Formula 3 and 4.

$$|F_1| = |F_2| = ... = |F_{t-1}| = \varepsilon, 1 \leq |F_t| \leq \varepsilon \qquad (1)$$

$$D = F_1 \cup F_2 \cup ... \cup F_t \tag{2}$$

where $|X|$ represent the number of elements contained in the list or set X.

Definition 2. *Parameter-(δ, β).* δ is the maximum number of documents containing a certain keyword w_j in a certain fragment F_i. And β is the maximum number of keywords contained in a certain fragment of the F.

$$\begin{cases} \delta = max\{|F_{v,j}| \mid 1 < v < t, 1 < j < m\} \\ \beta = max\{|W_v| \mid 1 < v < t, 1 < j < m\} \end{cases} \tag{3}$$

where $F_{v,j} \subset F_v$ is the document set containing the keywords w_j in fragment F_v, and $W_v \subset W$ is the keywords set contained in fragment F_v.

Definition 3. *Fragment-based Encrypted Inverted Indexes.* $\widetilde{I} = \{\widetilde{I}_1, \widetilde{I}_2, ..., \widetilde{I}_t\}$. Here, $\widetilde{I}_v \in \widetilde{I}$ is an encrypted inverted index corresponding to fragment F_v. Each row in $\widetilde{I}_v$ is $<tag_j, \widetilde{PL}_j>$, corresponding to a keyword w_j. tag_j is a hash-based message authentication code of w_j generated by key c, $tag_j = hash(c, w_j)$. $\widetilde{PL}_j$ is the encrypted posting list of w_j. To protect the private information of keyword frequencies, we make the index $\widetilde{I}_v$ corresponding to a fragment $F_v \in F$ has the same number of rows and the posting list in each row has the same number of posting. Thus, the generated indexes $\widetilde{I}$ indistinguishable. The construction of the index $\widetilde{I}_v$ is given as follows:

(1) For any $d_i \in F_{v,j}$, the corresponding posting $\widetilde{P}_{j,i} = < id(d_i), \widetilde{V}_{d_i} >$ is generated to form the $\widetilde{PL}_j$. Here, $id(d_i)$ is the id information of document d_i. If $|F_{v,j}| < \delta$, $\delta - |F_{v,j}|$ different artificial padding $\widetilde{P}_{j,s} = < id(d_s), \widetilde{V}_{d_s}' >$ are constructed to add to the $\widetilde{PL}_j$. d_s represents a randomly selected document that satisfies $d_s \in F_v - F_{v,j}$. V_{d_s}' represents a randomly generated m-dimensional vector, and the values of each dimension are as follows:

$$V_{d_s}'[k] = \begin{cases} 0, & k \neq j \\ rand(min\{V_{d_i}[k]\}), & k = j \wedge d_i \in F_{v,j} \end{cases} \tag{4}$$

(2) For any $w_j \in W_v$, the corresponding row $<tag_j, \widetilde{PL}_j>$ is generated according to the above steps to form the $\widetilde{I}_v$. If $|W_v| < \beta$, $\beta - |W_v|$ different artificial rows $<tag_s, \widetilde{PL}_s>$ are constructed to add to the $\widetilde{I}_v$. tag_s is generated by randomly selected keyword w_s, where $w_s \in W - W_v$. $\widetilde{PL}_s$ is composed of δ artificial padding generated by the above steps.

We take an example to explain the above definitions. We assume $D = \{d_i | i = 1, ..., 10\}$ and $W = \{w_1, w_2, w_3\}$. D is divided to $F_1 = \{d_1, d_2, d_3, d_4\}$, $F_2 = \{d_5, d_6, d_7, d_8\}$, $F_3 = \{d_9, d_{10}\}$. Then, $\delta = 3$ and $\beta = 3$ are calculated. Finally, the indexes $\widetilde{I} = \{\widetilde{I}_1, \widetilde{I}_2, \widetilde{I}_3\}$ are generated as shown in Fig. 3.

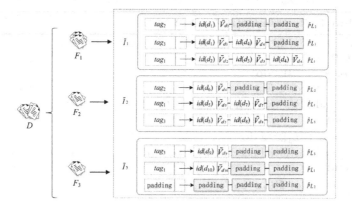

Fig. 3. Example of $\widetilde{I}$

3.2   Data Preprocessing and Outsourcing

The data preprocessing and outsourcing of PPTS are mainly performed by DO which include three algorithms: *GenKey*, *Setup* and *BuildIndex*.

$\boldsymbol{K \leftarrow GenKey(1^\lambda)}$**:** On input a security parameter λ, the key generation algorithm output the key K. DO randomly generate the key sk, $c \in \{0,1\}^\lambda$, an m-dimensional vector S and two $m \times m$ invertible matrices M_1, M_2. Finally, the key $K = (sk, c, S, M_1, M_2)$ is formed. K is shared between DO and DU but is private to CS.

$\boldsymbol{(\widetilde{V}, \widetilde{D}) \leftarrow Setup(D)}$**:** On input the document set D, this algorithm output the encrypted document vector set $\widetilde{V}$ and encrypted document set $\widetilde{D}$. DU encrypts document d_i into $\widetilde{d_i}$ using sk. Then DU generate document vector V_{d_i} according to VSM and TF-IDF models. The key S is used to split the document vector V_{d_i} into V'_{d_i} and V''_{d_i} according to the following formula, and then the reversible matrices M_1 and M_2 are used to encrypt V_{d_i} to $\widetilde{V_{d_i}} = (M_1^T V'_{d_i}, M_2^T V''_{d_i})$. Finally, the generated $\widetilde{d_i}$ and $\widetilde{V_{d_i}}$ are added to $\widetilde{D}$ and $\widetilde{V}$ respectively.

$$\begin{cases} V'_{d_i}[j] = V''_{d_i}[j] = V_{d_i}[j], & S[j] = 0 \\ V'_{d_i}[j] + V''_{d_i}[j] = D_i[j], & S[j] = 1 \end{cases} \tag{5}$$

$\boldsymbol{\widetilde{I} \leftarrow BuildIndex(K, D, \widetilde{V}, \varepsilon)}$**:** This algorithm is run by DU to generate encrypted indexes. Its inputs are the key K, the document set D, the encrypted document vector set $\widetilde{V}$, and the fragmentation parameter ε, and output the $\widetilde{I}$. Procedures of this algorithm is shown in Algorithm 1 where DO first divides D into fragments and then builds an encrypted index for each fragment. Since the operations on each fragment are exactly the same after fragmented, the data preprocessing stage can be executed in parallel. Finally, DO outsource the $\widetilde{D}$ and $\widetilde{I}$ to CS.

Algorithm 1. $BuildIndex(K, \widetilde{V}, D, \varepsilon)$

---

1   Calculate the value of Parameter-(δ, β);
2   **for** *each $F_v \in F$* **do**
3   **for** *each $w_j \in W_v$* **do**
4   **for** *each $d_i \in F_{v,j}$* **do**
5   add $\widetilde{P}_{j,i} = <id(d_i), \widetilde{V}_{d_i}>$ to $\widetilde{PL}_j$;
6   **end**
7   **while** $|\widetilde{PL}_j| < \delta$ **do**
8   add artificial padding $\widetilde{P}_{j,s} = <id(d_s), \widetilde{V}_{d_s}'>$ to $\widetilde{PL}_j$;
9   **end**
10   add $<tag_j, \widetilde{PL}_j>$ to $\widetilde{I}_v$;
11   **end**
12   **while** $|\widetilde{I}_v| < \beta$ **do**
13   add artificial row $<tag_s, \widetilde{PL}_s>$ to $\widetilde{I}_v$;
14   **end**
15   add $\widetilde{I}_v$ to $\widetilde{I}$;
16 **end**
17 **return** $\widetilde{I}$

3.3   Map-Reduce-Based Top-k Search

The Map-Reduce-based top-k search phase of PPTS is performed by DU and CS. DU generate a search trapdoor TD and submit it to CS. CS perform the *Map* operation according to TD to obtain the k documents most relevant to each fragment, and then perform *Reduce* operation to merge and rank the previously acquired documents to generate the final top-k results. This phase mainly contains two polynomial-time algorithms: *GenTrapdoor* and *Search*.

$TD \leftarrow GenTrapdoor(K, Q, k)$: This algorithm takes a plaintext query containing the key K, the search keyword set Q, and the number of documents to be returned k, and outputs the encrypted query as a trapdoor TD. Its goal is to protect the keyword information in the query from CS. The construction process of TD as the following steps:

(1) The query vector V_q is constructed according to Q. If $w_i \in Q$, the IDF of w_i is stored in $V_q[i]$, otherwise, the value of $V_q[i]$ is 0. Then, according to the following formula, V_q is split into two vectors V_q' and V_q''. Finally, V_q' and V_q'' are encrypted with reversible matrices M_1 and M_2 to obtain the encrypted query vector $\widetilde{V_q} = (M_1^{-1}V_q', M_2^{-1}V_q'')$.

$$\begin{cases} V_q'[j] + V_q''[j] = V_q[j], & S[j] = 0 \\ V_q'[j] = V_q''[j] = V_q[j], & S[j] = 1 \end{cases} \tag{6}$$

(2) The hash-based message authentication code tag_i of w_i is calculated and constitutes the set $T = \{tag_i \mid tag_i = hash(c, w_i) \wedge w_i \in Q\}$.

(3) Output $TD = (T, \widetilde{V}_q, k)$.

Algorithm 2. $Search.Map(\widetilde{I}_v, TD)$

---

1 **for** each $< tag_j, \widetilde{PL}_j > \in \widetilde{I}_v$ **do**
2   **if** $tag_j \in T$ **then**
3   **for** each $\widetilde{P}_{j,i} \in \widetilde{PL}_j$ **do**
4   **if** $Score(\widetilde{V}_{d_i}, \widetilde{V}_q) > minScore\{RS_i\}$ **then**
5   **if** $|RS_i| = k$ **then**
6   Delete the document with the lowest relevance score in RS_i;
7   **end**
8   add $< id(d_i), Score(\widetilde{V}_{d_i}, \widetilde{V}_q) >$ to RS_i;
9   **end**
10   **end**
11   **end**
12 **end**
13 **return** RS_i

$RS \leftarrow Search(\widetilde{I}, TD)$: When CS receives the trapdoor TD, it performs the top-k search in parallel on the basis of the indexes $\widetilde{I}$, and then returns the result encrypted documents. The standard Map-Reduce model is adopted to find the top-k relevant documents. In the Map stage, local top-k result is obtained in each fragment. In the Reduce phase, all local top-k results are merged to obtain the global top-k result is calculated. Detailed procedures are shown in Algorithms 2 and 3.

Algorithm 3. $Search.Reduce(RS_1, RS_2, ..., RS_t)$

---

1 **for** each RS_v **do**
2   **if** $RS_v.Score(\widetilde{V}_{d_i}, \widetilde{V}_q) > minScore\{RS\}$ **then**
3   **if** $|RS| = k$ **then**
4   Delete the document with the lowest relevance score in RS;
5   **end**
6   Obtain the $\widetilde{d}_i$ according to $id(d_i)$, and add it to the RS;
7   **end**
8 **end**
9 **return** RS

According to the structure of index and the top-k search algorithms, we have that each encrypted inverted index for a fragment is independent and the top-k search follows the Map-Reduce model. Thus, the proposed search scheme is scalable. It means that, when the volume of outsourced data grows, the search efficiency can be preserved by adding servers.

4   Security Analysis

This chapter mainly elaborates PPTS from two aspects of security and efficacy. Security is to analyze the confidentiality of documents, indexes, and trapdoors. Efficacy is to analyze the scalability of PPTS, and prove that it has the capability of parallel search and can store large document sets.

Theorem 1. *PPTS satisfies privacy requirements.*

Proof. First, the symmetric encryption algorithm is used to encrypt documents in PPTS, which can protect the privacy of documents when the key is not leaked. Second, the security of the $\widetilde{I}$ is guaranteed by random mapping of keywords, filling redundant values and matrix encryption. Because of the characteristics of the hash-based message authentication code, attackers cannot recover keyword information according to the codes. The document vectors and query vectors are encrypted by matrix encryption technology. Therefore, it can be fully proved that the indexes and trapdoors of PPTS are confidential. In addition, the problem of correlation between similar trapdoors can be solved by randomly adding redundant values to the search trapdoors.

Theorem 2. *PPTS has parallel execution capability.*

Proof. First, the indexes $\widetilde{I}$ is designed according to the parallel computing framework Map-Reduce, that is, both Map and Reduce phases can be executed in parallel. Second, HDFS as a distributed file system can store large data. On the Hadoop cluster, DO do not need to pay attention to the details of data storage and transmission. It only needs to submit the $\widetilde{D}$ and the $\widetilde{I}$ to Hadoop to provide a secure, stable and effective search service for DU, which has very high practicability. Also, the execution capability can be linearly improved by server expansions, providing almost unlimited processing power.

Theorem 3. *The accuracy and privacy of search are not affected by artificial paddings and rows.*

Proof. We assume that $\widetilde{P}_{j,s} = \, < id(d_s), \widetilde{V}_{d_s}{}' >$ is an artificial padding added to $\widetilde{PL}_j$ corresponding to the keyword w_j. $\forall d_i \in F_{v,j}$ satisfies the following equation:

$$Score(V_{d_s}{}', V_q) = V_{d_s} \cdot V_q = rand(min\{V_{d_i}[j]\}) \times V_q[j] < Score(V_{d_i}{}', V_q) \quad (7)$$

Therefore, the relevance score corresponding to the padding must be lower than any document in $F_{v,j}$. When searching, DU only focuses on the documents with the highest relevance score, so the added paddings will not affect the accuracy of the search results. By adding paddings and rows, the posting list corresponding to each keyword is equal in length and to each fragment equal in width. Therefore, it is impossible to judge whether it is artificial padding or row based on the length of the posting list. Because $d_s \notin F_{v,j} \wedge d_s \in F_v$, d_s has uniqueness and indistinguishability in posting list PL_j. As a result, the added paddings will not affect the privacy of the search results.

5   Performance Evaluation

To evaluate the performance of PPTS, we implement it on the Hadoop platform and compared time cost with SPTS. Here, SPTS is a sequential privacy-preserving top-k search scheme running on a single server. In other words, SPTS use one server to search each $\widetilde{I}_v \in \widetilde{I}$ sequentially. We extent the New York Times Dataset [19] to generate our experimental dataset which has 3,600,000 documents and 228,623 keywords are extracted. We implement the schemes using Java in Hadoop platform with three servers. Each server has 3.2 GHz, 8-core CPU, 16 G memory and 1 T hard disk. Default parameters are $n = 3{,}600{,}000, |Q| = 15$, $k = 5$, and $\varepsilon = 30{,}000$ which are the number of documents, search keywords, search documents and fragmentation parameter respectively.

In the following experiments, we evaluate the time cost of searches where one of the parameters n, k, and $|Q|$ changes and the other parameters adopt the default values. The results are shown in Figs. 4, 5 and 6.

Figures 4, 5 and 6 all show that the proposed PPTS outperforms SPTS in the time cost of ranked searches, and the former saves at least 80% of the time cost compared with the latter. The reason is that both PPTS and SPTS are based on the inverted index. In the inverted index, the number of candidate posting corresponding to search keywords is positively correlated with the number of documents and search keywords, but is not affected by the value of search documents. As the number of documents or search keywords increases, more resources are needed to calculate the relevance score. SPTS only use a single server with limited processing power, which can possibly reach its processing bottleneck and make the search speed slower and slower. However, PPTS use multiple servers to perform the search at the same time, and the task pressure is shared on multiple servers, so the time cost will not increase too much.

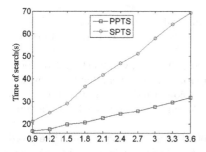

Fig. 4. Number of documents n ($\times 10^6$)

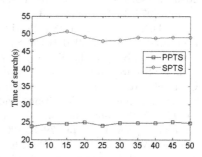

Fig. 5. Number of search documents k

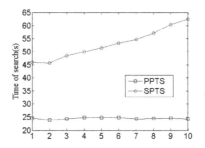

Fig. 6. Number of search keywords $|Q|$

6 Conclusion

In this paper, we propose a parallel privacy-preserving top-k search scheme over encrypted cloud data. In this scheme, the fragment-based encrypted inverted index is designed, which is indistinguishable and can be used for parallel searching. On the basis of such indexes, the Map-Reduce-based distributed computing framework is adopted and the parallel multi-keyword top-k search algorithms are proposed. Security analysis and experiment evaluation show that the proposed scheme is privacy-preserving, efficient and scalable.

References

1. González, L.M.V., Rodero-Merino, L., Caceres, J., Lindner, M.A.: A break in the clouds: towards a cloud definition. Comput. Commun. Rev. **39**(1), 50–55 (2008)
2. Kamara, S., Lauter, K.: Cryptographic cloud storage. In: Sion, R., et al. (eds.) FC 2010. LNCS, vol. 6054, pp. 136–149. Springer, Heidelberg (2010). https://doi.org/10.1007/978-3-642-14992-4_13
3. Song, D.X., Wagner, D.A., Perrig, A.: Practical techniques for searches on encrypted data. In: 2000 IEEE Symposium on Security and Privacy, Berkeley, California, USA, pp. 44–55, May 2000
4. Goh, E.: Secure indexes. IACR Cryptology ePrint Archive, vol. 2003, p. 216 (2003)
5. Curtmola, R., Garay, J.A., Kamara, S., Ostrovsky, R. : Searchable symmetric encryption: improved definitions and efficient constructions. In: Proceedings of the 13th ACM Conference on Computer and Communications Security, CCS 2006, Alexandria, VA, USA, pp. 79–88 (2006)
6. Cao, N., Wang, C., Li, M., Ren, K., Lou, W.: Privacy-preserving multi-keyword ranked search over encrypted cloud data. In: 30th IEEE International Conference on Computer Communications, Joint Conference of the IEEE Computer and Communications Societies, INFOCOM 2011, pp. 829–837, April 2011
7. Sun, W., Wang, B., Cao, N., Li, M., Lou, W.: Verifiable privacy-preserving multi-keyword text search in the cloud supporting similarity-based ranking. IEEE Trans. Parallel Distrib. Syst. **25**(11), 3025–3035 (2014)
8. Xia, Z., Wang, X., Sun, X., Wang, Q.: A secure and dynamic multi-keyword ranked search scheme over encrypted cloud data. IEEE Trans. Parallel Distrib. Syst. **27**(2), 340–352 (2016)

9. Jiang, X., Yu, J., Yan, J., Hao, R.: Enabling efficient and verifiable multi-keyword ranked search over encrypted cloud data. Inf. Sci. **403**, 22–41 (2017)
10. Chen, C., Zhu, X., Shen, P., Hu, J., Guo, S.: An efficient privacy-preserving ranked keyword search method. IEEE Trans. Parallel Distrib. Syst. **27**(4), 951–963 (2016)
11. Zhu, X., Dai, H., Yi, X., Yang, G., Li, X.: MUSE: an efficient and accurate verifiable privacy-preserving multikeyword text search over encrypted cloud data. Secur. Commun. Netw. **2017**, 1 923 476:1–1 923 476:17 (2017)
12. Fu, Z., Wu, X., Guan, C., Sun, X., Ren, K.: Toward efficient multi-keyword fuzzy search over encrypted outsourced data with accuracy improvement. IEEE Trans. Inf. Forensics Secur. **11**(12), 2706–2716 (2016)
13. Ge, X., Yu, J., Hu, C., Zhang, H., Hao, R.: Enabling efficient verifiable fuzzy keyword search over encrypted data in cloud computing. IEEE Access **6**, 45 725–45 739 (2018)
14. Guo, C., Zhuang, R., Chang, C., Yuan, Q.: Dynamic multi-keyword ranked search based on bloom filter over encrypted cloud data. IEEE Access **7**, 35 826–35 837 (2019)
15. Sun, W., et al.: Privacy-preserving multi-keyword text search in the cloud supporting similarity-based ranking. In: 8th ACM Symposium on Information, Computer and Communications Security, ASIA CCS 2013, Hangzhou, China, pp. 71–82, May 2013
16. Yang, Y., Zhan, Y., Liu, J., Liu, X., Yuan, F., Zhong, S.: Chinese multi-keyword fuzzy rank search over encrypted cloud data based on locality-sensitive hashing. J. Inf. Sci. Eng. **35**(1), 137–158 (2019)
17. Zhang, R., Xue, R., Yu, T., Liu, L.: Dynamic and efficient private keyword search over inverted index-based encrypted data. ACM Trans. Internet Technol. **16**(3), 21:1–21:20 (2016)
18. Wang, H., Dong, X., Cao, Z.: Secure and efficient encrypted keyword search for multi-user setting in cloud computing. Peer-To-Peer Netw. Appl. **12**(1), 32–42 (2019)
19. B. D: New York times dataset[db/ol] (2018). http://developer.nytimes.com/docs

N-Docker: A NVM-HDD Hybrid Docker Storage Framework to Improve Docker Performance

Lin Gu[1], Qizhi Tang[1], Song Wu[1]([✉]), Hai Jin[1], Yingxi Zhang[2], Guoqiang Shi[2], Tingyu Lin[2], and Jia Rao[3]

[1] National Engineering Research Center for Big Data Technology and System Services Computing Technology and System Lab, Cluster and Grid Computing Lab, School of Computer Science and Technology, Huazhong University of Science and Technology, Wuhan 430074, China
wusong@hust.edu.cn

[2] State Key Laboratory of Intelligent Manufacturing System Technology, Beijing 100854, China

[3] The University of Texas at Arlington, Arlington, TX 76019, USA

Abstract. Docker has been widely adopted in production environment, but unfortunately deployment and cold-start of container are limited by the low speed of disk. The emerging *non-volatile memory* (NVM) technology, which has high speed and can store data permanently, brings a new chance to accelerate the deployment and cold-start of container. However, it is expensive to replace the whole *hard disk driver* (HDD) with NVM. To achieve the fastest deployment and cold-start with lowest cost, we conduct in-depth analysis on the Top-134 images in Docker Hub and obtain two main insights as: (1) the storing latency of layered image has become the bottleneck of container deployment; (2) only a few image layers are required for container cold-start. Based on these two findings, we propose a NVM-HDD hybrid docker storage framework as N-Docker. It can effectively accelerate container cold-start by detecting the bottleneck layers as well as cold-start required layers and storing them into NVM for faster container startup with limited NVM capacity. Experimental results show that N-Docker can accelerate the container deployment by 1.21X and cold-start by 2.96X. Compared to NVM-Docker, which stores all images into NVM, N-Docker achieves the same performance improvements while reducing the usage of NVM by 88.22%.

Keywords: Container deployment · Container cold-start · Docker · Image · NVM

© IFIP International Federation for Information Processing 2019
Published by Springer Nature Switzerland AG 2019
X. Tang et al. (Eds.): NPC 2019, LNCS 11783, pp. 182–194, 2019.
https://doi.org/10.1007/978-3-030-30709-7_15

1   Introduction

Docker [2] is a lightweight virtualization system, with the advantages of continuous integration, version control, portability and fast migration, which has been widely used in industry. Different from virtual machines [5], containers share the operating system kernel with the underlying host, which enables rapid deployment with low performance overhead. Docker packages everything needed by an application as an image, including runtime tools, system tools, and system dependencies. Images are layered and read-only, and adopt copy-on-write to reduce the usage of storage space. Despite being lightweight, container's startup is much slower in practice due to deploying image and file-system provisioning bottlenecks. The startup of non-local containers includes two processes: deployment and cold-start.

The non-local containers images must be first downloaded from remote registry, then stored in local disk [9]. Note that the image downloading latency is usually determined by the image size, network dynamics and available bandwidth. To cope with slow downloading, a method based on peer-to-peer is adopted by some cluster management systems (such as Tupperware [12], Borg [15] etc.) to accelerate the distribution of package. Slacker [10] accelerates container deployment by lazily pulling image data when needed. Unfortunately, these pioneer work all focus on reducing the downloading latency, but fail to take the image storing latency into consideration. When the image layers are downloaded, they must be stored layer by layer sequentially into local disk. Note that the upper layer downloading may complete first but still have to wait for the lower ones, leading to a long storing latency after the network download is completed. Our analysis shows that with a network speed of 100 Mbps, image storing accounts for at least 23.5% of container deployment latency, and should be carefully analysed and studied to improve the overall deployment latency.

After successful deployed in local disks, the containers are ready to be launched to provide certain services with a cold-start latency. It widely agrees that containers are usually short-lived and dynamically activated/deactivated according to real time service demands such as serverless computing, hence the long cold-start latency severely hinders the service quality [7]. Moreover, the cold-start latency of launching multiple container simultaneously increases significantly with the container number. For example, launching 20 containers is about 7 times slower than launching 1 container [16]. According to the studies conducted by Google Borg [15], the median task cold-start latency is 25 s, and above 80% of the latency is caused by the slow I/O speed of local disk. To mitigate this problem, Akkus et al. [7] try to lower the function instances from container to separate process to share libraries with other functions of an application. Oakes et al. [13] create a cache of pre-warmed Python interpreters to speed-up the I/O process. However, they either weaken the function isolation by sharing libraries or are designed for specialized system with limited application scenarios.

Facing the above problems, it is desired to design a container acceleration framework to speed up both deploying and cold-start without loss of generality. Recently, the development of new hardware, i.e. non-volatile memory (NVM) [17], brings a new opportunity. NVM shows excellent characteristics of non-volatility, byte addressing, and superior reading performance. However, due to the high cost of NVM, its size is usually limited, which makes it impossible to store all images in NVM. To make full use of the limited NVM resource, and accelerate the deployment and cold-start of container, we should carefully select the image layers and schedule the NVM resource accordingly.

To address this issue, we conduct an in-depth analysis on bottlenecks of container deployment and cold-start on the Top-134 images downloaded more than 1 million times on the Docker Hub [3]. Based on our analysis, we propose a N-Docker framework to optimize the deployment and cold-start of container with limited NVM resource. The main contributions of this paper are as follows:

- We analysis the deployment and cold-start on the Top-134 images and find two key issues of container startup: (1) container deployment latency can be greatly reduced by improving the image layer storing; (2) container cold-start only requires a small part of files in the image. Based on these two findings, we discuss the opportunity and challenges of adapting NVM to speedup container startup.
- A N-Docker framework is designed to improve container deployment and cold-start. We focus on the storing layers and leverage NVM to accelerate its storing. Furthermore, we detect and write the hot image files required by container cold-start to NVM in order to reduce cold-start latency.
- We implement N-Docker, a NVM-HDD hybrid docker storage framework. The experimental results demonstrate that N-Docker achieves the same performance as NVM-Docker. Moreover, N-Docker can reduce the size usage of NVM by 88.22%. Compared to traditional Docker which stores all images in hard disk, N-Docker can speed up the deployment and cold-start of containers by 1.21X and 2.96X separately.

The rest of this paper is organized as follows. We discuss the design and implementation of N-Docker in Sect. 2. In Sect. 3 we evaluate the effectiveness of N-Docker as well as the overhead. In Sect. 4 we discuss the related work. Finally, Sect. 5 concludes this paper.

2   Design and Implementation

2.1   Opportunities and Challenges from Emerging NVM

Container technique has been widely used in the industry during the past few years [12,15]. In this section, we introduce the deployment and cold-start of container in detail and discuss the opportunities of NVM.

Time-Consuming Image Deploying. Before a container can be started up, its image has to be downloaded and stored from the Internet first. In detail, the data downloaded from the Internet is just some compressed files, which will be decompressed and stored as layered image later. We do a lot of research on the Top-134 images downloaded more than 1 million times on the Docker Hub and find that the average deployment latency of the Top-134 images is about 20.7 s, and find that storing latency accounts for 23.4% of deploying time. As a result, the storing latency should also be considered during container deployment acceleration.

Slow Container Cold-Start. Similarly, container cold-start is also slower in practical cases due to the poor performance of local disk I/O. We do comprehensive analysis and research on the Top-134 images and find out only a few base files in the large image, called *Hot Image File* (HIF), are required during the cold-start for different types of containers. The HIF usually includes bin files, system dependencies, application files, and execution engines, and will be access when a container is launched.

Now the server platform supports NVM in the form of NVDIMMs [6]. However, naively store the whole image into the NVM is not realistic in practical cases, since the price of NVM devices is relatively high. To this end, we must take the characteristics of container into consideration to accelerate container startup with limited NVM resource. Through the above analysis and discussion, we conclude that the deployment and cold-start of container are the main bottleneck of container startup. In this section, we introduce N-Docker, a NVM-HDD hybrid docker storage framework, to speed up its deployment and cold-start.

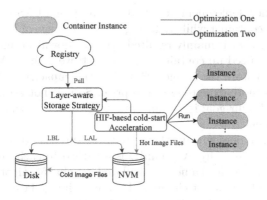

Fig. 1. N-Docker overview

2.2   Overview

According to the previous researches of the process on container deployment, analysis of the files used for container cold-start and the characteristics of NVM, we design N-Docker based on the following three objectives:

- **Container deployment acceleration:** In general, we can accelerate the container deployment by storing the images in NVM. However, considering the capacity limitation of NVM devices, we design a container deployment strategy to store the bottleneck layers in NVM instead of the entire images in NVM during the process of container deployment.
- **Container cold-start improvement:** Similarly, to improve the container cold-start via NVM, we also need detect and store the HIF in NVM, achieving fast container cold-start and high NVM resource utility at the same time.
- **Generality and transparency:** In terms of generality, N-Docker should support a wide range of workflows to accelerate deployment and cold-start. As for transparency, N-Docker should support these workflows without modifying the application or weaken the isolation.

Figure 1 describes the overview of the N-Docker architecture. It is clear that N-Docker has two core components. According to the finding that storing image is one of the reasons for the slow deployment of containers, we design *Layer-aware Storage Strategy* (LASS) to store partial image in NVM during the deployment of container. Based on the finding that the cold-start of container only needs Hot Image Files, we propose the *HIF-based Cold-start Acceleration* (HBCSA) method to acquire HIF, store them in NVM, and write other cold image files back to hard disk.

2.3   Layer-Aware Storage Strategy

To achieve container deployment acceleration, we design LASS which speeds up the deployment of container while reducing the space usage of NVM. The latency of container deployment is mainly resulted from the download image and storing image. In order to speed up container deployment, we take NVM instead of the traditional disks to store the images. However, the capacity of NVM is usually limited, since it is more expensive in the price. So it is not economical to store all images in NVM. Our goal is to minimize usage of NVM while enabling rapid container deployment.

During the image deployment, there are three threads downloading different layers of one image parallelly. A ChainID is attached to each downloaded layer to identify a layer, and its value is calculated by sha256 algorithm according to layer's diffID and its parent chainID. Therefore, image layers must be stored layer by layer sequentially into local disk. The different layer sizes usually lead to different download latency. That is, one layer may still in download while the others have finished. We refer to the layer being downloaded lastly as *Last Downloading Layer* (LDL), which is usually the largest layer. As shown in Fig. 3, the lower image layers of LDL is called *layers below LDL* (LBL) and its status is *Pull complete*, which has already been downloaded and stored in the hard disk. The higher image layer above LDL is called *Layers above LDL* (LAL) and its status is *Download Complete*, which has been downloaded while not yet stored in the hard disk. The storing of LAL will be postponed until LDL has been downloaded and stored due to layer sequential storing. Once the LDL is downloaded, LAL and LDL will be stored together, which may lead to high latency of intensive writes after the network download is completed. To address this issue, we design LASS which can identify LDL and only store LDL and LAL in NVM.

Layer-Aware Storage Strategy Design. To take advantages of fast I/O speed of NVM with a constrained capacity, we need design LASS to determine the layers that should be stored in NVM and disk, towards the goal of fastest storing and fewest NVM usage. Hereafter, we investigate different schemes and find the optimal strategy. To answer this question, we use a *Boundary Layer* (BL) to divide the layers of one image into two parts, namely L_n (above BL) and L_d (below BL) with the sizes of S_n and S_d, to be stored in NVM and local disk, respectively. BL stores in NVM and its size is S_{bl}. We suppose that an image has N layers, LDL is the K_{th} layer and BL is the M_{th} layer. In order to find the optimal layer aware storage strategy, we introduce two indicators as the criteria to evaluate the performance of different strategies, namely T_{Total} (the total latency of container deployment) and U_{Total} (the total NVM usage of container deployment). T_{Total} and U_{Total} are calculated according to the equations as follows.

$$T_{Total} = T_D + T_{Disk} + T_{NVM} \tag{1}$$

As shown in Eq. 1, T_{Total} consists of three parts. T_D represents the latency caused by the download image. T_{Disk} and T_{NVM} are the latencies storing image in hard disk and storing image in NVM after the network download is completed, separately.

$$T_{Disk} = \begin{cases} 0 & M \leq K \\ \alpha * (S_{LDL} + S_{LAL} - S_n - S_{bl}) & M > K \end{cases} \tag{2}$$

T_{Disk} is equal to the total image size written to the hard disk after the network download is finished divided by the hard disk write speed. As shown in Eq. 2, α is the reciprocal of the hard disk write speed. When the network download is completed, LBL has been stored. Therefore, if BL is LBL or LDL, T_{Disk} is 0. If BL is LAL, layers between LDL and BL store in disk after network download is completed.

$$T_{NVM} = \begin{cases} \beta * (S_{LDL} + S_{LAL}) & M \leq K \\ \beta * (S_{bl} + S_n) & M > K \end{cases} \tag{3}$$

T_{NVM} is equal to the total image size written to NVM after the network download is completed divided by the NVM write speed. As shown in Eq. 3, β is the reciprocal of NVM write speed. After network download is completed, if BL is LBL or LDL, layers between LDL and the highest layer store in NVM. If BL is LAL, layers between BL and the highest layer store in NVM.

$$T_{Total} = \begin{cases} T_D + \beta * (S_{LDL} + S_{LAL}) & M \leq K \\ T_D + \alpha * (S_{LDL} + S_{LAL} - S_{bl} - S_n) + \beta * (S_{bl} + S_n) & M > K \end{cases} \tag{4}$$

$$U_{Total} = S_{bl} + S_n \tag{5}$$

Combining 1, 2, and 3, we can easily get Eq. 4. Since the write speed of NVM is several orders of magnitude faster than that of hard disks, it is assumed that $\alpha \gg \beta$. Equation 5 shows that U_{Total} is the space usage of NVM, which is another performance indicator. Because we only care about latency caused by storing image, we set T_D as a constant. Our goal is to minimize U_{Total} on the premise of minimizing T_{Total}. According to the location of boundary layers, we design the following three strategies as shown in Fig. 2.

Strategy 1 set the Boundary Layer as the LBL. At this time, $M < K$ and $S_{LDL} + S_{LAL} < S_{bl} + S_n$. The result is shown in Eq. 6. T_{Total} is the sum of T_D and the latency caused by storing S_{LAL} to NVM, and U_{Total} is S_n.

$$\begin{cases} T_{Total} = T_D + \beta * (S_{LDL} + S_{LAL}) \\ U_{Total} = S_{bl} + S_n \end{cases} \tag{6}$$

Strategy 2 takes LAL as the Boundary Layer. At this time, $M > K$. The result is shown in Eq. 7. Because $\alpha \gg \beta$, the delay T_{Total} of strategy 2 is higher than that of strategy 1, so strategy 1 is better than strategy 2.

$$\begin{cases} T_{Total} = T_D + \alpha * (S_{LDL} + S_{LAL} - S_{bl} - S_n) + \beta * (S_{bl} + S_n) \\ U_{Total} = S_{bl} + S_n \end{cases} \tag{7}$$

Strategy 3 selects the LDL as the Boundary Layer. At this time, $M = K$, $S_{bl} + S_n = S_{LDL} + S_{LAL}$. The result is shown in Eq. 8. The strategy 3's T_{Total} is the same as the strategy 1. In strategy 1, $S_{LDL} + S_{LAL} < S_{bl} + S_n$. So the U_{Total} of strategy 3 is smaller than strategy 1, strategy 3 is better than strategy 1.

$$\begin{cases} T_{Total} = T_D + \beta * (S_{LDL} + S_{LAL}) \\ U_{Total} = S_{LDL} + S_{LAL} \end{cases} \tag{8}$$

It is easy to conclude that strategy 3 is the best choice with the lowest latency. On the premise of enabling rapid deployment of containers, the usage of NVM is minimized. Therefore, we adopt strategy 3 and set the Boundary Layer as LDL, as shown in Fig. 3. While LBL are storing, other image layers are also being downloaded from the network. We chose to store LBL in disk to reduce the usage of NVM. When LAL and LDL are being stored, the network download process has ended. At this time, the latency of container deployment depends entirely on the storing LDL and LAL. We store LDL and LAL in NVM. As a result, containers deployment is significantly accelerated.

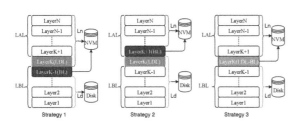

Fig. 2. Strategy overview

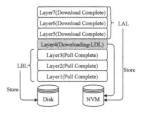

Fig. 3. Image storage

2.4   HIF-Based Cold-Start Acceleration

To achieve fast cold-start, we propose HBCSA to speed up the cold-start of containers while reducing the usage of NVM. Note that containers cold-start only requires HIFs, hence only storing the HIFs in NVM devices during cold-start can speedup the cold-start. To accelerate the cold-start of container for the first time, we execute static analysis to identify image layers including HIFs, and store them in NVM during the deployment of container. The other is dynamic analysis. HIFs obtained by static analysis are redundant. Therefore, we execute dynamic analysis to obtain accurate HIFs during the cold-start of container. Static and dynamic analysis are detailed separately as follows.

Static Analysis. In the process of deploying the image, if the layer contains HIFs, the whole layer will be stored in NVM to obtain some HIFs initially. Dockerfile consists of a series of commands which can be obtained by a simple "string parsing" method. From Sect. 2.1, we can know that the HIFs include Bin files, system dependencies, application files, and execution engines. Bin files and system dependencies account for a small proportion of the total HIFs. And the image layers containing bin files or system dependencies are generally large. So an image layer that contains only bin files or system dependencies is stored in Disk without wasting NVM resources. For an image layer containing the execution engine or application files, we choose to store it in NVM as a coarse-grained HIFs.

Dynamic Analysis. Once the container cold-start is finished, the application files in need will be loaded into memory. Dynamic analysis mainly analyzes the necessary files and file dependencies in the image by tracking system calls, changes of files or directories, and running of processes. These files are HIFs. In order to improve the utilization rate of NVM, we only store HIFs in NVM, with other image files brushed back to the hard disk. In this way, the utilization rate of NVM is greatly improved, and the cold-start speed of the container is also accelerated. Compared with the traditional architecture, HIFs will not be replaced back to disk due to memory collection in the multi-container scenario with the same host. In this way, container running reduce the disk I/O overhead caused by missing page interruptions. When a container is suspended for a period of time or restarted, it can start running faster by reducing I/O latency caused by page missing interruptions.

3   Evaluation

We implement N-Docker, a NVM-HDD Hybrid Docker Storage Framework to accelerate container deployment and cold-start. In order to evaluate the performance of N-Docker, we conduct a comparative experiment between N-Docker and native Docker, and a comparative experiment between N-Docker and NVM-Docker. Our experiments are based on 134 images in Docker hub, which are downloaded more than 1 million times.

3.1   Experiment Setup

Environment. Table 1 provides a detailed description of memory configuration. We simulate NVM as a fast block device [4] and install ext4 with DAX (direct access) [1] on it. The machines interconnect with each other in 1 Gbps network. We implement N-Docker based on Ubuntu 16.04 and Docker 18.06-ce.

Table 1. Memory configuration

	DRAM	NVM
Capacity	4G	4G
Channels	1	2
Bandwidth	8 GB/s	3.6 GB/s (Read)
		1.3 GB/s (Write)
Read/Write Latency (Normalized to DRAM)	1	4.4x (Read)
	1	12x (Write)

3.2   Deployment

N-Docker divides the image into two parts, one of which is stored in NVM and the other is stored in Disk. In this section, to compare the performance of N-Docker with that of NVM-Docker, we deploy Top-134 containers through N-Docker and NVM-Docker respectively. The experimental results of container deployment latency are shown in Fig. 4. As can be seen from the figure, the deployment latency of NVM-Docker container in each category is larger than that of N-Docker by more than 97%. Therefore, it can be concluded that N-Docker divides the image into two parts without incurring additional latency. The space usages of NVM of N-Docker and NVM-Docker are shown in Fig. 5. With regards to the category of distro, N-docker uses the same space size of NVM as NVM-Docker. The main reason is that the distro category is the basic image, and the only one layer or the first layer accounts for most of the entire image size. In this case, we store the entire image in NVM. The category of web fwk, which has the most decrease in NVM usage, has a 38.5% decrease in NVM usage. The reason for it is that the LDL of this category of image is located further back in the image layer, and more image layers are stored in DISK. In addition, N-Docker's container deployment is almost as fast as NVM-Docker. On average, N-Docker's NVM usage is 28.53% less than NVM-Docker.

In Sect. 2, we have seen the latency of various container deployments, with downloading latency accounting for 76.6% and storing latency accounting for 23.4%. In this section, we evaluate the performance of N-Docker in container deployment by deploying Top-134 containers separately through N-Docker and Docker. Experimental results are shown in Fig. 4. The most significant drop in container deployment latency is 26.7% for the category of distro, as the entire image of distro is stored in NVM. The percentage of image layer stored in NVM is the highest in all categories, and the benefits brought by accelerating container deployment through NVM is the largest. The lowest reduction in container deployment latency is 19.3% for the category of web fwk, since the percentage of image layer stored in NVM by web fwk is the lowest in all categories, and the benefits of accelerating container deployment through NVM is the least. On average, N-Docker's container deployment latency is 21.14% lower than Docker's. We speed up the entire container deployment process by reducing the latency of storing images. So we evaluate the latency caused by NVM-Docker, N-Docker and Docker in the process of storing images. The results are shown in Fig. 6. On average, the latency of the N-Docker storing image after the network download is finished is reduced by 90.3% compared to Docker, and is comparable to NVM-Docker.

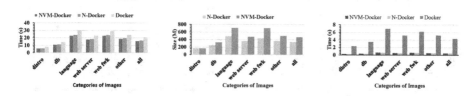

Fig. 4. Deployment time   **Fig. 5.** NVM usage in docker deployment   **Fig. 6.** Storing image time

3.3 Cold-Start

We store Hot Image Files in NVM to speed up the cold-start of the container. In order to verify that the Hot Image Files selected by our scheme is indeed the file necessary for container cold-start, we conduct the experiments on the Top-134 containers's cold-start through N-Docker and NVM-Docker. The result of cold-start latency is shown in Fig. 7, which demonstrates that N-Docker's cold-start latency is only 2% slower than NVM-Docker. The space usage of NVM is shown in Fig. 8. As can be seen from the figure, the largest reduction in NVM usage is 97.12 % for the category of distro, as the distro class is the basic image. The vast majority of such images are auxiliary tools, package managers, and dependencies. The files needed for by the category of distro are very few. The minimum reduction in NVM usage is 70.12% for the category of web server. This type of container contains more executable files, configuration files and the underlying execution engine. Taking the JVM as an example of execution engine, common versions of JVM exceed 100M, which makes Hot Image File larger. In summary, the Hot Image File used by N-Docker contains almost all the files necessary for container cold-start, and the NVM's usage of N-Docker is 88.22% less than NVM-Docker.

In order to evaluate the cold-start performance of N-Docker, we compare the cold-start latency of containers by N-Docker and Docker. As shown in Fig. 7, the maximum reduction of cold-start delay is 76.1% for distro container. The reason for it is that containers of the distro category are the simplest, which requires only a small number of files and then builds an independent execution environment. The cold-start latency of containers in the distro category is mainly resulted from the overhead of I/O. Containers of the distro category get higher promotion by using NVM to store Hot Image Files to speed up container cold-start. The minimum reduction in container cold-start latency is 62.6% in the category of web server, as containers in the category of web servers are the most complex. Such containers's cold-start requires not only building an independent execution environment, but also starting the server's daemon process. Therefore, accelerating container cold-start by taking NVM devices to store Hot Image Files gets the least benefits. On average, N-Docker can reduce the latency of the containers's cold-start by 33.8%, compared to that of Docker.

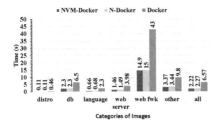

Fig. 7. Cold-start time   **Fig. 8.** NVM usage

4 Related Work

This research work is to accelerate container deployment and cold-start based on emerging NVM. The slow deployment and cold-start of containers has also been widely discussed by other researchers.

Deployment: Some cluster management systems, such as Tupperware [12], Borg [15], use peer-to-peer technique to reduce the load on the central repository and speed up packet distribution. However, they are not applicable to Docker images. Slacker [10] accelerates container deployment by reducing network I/O, which lazily pulls image data when needed. However, slacker needs a longer time to build image and a greater demand for storage in the registry. Cider [9] changes the working node's local Docker storage to an all-nodes-sharing network storage, allowing image data to be loaded on demand when deploying containers.

Cold-start: CNTR [14] divides the traditional image into two parts: the "fat" image contains complete functions, while the "slim" image contains only the core files needed by common user-case. CNTR reduces image size, which makes Docker lighter. However, CNTR incurs overhead for some benchmarks. Uniker-nel [11] uses the library OS [8] to screen out the required operating system components to construct a lighter-weight executable application operating system. But Unikernels cannot debug and require static linking tools in the library

OS. SAND [7] weakens the function instances from container-level isolation to separate process-level to share libraries, which only require to be loaded into container once, with other functions of an application. SOCK [13] create a cache of pre-warmed Python interpreters to avoid that Python runtime is initialized repeatedly.

Existing work does not consider storing image during container deployment, weaken function's isolation or cannot be applied to general containers for cold-start acceleration.

5 Conclusion

Rapid deployment and cold-start of container are very important, such as in the serverless computing scenario. To achieve this goal, we leverage the emerging NVM device and design N-Docker, a NVM-HDD hybrid docker storage framework. N-Docker stores LAL and LDL in NVM during container deployment and Hot Image Files in NVM during container cold-start. Through extensive experiments, we validate the efficiency of N-Docker by the fact that it can accelerate the median container deployment by 1.21X and cold-start by 2.96X with very few NVM. Compared to NVM-Docker, which stores all images in NVM, the proposed N-Docker achieves the same performance improvements while reducing the usage of NVM by 88.22%.

References

1. Add support for NV-DIMMs to ext4. https://lwn.net/Articles/613384/
2. Docker. https://www.docker.com/
3. Docker hub. https://hub.docker.com/u/library/
4. Emulate persistent memory. http://pmem.io/2016/02/22/pm-emulation.html
5. Linux kernel virtual machine. https://www.linux-kvm.org/page/MainPage
6. Nvdimm. https://www.micron.com/products/dram-modules/nvdimm
7. Akkus, I.E., et al.: SAND: towards high-performance serverless computing. In: Proceedings of the 2018 USENIX Annual Technical Conference, pp. 923–935 (2018)
8. Belay, A., Bittau, A., Mashtizadeh, A.J., Terei, D., Mazières, D., Kozyrakis, C.: Dune: Safe user-level access to privileged CPU features. In: Proceedings of the 10th USENIX Symposium on Operating Systems Design and Implementation, pp. 335–348. USENIX Association (2012)
9. Du, L., Wo, T., Yang, R., Hu, C.: Cider: a rapid docker container deployment system through sharing network storage. In: Proceedings of 19th International Conference on High Performance Computing and Communications, pp. 332–339. IEEE (2017)
10. Harter, T., Salmon, B., Liu, R., Arpaci-Dusseau, A.C., Arpaci-Dusseau, R.H.: Slacker: fast distribution with lazy docker containers. In: Proceedings of the 14th USENIX Conference on File and Storage Technologies, pp. 181–195. USENIX Association (2016)
11. Madhavapeddy, A., Scott, D.J.: Unikernels: the rise of the virtual library operating system. Commun. ACM 57(1), 61–69 (2014)
12. Narayanan, A.: Tupperware: containerized deployment at facebook (2014)

13. Oakes, E., et al.: Sock: rapid task provisioning with serverless-optimized containers. In: Proceedings of the 2018 USENIX Annual Technical Conference, pp. 57–70. USENIX Association (2018)
14. Thalheim, J., Bhatotia, P., Fonseca, P., Kasikci, B.: CNTR: lightweight OS containers. In: Proceedings of the 2018 USENIX Annual Technical Conference, pp. 199–212. USENIX Association (2018)
15. Verma, A., Pedrosa, L., Korupolu, M., Oppenheimer, D., Tune, E., Wilkes, J.: Large-scale cluster management at Google with Borg. In: Proceedings of the 10th European Conference on Computer Systems, p. 18. ACM (2015)
16. Wang, L., Li, M., Zhang, Y., Ristenpart, T., Swift, M.: Peeking behind the curtains of serverless platforms. In: Proceedings of the 2018 USENIX Annual Technical Conference, pp. 133–146. USENIX Association (2018)
17. Xu, J., et al.: NOVA-fortis: a fault-tolerant non-volatile main memory file system. In: Proceedings of the 26th Symposium on Operating Systems Principles, pp. 478–496. ACM (2017)

HPC

MMSR: A Multi-model Super Resolution Framework

Ninghui Yuan, Zhihao Zhu, Xinzhou Wu, and Li Shen[✉]

School of Computer, National University of Defense Technology,
Changsha 410073, Hunan, China
lishen@nudt.edu.cn

Abstract. *Single image super-resolution (SISR)*, as an important image processing method, has received great attentions from both industry and academia. Currently, most super-resolution image reconstruction approaches are based on the deep-learning techniques and they usually focus on the design and optimization of different network models. But they usually ignore the differences among image texture features and use the same model to train all the input images, which greatly influence the training efficiency. In this paper, we try to build a framework to improve the training efficiency through specifying an appropriate model for each type of images according to their texture characteristics, and we propose *MMSR*, a multi-model super resolution framework. In this framework, all input images are classified by an approach called *TVAT (Total Variance above the Threshold)*. Experimental results indicate that our MMSR framework brings a 66.7% performance speedup on average without influencing the accuracy of the results HR images. Moreover, MMSR framework exhibits good scalability.

Keywords: Super resolution · Multi-model · General framework · Classification

1 Introduction

Super Resolution (SR) technique is used to recover a super-resolution[1] image from a single (or a series of) low-resolution image(s). This technique has been widely used in the fields including remote sensing, video, medicine and public security, etc. In recent years, with the wide application of deep learning, more and more researches focus on the study of *single image super resolution (SISR)*.

Since SRCNN [1] is proposed by Dong et al., deep convolution neural work has been the basis of other researches of super resolution. This work starts the deep-learning-based super resolution studies. VDSR [4] is also a revolutionary model in the development of super resolution, in which the residual block is firstly proposed and

---

[1] To distinguish between the output images and the reference images, the output images are called SR (Super-Resolution) images and the reference images are called HR (High-Resolution) images in this paper.

© IFIP International Federation for Information Processing 2019
Published by Springer Nature Switzerland AG 2019
X. Tang et al. (Eds.): NPC 2019, LNCS 11783, pp. 197–208, 2019.
https://doi.org/10.1007/978-3-030-30709-7_16

used in a deep network. SRGAN [8], proposed by Christian Ledig et al., makes use of the generative adversarial network (GAN) [13] in super resolution for the first time.

There are still several problems and challenges in current super resolution studies. Firstly, researchers usually use the same model to train and reconstruct all the images, and they do not pay any attention to the differences of images features. For example, some images are smooth while other images have more textures. In general, for images with relatively simple texture features, a simple network model is enough to obtain satisfactory results, with a relatively short time overhead. Therefore, using the same model to train all the images will usually increase the time overheads and waste some computation resources. Secondly, researchers pay all their attentions on the quality of the result SR images, and they usually ignore the training or reconstruction efficiency. In fact, in particular scenarios, the efficiency is significant as well, such as scenarios having high real-time requirements. Thirdly, there is not a satisfactory criterion that can totally fit how the human eyes feel. MSE-based criteria usually make the output images too smooth, their visual results are usually not as good as expected.

In this paper, we focused on how to solve the first two problems. We found that images in the training dataset usually have different texture features. Some images have simple textures and others have complex textures. And we found that for images with different texture features, the most appropriate models are usually different. According to these observations, we proposed a *multi-model super resolution (MMSR)* framework. MMSR can choose a suitable network model for each image for training. MMSR shortens the training time efficiently without decreasing the quality of reconstruction image. We implement a MMSR framework based on SRGAN [8], and experimental results indicate that using *DIV_2K* as the training set, MMSR can reduce 40% training time on average. Moreover, the MMSR framework shows good stability. The main contributions of this paper are as follows:

(1) We proposed *MMSR*, a multi-model super resolution framework. This framework can choose a suitable model according to the texture characteristics of the input images. Therefore, it can improve the training efficiency without influencing the quality of the output SR images.

(2) We proposed *TVAT (Total Variance above the Threshold)*, a method to classify the training images. This approach can be used to describe the complexity of the image texture, and it does not introduce extra computational overheads. Moreover, since points with low pixel variations have almost no effect on the calculation of image texture, they could be removed to improve the accuracy of classification.

The rest of this paper is organized as follows. Section 2 lists some related works. Section 3 introduces our MMSR framework and the image classification method in detail. In Sect. 4, the performance of MMSR is evaluated and experimental results are given. And finally, in Sect. 5, some conclusions are given.

2 Related Works

In recent years, deep learning has been applied in many areas of image processing and analyzing, including super resolution [1–9]. Reference [1] is a pioneer work that brought super resolution into the deep learning area, in which the authors proposed a simple three-layer convolutional neural network called SRCNN and each layer sequentially deals with feature extraction, non-linear mapping, and reconstruction. The input of SRCNN uses an extra bicubic interpolation to enlarge the resolution of image. But this approach lacks enough high-frequency information and introduces some extra computations. Their later work, FSRCNN [2], removes the bicubic process and adds a deconvolution layer for reconstruction. VDSR [4] is another revolutionary work in the development of super resolution techniques, because the residual blocks are first used in its deep network. Almost all the successive researches on super resolution use residual blocks in their network models. SRGAN [8], proposed by Christian Ledig et al., makes use of (GAN) in super resolution for the first time.

Before the GAN network is used to solve the super resolution problem, the mean square error is often used as a loss function when training the network. Although a high peak signal-to-noise ratio can be obtained in this way, the reconstructed images lose some high-frequency details, which makes people hardly have a good visual experience. Figure 1 [8] describes the whole process of SRGAN, which consists of a generation phase using the Generator Network and an adversary phase using the Discriminator Network. In the last layer of the discriminator network, SRGAN uses perceptual loss to guarantee the quality of the output images. Perceptual loss describes the differences between the generated SR images and the reference HR images. If the perceptual loss of a SR image is larger than the threshold, the SR image will be regenerated.

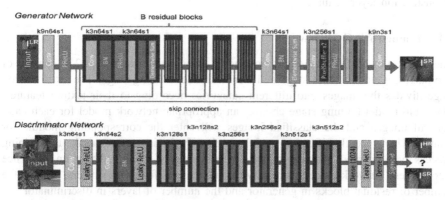

Fig. 1. Architecture of SRGAN with kernel sizes (k), numbers of feature maps (n) and stride (s) specified for each convolutional layer [8].

Most current super resolution approaches using deep learning techniques focus on the optimization of network models as well as the quality of the output SR images.

They did not care much about the efficiency of training, which may cause great waste of computation resources. Therefore, we propose a multi-model SR framework to improve the efficiency of training. Our framework is based on SRGAN, because it is a widely used in current super resolution studies and its reconstruction effect is better than other models.

Another problem concerned by researchers is how to evaluate the quality of output SR images. There are generally two categories of metrics. The first one describes the quality in terms of pixel features, such as MSE (Mean Square Error), PSNR (Peak-Signal to Noise Ratio), SSIM (Structure Similarity), etc. However, under the guidance of such metrics, the texture features of images are usually ignored and the output images tend to be too smooth or too fuzzy. The other one is based on the visual effect of human eyes, such as NIQE (Natural Image Quality Evaluator) [10] and PI (perceptual index). Sometimes the output images are shown and judged by the naked eyes. Obviously, the sharper and the more natural an image is, the better NIQE or PI value it can gets. In recent researches on super resolution, the second category of metrics gradually become the mainstream choice. Therefore, in this paper, we choose PI as the image quality metric. The PI value is calculated using the NIQE method [10].

3   MMSR Framework

In this section, we will first introduce our MMSR framework. The MMSR is composed of a training module and a reconstruction module. The training module trains the models with a train image set and the reconstruction module recovery the LR images to SR images. MMSR has good versatility and different deep learning network models can be integrated into this framework. Then, we will introduce the image classification method, the structure of the multi-model training module and the design of the reconstruction layer in turn.

3.1   Framework Overview

The first part of MMSR is the training module, which is shown in Fig. 2. It consists of two stages, a classification stage and a multi-model training stage. The classification stage divides the images into different categories according to their texture features. The multi-model training stage chooses an appropriate network model for each category of images and the classified images will enter the corresponding module for training. Using the classification module to classify the images can make the training process more targeted, and also improve the efficiency of training. The main difference among these network models is mainly that they use different parameters, such as the number of residual blocks in generator and the number of layers in discriminator.

The second part of MMSR is the reconstruction framework, which is shown in Fig. 3. It consists of four main parts, i.e. the segmentation layer, the classification module, the multi-model training module and the reconstruction module. The classification module and the multi-model training module are the same as those in Fig. 2. The segmentation layer is used to divide the input LR images to be reconstructed into a group of fragments. Then, these fragments enter the classification module and are

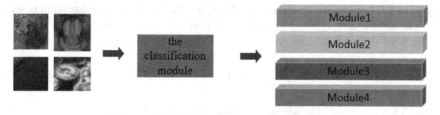

Fig. 2. The architecture of training module.

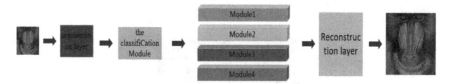

Fig. 3. The architecture of reconstruction module.

classified according to their texture features. After the classification (and network model assignment), these fragments will enter the corresponding modules for reconstruction. And finally, the reconstructed fragments are assembled by the reconstruction layer into a complete SR image.

3.2 Image Classification

In general, different images have different texture features. At present, most of the deep-learning-based SR approaches do not take the influence of the characteristics of the image on the training or the reconstruction process into account. This paper proposes a classification method to divide the images into several categories based on their texture features with low time overheads.

3.2.1 Total Variance Above the Threshold

We tested some images in order to observe the training methods of different fragments and the features hidden in them. We found that for most images, the more complex the texture of an image is, the longer the training time it requires. Therefore, we try to propose a suitable method to describe the texture feature of an image. The simplest way to describe the texture complexity is usually based on the variance of the whole image. However, we have found that this method does not work well, because the variance of some images with relatively uniform texture is large, although the variance of each point is relatively small. We found the training of these images does not require a very deep network model, but the method to assign the network models to image categories requires some training. Therefore, we consider using an innovative method to describe the variance of the pixel variation between each pixel and its 8 neighbor pixels, as shown in Fig. 4. In this paper, it is called the variance of single pixel (VSP). All VSPs in an image larger than a threshold are added together to obtain the total variance above the threshold (TVAT). The threshold is chosen through tests. We set the threshold to all integers in 0–25 to test the classification effect. We find that when the threshold is set to 5, we can get the best effect, so we choose to set the threshold to 5.

The VSP value of the i-th pixel can be calculated as follows.

$$VSP_i = \sum_{j=1}^{8} \left((Ri, Gi, Bi) - (Rj, Gj, Bj)\right)^2 \tag{1}$$

The TVAT value of the whole fragment can be calculated as follows.

$$TVAT = \left(\sum_{i=1}^{n} VSP_i * judge_{index_i}\right)/(cols * rows) \tag{2}$$

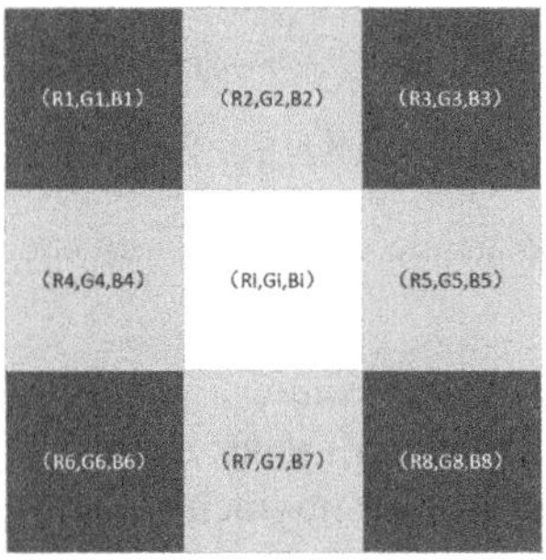

The cells of the 3×3 neighborhood are labeled (R1,G1,B1), (R2,G2,B2), (R3,G3,B3), (R4,G4,B4), (Ri,Gi,Bi), (R5,G5,B5), (R6,G6,B6), (R7,G7,B7), (R8,G8,B8).

Fig. 4. The VSP of a pixel in the 3×3 Neighborhood.

Here *cols* and *rows* represent the number of columns and rows of the fragment respectively, and $judge_{index_i}$ is a step function, which is calculated as follows:

$$judge\_index_i = \begin{cases} 0, & VSP_i < threshold \\ 1, & VSP_i \geq threshold \end{cases} \tag{3}$$

3.2.2   TVAT Values

In this work, TVAT values are used to guide the image classification. In Table 1, *X-Y* means the GAN model has *X* residual blocks in generator and *Y* layers in discriminator. For example, *16-8* means that the GAN model has 16 residual blocks in generator and 8 layers in discriminator. We can find that for most images, the larger the TVAT value is, the more complicated an image is. However, when the depth of the network increases, the results do not always get better.

We randomly selected 80 image fragments from DIV_2K image set to test the recovery quality of these images in different models. We calculated the TVAT of images and found the following observations, as shown in Fig. 5: when the value of TVAT is relatively small (i.e. between 0 and 2), a 4-6 model can get the best performance. When the TVAT value is between 2 and 4, the 2-2 and 16-8 models have better performance. When the value of TVAT is large than 4, the 4-2 and 16-8 models perform best. Therefore, in this paper we classify images according to their TVAT values.

Table 1. The relationship between TVAT and the number of residual blocks in generator and the number of layers in discriminator in GAN.

Image number	TVAT	Perceptual index					
		2-2	4-2	4-6	8-6	16-8	Best
1	0.22316	10.7709	10.7093	8.9474	10.6772	13.4531	4-6
2	0.14125	10.7286	10.2412	9.8015	10.6265	11.0304	4-6
3	5.84019	6.8456	6.7848	6.9019	6.8478	6.4810	16-8
4	0.36391	14.9585	10.4451	14.3677	12.9602	11.1268	4-2
5	1.46280	7.0466	6.9128	7.6555	7.4745	7.7626	4-2
6	4.54511	5.8232	5.7170	6.0661	5.4763	5.3578	16-8
7	4.15810	6.4053	6.4270	6.8263	6.4317	6.6969	2-2
8	2.50022	6.8408	6.9172	6.8703	6.7201	7.0015	8-6
9	0.05513	12.8878	9.0682	14.9559	12.1362	14.2192	4-2
10	3.30724	6.6211	6.2182	6.9568	6.1581	6.1914	16-8
11	3.01382	7.2778	6.9914	6.6588	6.6947	6.5318	16-8
12	0.06423	11.7144	10.3977	10.3883	11.1998	11.4946	4-6
13	4.39549	6.6559	6.5765	6.7881	6.6619	5.8498	16-8
14	0.27429	11.2172	10.5885	9.8657	12.9558	10.6933	4-6
15	0.32317	9.6611	9.0914	8.961	9.6730	9.5313	4-6
16	5.95909	7.7423	7.5606	7.6745	7.7127	7.1994	16-8

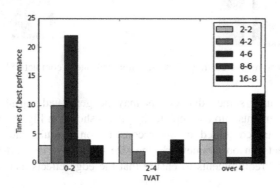

Fig. 5. Different network models are suitable for different TVAT values

3.3 Multi-model Training Module

After image classification, we need to use different models to train each class of images, as shown in Fig. 6. We deploy different training models on different GPU nodes. These models may be completely different kinds of deep learning network models, or the same kind of models with different depths. This paper chooses the second way because no matter for simple texture images or complex images, the recovery quality of SRGAN is better than previous works.

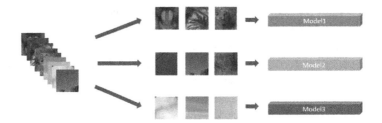

Fig. 6. Image fragments are input into different models based on their texture features.

3.4   Reconstruction Layer

As shown in Fig. 7, the reconstruction layer is used to recovery fragments into a complete SR image. Since different fragments will enter different models for training after an image is segmented, the recovery time of a set of images maybe different. After all the fragments reach the reconstruction module, the reconstruction module combines them into a complete SR image.

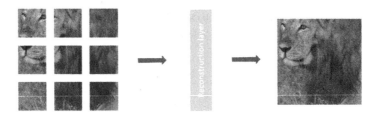

Fig. 7. The recovered fragments are combined into a complete SR image.

At the same time, some edge effects may be generated during the process of assembling the fragments into a complete image. As shown in Fig. 8, some overlapped image fragments are combined in the reconstruction module and the overlapping reduces edge effects. In our framework, the size of the overlap part can be adjusted according to users' requirements to ensure that the edge effects can be eliminated as possible.

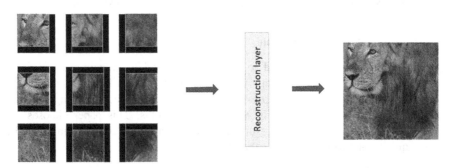

Fig. 8. The overlapped fragments are combined into a complete SR image.

4   Experiment Results

4.1   Environment Setup

We construct a cluster which consists of 4 CPU-GPU heterogeneous nodes to evaluate the performance and scalability of our MMSR framework. The main system parameters of each node are listed in Table 2.

Table 2. System parameters of each computation node.

HW/SW module	Description
CPU	Intel® Xeon® E5-2660 v3 @2.6 GHz x 2
GPU	NVIDIA Tesla K80 x 2
Memory	64 GB
OS	Linux CentOS 7.4
Development Environment	Anaconda 3, Pytorch 1.0

In this work, *DIV_2K* image set is used as both train set and test set. As a widely used image quality metric, the *perceptual index (PI)* value is used by us to compare different SR frameworks or models. The PI value can be calculated using following formula:

$$Preceptual\ index = 12((10 - Ma) + NIQE) \tag{4}$$

In formula (4), *NIQE* (Natural Image Quality Evaluator) is based on the construction of a "quality aware" collection of statistical features based on a simple and successful space domain natural scene statistic (NSS) model. And Ma is an effective and efficient metric to assess the quality of super-resolution images based on human perception, it uses three types of low-level statistical features in both spatial and frequency domains to quantify super-resolved artifacts, and learn a two-stage regression model to predict the quality scores of super-resolution images. Figure 9 shows that a lower perceptual index indicates better perceptual quality. We can see that mathematically that distortion and perceptual quality are at odds with each other [10–12].

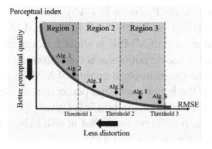

Fig. 9. The relationship between perceptual quality and distortion of images

The advantage of PI value is that different from the traditional image quality metrics. It can better match the senses of the human eye. Moreover, the GAN model itself tries to improve the sensory level of SR images. Therefore, using PI value as the metric can directly reflect the advantages of our MMSR framework.

4.2 Experiment Details

Since we need to classify the images into different parametric models for learning, we tested the training under different parameters, using the SRGAN model. Firstly, we input different texture complexity images into different models for training. We noticed that different types of images have different training effects under different model. In other words, in a limited training time, the training effect and the depth of the model are not necessarily positively correlated.

We choose the Python language to implement the framework. The classification module can divide the image into suitable block. In this experiment, we divide images into three types according to the TVAT value, and the images are sent to different GPU nodes to train. Finally, the whole image is merged into SR image. Our training time is shortened compared with 16-8 SRGAN [8] (i.e. standard SRGAN). The reconstruction effect of model trained by MMSR will not reduce obviously for most of the pictures, and the reconstruction effect of some pictures even increase. The training time of these three methods are listed in Table 3.

Table 3. Training time of different methods.

Method	Training time (s)	Average time (s)
SRGAN (one GPU node)	16415.600 17680.502 17417.912	17171.338
MMSR (one GPU node)	10683.279 10592.480 10598.944	10624.901
MMSR (three GPU nodes)	5921.761 5872.322 5890.451	5894.845

We trained all networks on PyTorch [14–16], which is an open source Python machine learning library based on Torch, used in the field of artificial intelligence. It can be seen the acceleration ratio of MMSR is about 1.62 on one GPU node. And when we use three GPU nodes for acceleration, the acceleration ratio of MMSR is about 2.9.

Figure 10 compares the reconstruction quality of MMSR with other methods. The smaller the value of PI is, the better the visual perception of the result image is, so we can find that the effect of the bicubic method is relatively poor, and the reconstruction effect of MMSR is not much different from that of SRGAN, and even achieves better results for some images (such as 4, 9 and 13).

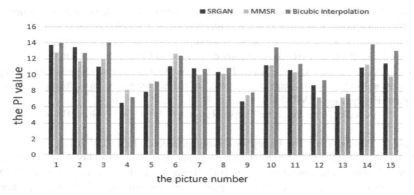

Fig. 10. The comparison of reconstruction quality of different approaches.

5 Conclusions

This paper proposes MMSR, a general multi-model framework for super-resolution image reconstruction. The highlight of our work is to build a general-purpose framework to improve the training or reconstruction efficiency of SR. To implement this framework, we propose a classification method based on experiments (TVAT) to classify the training set. This classification method can divide images into several categories according to their texture characteristics, and we input the images into the most suitable model to train. Experimental results show that the proposed framework is efficient to train the models and do not have too much impact on the training effect. Moreover, because we can use different models in MMSR, our framework has wide applicability.

References

1. Dong, C., Loy, C.C., He, K., Tang, X.: Learning a deep convolutional network for image super-resolution. In: Fleet, D., Pajdla, T., Schiele, B., Tuytelaars, T. (eds.) ECCV 2014. LNCS, vol. 8692, pp. 184–199. Springer, Cham (2014). https://doi.org/10.1007/978-3-319-10593-2_13
2. Dong, C., Loy, C.C., Tang, X.: Accelerating the super-resolution convolutional neural network. In: Leibe, B., Matas, J., Sebe, N., Welling, M. (eds.) ECCV 2016. LNCS, vol. 9906, pp. 391–407. Springer, Cham (2016). https://doi.org/10.1007/978-3-319-46475-6_25
3. Shi, W., et al.: Real-time single image and video super-resolution using an efficient sub-pixel convolutional neural network. In: CVPR (2016)
4. Kim, J., Lee, J.K., Lee, K.M.: Deeply-recursive convolutional network for image super-resolution. IEEE Conf. Comput. Vis. Pattern Recognit. (CVPR) 3(6), 8 (2016)
5. Mao, X.-J., Shen, C., Yang, Y.-B.: Image restoration using convolutional auto-encoders with symmetric skip connections. In: The Annual Conference on Neural Information Processing Systems (NIPS), August 2016
6. Tai, Y., Yang, J., Liu, X.: Image super-resolution via deep recursive residual network. In: IEEE Conference on Computer Vision and Pattern Recognition (CVPR) (2017)

7. Tong, T., Li, G., Liu, X., Gao, Q.: Image super-resolution using dense skip connections. In: IEEE International Conference on Computer Vision (ICCV) (2017)

8. Ledig, C., et al.: Photo-realistic single image super-resolution using a generative adversarial network. In: IEEE Conference on Computer Vision and Pattern Recognition (CVPR) (2017)

9. Wang, X., Yu, K., Dong, C., Change Loy, C.: Recovering realistic texture in image super-resolution by deep spatial feature transform. In: IEEE Conference on Computer Vision and Pattern Recognition (CVPR) (2018)

10. The Pirm Challenge on Perceptual Super Resolution. https://www.pirm2018.org/PIRM-SR.html

11. Blau, Y., Michaeli, T.: The Perception-distortion tradeoff. In: ECCV 2018

12. Wang, Z., Bovik, A.C., Sheikh, H.R., Simoncelli, E.P: Image Quality Assessment: From Error Visibility to Structural Similarity. IEEE Trans. Image Process. **13**(4), 600–612 (2004)

13. Goodfellow, I., et al.: Generative adversarial nets. In: Advances in Neural Information Processing Systems (NIPS), pp. 2672–2680, March 2014

14. Yegulalp, S.: Facebook brings GPU-powered machine learning to Python. InfoWorld, 19 January 2017

15. Lorica, B.: Why AI and machine learning researchers are beginning to embrace PyTorch. O'Reilly Media, 3 August 2017

16. Ketkar, N.: Deep Learning with Python, pp. 195–208. Apress, Berkeley (2017)

HiPower: A High-Performance RDMA Acceleration Solution for Distributed Transaction Processing

Runhua Zhang[1,2,3], Yang Cheng[2], Jinkun Geng[2], Shuai Wang[2], Kaihui Gao[2], and Guowei Shen[1,3(✉)]

[1] Department of Computer Science and Technology, Guizhou University, Guiyang, Guizhou, China
`gwshen@gzu.edu.cn`
[2] Department of Computer Science and Technology, Tsinghua University, Beijing, China
{`zhangrh18,cheng-y16,s-wang17,gkh18`}`@mails.tsinghua.edu.cn`,
`steam1994@163.com`
[3] CETC Big Data Research Institute Co. Ltd., Chengdu, China

Abstract. The increasing complex tasks and growing size of data have necessitated the application of distributed transaction processing (DTP), which decouples tasks and data among multiple nodes for jointly processing. However, compared with the revolutionary development of computation power, the network capability falls relatively behind, leaving communication as a more distinct bottleneck. This paper focuses on the recent emerging RDMA technology, which can greatly improve communication performance but cannot be well exploited in many cases due to improper interactive design between the requester and responder. Our research finds that the typical implementation of confirming per work request (CPWR) triggers considerable CPU involvement, which further degrades the overall performance of RDMA communication. Targeting at this, we propose HiPower, which leverages a batched confirmation scheme with lower CPU utilization, to improve high-frequency communication efficiency. Our experiments show that, compared with CPWR, HiPower can improve the communication efficiency by up to 75% and reduce CPU cost by up to 79%, which speeds up the overall FCT (Flow Completion Time) by up to 14% on real workflow (Resnet-152).

Keywords: RDMA · Distributed transaction processing · Batched confirmation · One-by-one confirmation

1 Introduction

Distributed transaction processing (DTP) has become a practical problem with extensive study in system building. Generally speaking, performance optimization for DTP systems can be summarized as computation acceleration and communication acceleration. In the past decades, the computation power has been

© IFIP International Federation for Information Processing 2019
Published by Springer Nature Switzerland AG 2019
X. Tang et al. (Eds.): NPC 2019, LNCS 11783, pp. 209–221, 2019.
https://doi.org/10.1007/978-3-030-30709-7_17

greatly improved due to the rapid development of hardware accelerators, such as GPUs and TPUs [16]. Whereas the communication capability, although also making some progress, cannot match the speed of computation enhancement and is left as the major bottleneck in DTP. A series of related works have shown that network performance now has a substantial impact on the efficiency of DTP [2,15,16]. Considering this, the recent emerging RDMA technology is widely concerned and is believed to remedy the communication deficiency in DTP.

There are two types of RDMA operations (i.e., one-sided RDMA and two-sided RDMA) widely used today. Between them, the former one is more pursued in DTP scenarios, because it directly accesses the memory of remote server without involving kernel and remote CPUs, which has been widely implemented in some applications like the Key-Value system [5,17], etc. However, pure one-sided RDMA is not suitable for the distributed applications that the receiver needs to perceive data. Some current RDMA designs use the native `RDMA_WRITE_WITH_IMM` operation to ensure that the receiver can perceive data, which also have some distinct drawbacks, and we argue as follows:

First, one-by-one message confirmation triggers much CPU overhead. Concretely speaking, the `RDMA_WRITE_WITH_IMM` operation first writes user-data to the remote memory in a context-oblivious way and then uses the immediate value to notify the receiver. The receiver can get the address of user-data written before from the immediate value. It provides higher communication performance than two-sided RDMA and TCP/IP. However, CPU is involved in the confirmation of each message, which incurs much overhead and affects the overall communication performance, especially when CPU resource is scarce or burnt for other processing logic. The performance drawbacks can become more distinct in many-to-one primitives like *gather* or high-frequency one-to-one transmission scenarios.

Second, one-by-one recycling reusable memory is an inefficient way with high-concurrent communication flows. In order to support concurrent operations and avoid the expensive cost of temporary registration[1], both sender and receiver register multiple MRs for data transmission. The receiver sends feedback to the sender after getting the user-data, which means the memory can be reused. Usually, the feedback can be achieved by two-sided RDMA. However, too many send/receive operations cause unacceptable communication overhead on both sides.

Targeting at the drawbacks, we design HiPower, which uses a batched confirmation mechanism to improve communication performance and reduce CPU utilization. Compared with existing works, HiPower enjoys three main advantages to achieve faster communication.

(1) **More efficient message confirmation**. HiPower allows the receiver to perceive messages through a well-designed bitmap and reduce the number of RECV operations, thereby reducing CPU involvement and improving communication efficiency.

---

[1] The cost includes the time to register MRs and exchange necessary information of MRs.

(2) **More economic reusability and better quality of service (QoS).** HiPower also uses the bitmap flag to represent batched reusable MRs, which can effectively improve the memory utilization, as well as improve the concurrency. In addition, HiPower guarantees better quality of service through pre-registered reusable memory pools.

(3) **Strong compatibility and usability.** HiPower is implemented as a middle layer between distributed applications and RDMA communication libraries, thus it keeps transparency to the upper layer and provides performance boosts for various DTP applications.

The rest of this paper is organized as follows: Sect. 2 briefly introduces the background and motivation of our work. Section 3 describes the design details of HiPower. Section 4 shows the results of the experiment and proves the outperformance of HiPower. Section 5 includes the related work and, Sect. 6 concludes the paper.

2   Background and Motivation

2.1   Typical Communication Pattern in DTP

DTP partitions a task among multiple servers for data processing and synchronization. Figure 1 illustrates the typical DTP architectures in practice, including Parameter Server (PS)-based, Ring-based, Map/Reduce-based, etc. Recent studies have shown that the performance of a single GPU has improved by 35× compared to that in previous years [16], but communication can hardly match the speed. Among the typical architectures for DTP tasks, the communication bottleneck is becoming more distinct. In order to mitigate the communication bottleneck, high-performance communication methods represented by RDMA are gaining more attention in the field of DTP. However, RDMA suffers from a couple of essential drawbacks, and requires us to carefully design to well embrace communication capabilities [3,10,11,19].

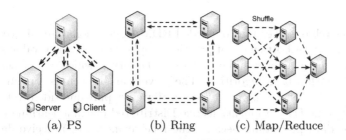

Fig. 1. Distributed transaction processing architecture

2.2   Background on RDMA

RDMA is well known for its zero-copy and kernel-bypass features, when compared with the TCP counterpart. Traditional TCP cannot serve high-speed dataflow well since it involves complicated kernel processing and at least two copies. In contrast, RDMA sinks the protocol stack to hardware and thus avoids the overhead of context switching. RDMA supports both one-sided and two-sided communication operations. As for one-sided RDMA, user-level applications can directly access the memory of a remote node. Note that the remote node is unaware and do not need CPU to involve. RDMA_WRITE and RDMA_READ are two typical one-sided operations. As for two-sided RDMA, operations must appear in pairs. More specifically, the RDMA_RECV operation should be prepared before launching an RDMA_SEND operation. One-sided operations achieve higher performance than two-sided operations [11].

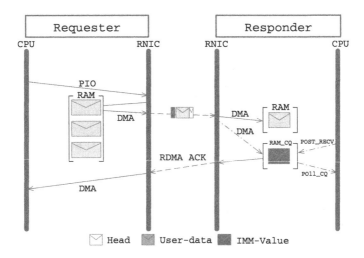

Fig. 2. The whole process of RDMA_WRITE_WITH_IMM operation.

Our research observes that one-sided RDMA operations can effectively solve the communication bottleneck of practical applications like key-value store systems. For example, a client submits data to the server, which can be achieved by RDMA_WRITE with a hash algorithm. The server can be seen as a memory-based database and don't need to perceive data. However, pure one-sided RDMA is not suitable for some DTP applications like Distributed machine learning (DML) or Distributed data processing because the receiver needs to perceive data. Most of current designs use RDMA_WRITE_WITH_IMM operations for data transfer and notification. The RDMA_WRITE_WITH_IMM operation first writes user-data to the remote memory in an unaware way and then uses the immediate value to notify the receiver. The receiver can obtain the storage address of user-data from the immediate value. It provides higher communication performance than two-sided RDMA and TCP/IP. Figure 2 presents the workflow of this operation.

However, in order to receive the immediate value, the receiver needs to prepare the RDMA_RECV event in advance as the two-sided RDMA. It means that the CPU needs to participate in the confirmation of each message, which triggers much CPU overhead and affects the overall communication performance to some extent. It becomes more serious in many-to-one primitives (e.g. Gather) or high-frequency one-to-one transmission scenarios. High CPU overhead is unacceptable to us for the reason that a large number of distributed applications are deployed in the cloud, where strong hardware is provided to speed up computation. Recent studies have shown that the CPU is a precious commodity in cloud service [13]. It's necessary to reduce CPU participation. In addition, in order to support concurrent operations and avoid the expensive cost of temporary registration, both sender and receiver register multi-MRs for data transmission. The receiver sends feedback to the sender after getting the user-data, which means the memory can be reused. Usually, the feedback can be achieved by two-sided RDMA. However, too many send/receive operations incur significant communication overhead on both sides.

3   Design for HiPower

In HiPower, we focus on improving the communication performance in DTP based on one-sided RDMA_WRITE operation, which has been proved higher efficiency than the two-sided RDMA_WRITE operation in previous works [5,11,17].

3.1   HiPower Overview

There are two roles involved in HiPower: the requester and the responder. The requester acts as a data generator, which will transport user-data to the responder. The responder acts as a data consumer and will process the user-data it received. The architecture of HiPower is shown in Fig. 3.

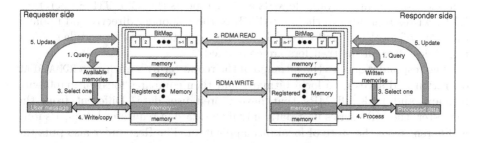

Fig. 3. The architecture of HiPower

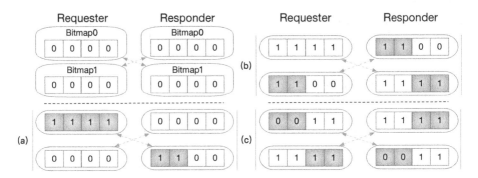

Fig. 4. The design of bitmap.

Both the requester and the responder first perform initialization operations and then take further actions according to their business logic. At the initialization stage, both requester and responder register a number of Memory Regions (MR). Every MR in each side is associated with a mirrored one in the peer's side with a unique id, respectively. These MRs can be registered in the CPU or GPU memory, which we call MRs-Pool. As shown in Fig. 4, we use an extra MR called bitmap0 on each side to record the transmission state of local MRs. Meanwhile, we use an extra MR called bitmap1 on each side to record the transmission state of the peer's MRs. Each item in the bitmap is associated with one local memory. There are two different states for each item. The state of each item will be converted once the related memory is changed.

The requester launches the RDMA_READ operation to query the state of remote bitmap0 and saves it to local bitmap1 and then compares local bitmap0 with bitmap1 to collect all available MRs. For the requester, the MR is available only if the state of local bitmap0's item is same as the state of local bitmap1's item. To limit the frequency of querying event and better overlap computation and communication, we put the reusable MRs of the responder into a pool and perform the RDMA_READ operation again when the number of reusable MRs in the pool below a certain threshold (e.g., 40% of the total MRs). The requester can select one or more available MRs and then posts the RDMA_WRITE operation. RNIC writes user-data to the related MR of the responder directly and updates the state of local bitmap0's item. Last, the requester sends an end-flag to the responder after all the transfers are completed.

The responder queries the bitmap0 of the requester by launching a RDMA_READ operation and saves it to local bitmap1 and then compares local bitmap0 with bitmap1 and finds all MRs that have stored user-data. Different from the requester, the MR carries user-data only if the state of requester bitmap1's item is different from the state of local bitmap0's item. The Responder also puts the MRs into a pool and then processes them according to its business logic.

3.2   Remarkable Advantages of HiPower

The bitmap design in HiPower enjoys three remarkable advantages compared to the baseline solutions, which we summarize as follows.

Batched Confirmation and Recycle Mechanism. In order to perceive the user-data, RNIC needs to confirm each packet and Requester needs to recycle reusable memory, which incurs a large amount of CPU utilization and significantly affects the communication efficiency. HiPower mitigates this problem with batched confirmation and recycles mechanism, which can perceive multiple data-MRs and obtain multiple reusable MRs with the bidirectional RDMA_READ operation. The reusable MRs will be put into a pool, which can be directly obtained from the pool when the next RDMA_WRITE operation is performed. The Responder also puts the MRs into a pool and then processes them according to its business logic.

Strong Quality of Service. Native RDMA lacks quality of service (QoS). For example, as for many-to-one communication under the PS architecture, both parties need to temporarily register memory and exchange memory information for further communication. At this point, the parameter server needs to temporarily register lots of MRs for multiple senders. If the available physical memory is insufficient, registration operations will fail. The application will not be aware of the physical memory state in time when the memory resource is released. HiPower guarantees QoS of the entire system by pre-registering memory and reusing the MRs more efficiently. As for larger DTP applications, the amount of MRs can be allocated according to the size of actual physical memory, business requirements and the ratio of sender/receiver.

Low Consumption for Memory. Bitmap0 of requester and responder will be modified and read at the same time. Our design avoids mutex locks and further reduced system overhead. To further mitigate the effects of query overhead, each bitmap occupies only one MR, and the size of each item is only 1 bit. Therefore, the time complexity of single processing is $O(C)$ and the total time complexity is $O(Cn)^2$, which is acceptable for most practical DTP cases.

4   Implementation and Evaluation

4.1   Experiment Setting

We deployed the comparative experiment on 5 servers. Each server is equipped with a Mellanox ConnectX-3 40 Gbps NIC, two Intel Xeon E5 CPUs (each CPU has 16 physical cores) and 64 GB DRAM. We implement both HiPower and

---

[2] C is a constant, which denotes the number of MRs pre-registered. n denotes the number of executions of RDMA_READ.

Vanilla with 4500 lines of C++ codes and run the prototypes in Ubuntu 16.04. More specifically, as for Vanilla, we use RDMA_WRITE_WITH_IMM to transfer user-data and the responder confirms them in turn. Then, we launch two-sided RDMA to feedback the message of reusable memory. As for HiPower, we use RDMA_WRITE to transfer user-data. Next, we use RDMA_READ with the bitmap to confirm user-data and recycle reusable memory. For fairness, we use the same size memory pool for HiPower and Vanilla. The threshold is set to 50% of the total MRs.

We conduct our experiments with three commonly-used types of communication primitives: one-to-one, one-to-many (broadcast), many-to-one (gather). As for broadcast and gather primitives, we choose one server as the master and the other four servers as workers. We evaluate the performance of the two prototypes using CPU-based and GPUDirect RDMA-based memory transport and take throughput, latency and CPU utilization as our metrics to evaluate the performance of both prototypes. We also use a practical DML application (i.e. ResNet-152 model training) as the benchmark to further compare the performance between HiPower and baseline solutions.

4.2   Experiment Result and Analysis

One-to-One Communication Pattern. One-to-One communication plays a vital role in ring-based architecture. In other words, each pair of adjacent servers in a ring-based architecture can be considered as one-to-one communication.

The comparison of CPU-based and GDR-based throughput performance can be illustrated in Fig. 5(a) and (d). The experimental results show that our batched confirmation strategy makes HiPower achieve a higher throughput performance than Vanilla, especially for data sizes below 4 KB. We calculate the average throughput of HiPower and Vanilla in one million iterations. In CPU-based transmission scenario, as for 512B and 1 KB packets, vanilla's throughput is 11.5 Gbps and 23.3 Gbps respectively, while HiPower's throughput is 20.1 Gbps and 29.9 Gbps respectively. Test performance based on GDR transmission scenarios is slightly inferior to the former, which improve throughput performance by 11.2% and 43% respectively. As for data sizes above 4 KB, the throughput performance of HiPower and Vanilla is basically the same. However, in most cases, HiPower's requester and responder CPU utilization are significantly lower than that of vanilla. We will discuss it in detail later. In addition, we use qperf [1] to measure the performance of TCP/IP on RNIC/40 Gbps. The throughput performance of TCP/IP is mostly lower than HiPower and Vanilla.

The comparison of CPU-based and GDR-based latency performance can be illustrated in Fig. 5(b) and (e). TCP/IP's latency is always higher than both HiPower and Vanilla. As for both CPU-based and GPU-based transmission scenario, HiPower has lower latency for packets below 4 KB than Vanilla. As the packet size grows, HiPower and Vanilla's latency continues to shrink, but HiPower's latency is never higher than vanilla.

Figure 5(c) and (f) take a closer look at the requester/CPU utilization comparison and responder/CPU utilization comparison of HiPower and Vanilla. Both of them contain two transmission scenarios. As for requester, HiPower has lower

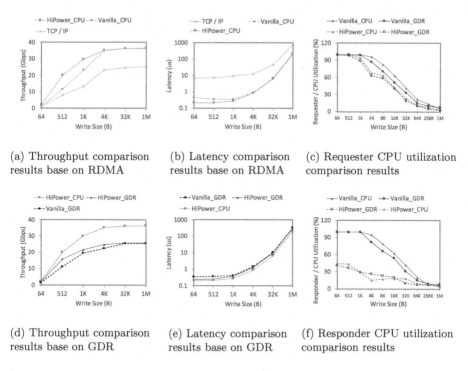

(a) Throughput comparison results base on RDMA

(b) Latency comparison results base on RDMA

(c) Requester CPU utilization comparison results

(d) Throughput comparison results base on GDR

(e) Latency comparison results base on GDR

(f) Responder CPU utilization comparison results

Fig. 5. One-to-one performance comparison.

CPU utilization than Vanilla. Among them, Vanilla's CPU utilization includes RDMA_WRITE_WITH_IMM operation and RDMA_RECV operation. HiPower's CPU utilization includes RDMA_WRITE and RDMA_READ operations. As for data sizes below 4 KB, HiPower launches more WRITE operations than Vanilla, but the CPU utilization is lower because Vanilla needs to continuously perform RDMA_RECV operations to obtain reusable memory messages. On the contrary, HiPower only needs one RDMA_READ operation to get multiple reusable memory, which greatly reduces CPU overhead. As for the data size between 4 KB and 256 KB, it can be further verified. The throughput performance of HiPower and Vanilla are basically the same, and the extra CPU utilization comes from inefficient recycling scheme. HiPower can reduce the frequency of RDMA_READ operations so that each reclaim can acquire multiple reusable memories while ensuring that the sender has enough reusable memory. As for the data size above 256 KB, Vanilla's RDMA_RECV operation slows down, so their CPU utilization is basically the same. As for responder, Vanilla's CPU utilization includes RDMA_RECV operation and RDMA_SEND operation while HiPower only has the RDMA_READ operation, HiPower saves half of the CPU utilization compared to Vanilla in CPU-based and GDR-based transmission scenarios. The main reason is that HiPower avoids lots of RDMA_SEND and RDMA_RECV (Used to receive imm_value) operations.

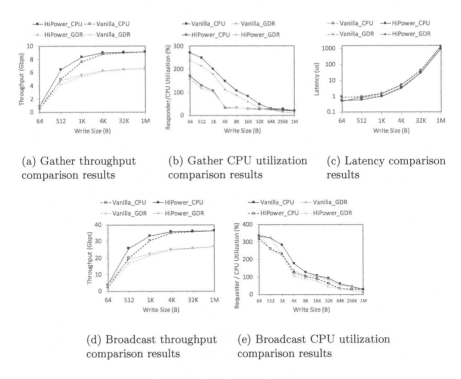

(a) Gather throughput comparison results

(b) Gather CPU utilization comparison results

(c) Latency comparison results

(d) Broadcast throughput comparison results

(e) Broadcast CPU utilization comparison results

Fig. 6. Broadcast and gather performance comparison.

Incast Communication Pattern. Incast communication mainly includes broadcast and gather. Map operations in the Map/reduce architecture and parameter update operations in the PS architecture can be viewed as broadcast [7]. Parameter syncing operations in the PS architecture can be viewed as gather.

We further use the incast pattern to compare the communication performance of the two prototypes, and the throughput performance comparison is illustrated in Fig. 6(a) and (d) respectively. In the experiment, we found that the throughput of each worker in the gather scenario is close to 1/n of the master server's total throughput in the broadcast scenario (where n represents the number of workers) and never exceeds the total throughput. The total throughput in the broadcast scenario is slightly higher than the one-to-one scenario. As for broadcast and gather, we try to send packets in a faster way, which trigger the PFC pause frame[3] and the throughput is no longer increasing. One major reason is that the Receiver queue buffer has reached the upper limit. This phenomenon is unavoidable when the rate of packet delivery is too fast. In this paper, we are more concerned about the maximum performance that two prototypes

---

[3] PFC (Priority-based Flow Control), priority-based flow control. The upstream device is notified to suspend the delivery by sending a Pause frame to prevent the buffer from overflowing.

can achieve in the incast pattern. Experimental results show that the batched confirmation scheme can detect user-data and reusable memory more quickly. As for broadcast, HiPower improves the throughput performance of 8.9%–70% and 5.9%–75% compared to Vanilla in CPU-based and GDR-based scenarios respectively. As for gather, HiPower improves the throughput performance of 7.8%–68% and 5.2%–72% compared to Vanilla in CPU-based and GDR-based scenarios respectively.

The latency comparison is illustrated in Fig. 6(c). In our experiments, we found that the latency for each stream (four in total) in the gather scenario is higher than the latency of each worker in the broadcast scenario (four in total). When transmitting packets over 4 KB, the latency is greatly increased, and each stream's latency is about four times of one-to-one. Similar to one-to-one communication Pattern, HiPower has lower latency for packets below 4 KB. As the packet size grows, the latency gap between HiPower and Vanilla shrinks, but HiPower's latency is never higher than vanilla.

For incast communication pattern, we pay more attention to the CPU utilization of the master server in both the gather and broadcast scenario, which are illustrated in Fig. 6(b) and (e). As for the gather server, HiPower can save up to 79% CPU utilization than Vanilla in both the CPU-based and GPU-based scenarios. When transferring large size packets, HiPower's CPU utilization is comparable to that of Vanilla. As for the broadcast server, compared with Vanilla, HiPower can save 20% and 15% CPU utilization in CPU-based and GPU-based scenarios respectively.

Case Study on ResNet-152. We conduct another experiment with a DML application (i.e. Resnet-152) to evaluate how HiPower performs on real workflow. We transport traffic flow of training ResNet [9] among the distributed cluster.

ResNet is a classic deep learning model and has won the champaign of ILSVRC2016, it inspires the model designer exploring deeper neural network for higher accuracy. How to design a DL model for high accuracy is beyond our discussing in this scope and e only focus on how to accelerate the training phase among the distributed cluster.

We choose ResNet-152 as our benchmark application, which contains 152 layers and 60 million parameters (organized in 514 blocks) in all. Most of these blocks are less than 4 KB and the maximum one is no more than 8 MB. To remove the bottleneck of the network, the 60 million parameters (i.e. 240 MB) must be synchronized within 0.15 s each iteration[4]. Bandwidth consumption will increase when training the ResNet-152 model with more nodes or by the smaller batch size, which brings a heavier burden to the network.

We compare the performance of our proposed prototype with the Vanilla and find that the FCT (Flow completion time) of HiPower is about 64 ms, FCT of Vanilla is about 73 ms. The HiPower shows 14% gains.

---

[4] It is derived from NVcaffe rc 0.17 with GPU supported. The input size is 224 * 224 * 224 * 3 and the batch size is 16. It is trained by P100. The forward stage cost 0.1 s, and backward cost about 0.15 s.

5   Related Work

Optimizations of RDMA Transmission. RDMA is a popular hardware-based solution. It aims to provide high-performance transport services but requires careful configuring and deep understanding. Recent study [4,12] gives a comprehensive analysis of RDMA verbs from a low-level perspective (e.g. PCIe, NIC and etc.) and offers guidelines on how to use RDMA verbs efficiently. Frey, et.al. [6] have also explored the hidden cost in using the RDMA verbs from the hardware-resources view. Taking task's workflow and RDMA verbs into consideration has been proved to be effective in many scenarios (e.g. Key-Value Stores, File System and etc.). Pilaf [17] optimizes the `get` operation using multiple `RDMA_READ` commands at the client side. Recently, RDMA has also been used in distributed machine learning. Yi, et.al. [2] focus on improving transport performance in distributed machine learning. They implemented a zero-copy RDMA-based transport service. Many works are explored to find the hidden obstacles in the application of RDMA such as PFC issues [8] etc. RDMA optimizations have brought benefits to computer systems, and this motivates us to start rethinking the design with RDMA in building high-performance transport services.

Mitigation of Communication Overheads in DTP. Distributed transaction processing has become the standard practice and there have been extensive studies focusing on mitigating the communication overheads in DTP. Communication compression technique can be well incorporated in the DTP process. Since most of the data transmitted by DTP are compressible numbers, such as zero, small integers and 32-bit foats with high precision, the communication costs can be reduced by using compression algorithms. Li et al. [14] compress the sparse matrix by eliminating most zeros values and Wei et al. [18] used a 16-bit float to replace 32-bit float value to improve the utilization of bandwidth. The effectiveness of these compression-based solutions are also demonstrated in the recent works for mitigation of communication overheads in DTP.

6   Conclusion

This paper proposes HiPower, which is a novel RDMA-accelerated solution for distributed transaction processing (DTP). Compared with existing works, HiPower leverages an elaborate bitmap design to execute batched transmission and confirmation, thus can efficiently improve communication efficiency and reduce CPU utilization for DTP tasks. HiPower can adjust dynamically to fit more complicated scenarios such as CPU-based computing and GPU-based computing. Our evaluations prove the effectiveness of DTP and the experimental results show that HiPower can achieve higher throughput performance and lower latency while consuming lower CPU utilization. Besides, HiPower can also reduce 14% FCT in ResNet-152 training compared to existing works (CPWR), which implies great potential performance gains to more practical applications.

Acknowledgement. This work is supported by the National Natural Science Foundation of China (No. 61802081), the Guizhou Provincial Natural Science Foundation (No. 20161052, No. 20183001).

References

1. qperf - measure RDMA and IP performance. Technical report, Johann George (2009). https://linux.die.net/man/1/qperf
2. How to compile, use and configure rdma-enabled tensorflow. Technical report, HKUST and Tensorflow community (2018). https://github.com/tensorflow/tensorflow/tree/master/tensorflow/contrib/verbs
3. Chen, H., et al.: Fast in-memory transaction processing using RDMA and HTM. In: TOCS 2017 (2017)
4. Dragojevic, A., Narayanan, D., Castro, M.: RDMA reads: to use or not to use? IEEE Data Eng. Bull. (2017)
5. Dragojević, A., Narayanan, D., Hodson, O., et al.: FaRM: fast remote memory. In: NSDI 2014 (2014)
6. Frey, P.W., Alonso, G.: Minimizing the hidden cost of RDMA. In: 2009 29th IEEE International Conference on Distributed Computing Systems (2009)
7. Geng, J.: CODE: incorporating correlation and dependency for task scheduling in data center. In: ISPA 2017 (2017)
8. Guo, C., et al.: RDMA over commodity ethernet at scale. In: SIGCOMM 2016 (2016)
9. He, K., Zhang, X., Ren, S., Sun, J.: Deep residual learning for image recognition. In: CVPR 2016 (2016)
10. Kalia, A., Kaminsky, M., Andersen, D.G.: FaSST: fast, scalable and simple distributed transactions with two-sided (RDMA) datagram RPCs. In: OSDI 2016 (2016)
11. Kalia, A., Kaminsky, M., Andersen, D.G.: Using RDMA efficiently for key-value services. In: SIGCOMM 2015 (2015)
12. Kaminsky, A.K.M., Andersen, D.G.: Design guidelines for high performance RDMA systems. In: ATC 2016 (2016)
13. Kim, D., et al.: HyperLoop: group-based NIC-offloading to accelerate replicated transactions in multi-tenant storage systems. In: SIGCOMM 2018 (2018)
14. Li, M., Andersen, D.G., Smola, A.J., Yu, K.: Communication efficient distributed machine learning with the parameter server. In: Advances in Neural Information Processing Systems, pp. 19–27 (2014)
15. Lu, X., Rahman, M.W.U., Islam, N., Shankar, D., Panda, D.K.: Accelerating spark with RDMA for big data processing: early experiences. In: Hot Interconnects 2014 (2014)
16. Luo, L., Nelson, J., Ceze, L., Phanishayee, A., Krishnamurthy, A.: Parameter hub: a rack-scale parameter server for distributed deep neural network training. In: SOCC 2018 (2018)
17. Mitchell, C., Geng, Y., Li, J.: Using one-sided {RDMA} reads to build a fast, CPU-efficient key-value store. In: ATC 2013 (2013)
18. Wei, J., et al.: Managed communication and consistency for fast data-parallel iterative analytics. In: Proceedings of the Sixth ACM Symposium on Cloud Computing
19. Wei, X., Dong, Z., Chen, R., Chen, H.: Deconstructing RDMA-enabled distributed transactions: hybrid is better! In: OSDI 2018 (2018)

Emerging Topics

LDAPRoam: A Generic Solution for Both Web-Based and Non-Web-Based Federate Access

Qi Feng(iD) and Wei Peng$^{(\boxtimes)}$(iD)

East China Normal University, 3663 N. Zhongshan Road, Shanghai, China
{qfeng,wpeng}@admin.ecnu.edu.cn

Abstract. Identity federation technology has been widely used in recent years. But the solution for federate access is totally different between the Web-Based and Non-Web-Based scenarios. Furthermore, it is highly limited for lack of support from Non-Web-Based scenarios now. This paper proposes a generic federate access solution based on LDAP roaming, which can provide reliable identity roaming in any internet service. To service providers, our solution is transparent and looks like a LDAP. The paper first presents the difficulties in realizing LDAP roaming and discusses offers solutions to the implementation of LDAP roaming. Then it evaluates the easy integration and usability of LDAP roaming. Finally it compares the Generic Solution with the existing federal access solution.

Keywords: Identity federation · Non-Web-Based · LDAP · SAML · Eduroam

1 Introduction

A consensus of resource sharing based on identity federation has been gradually reached [1]. However, the existing solutions for federate access, such as SAML [2] based on Web-based, cannot be applied under non-Web-Based scenarios. Although SAML can use the ECP mode to support applications on Non Browser, it is still limited for reliance on session and working on HTTP. The same is true of the case of the AAA-based identity federation, for example, eduroam [3], which does not depend on web, still asks for EAP to send authentication and accounting messages. Due to the lack of attributes, AAA-Based identity federation only appears in eduroam, but is hardly applicable under non-Web-Based scenarios, such as the console access (e.g., via SSH [4]) common in HPE.

Besides, the user experience of the two federate accesses is completely different. In SAML, users need to select their own home IdP on discovery service and input the username and password to finish the access, while in eduroam they should take user@domain as the username. In fact, the federations based on the two federate accesses are also different. In China, SAML Federation—CARSI contains 77 IdP members, almost comes from school libraries, while eduroam contains 235 IdP members, almost all of whom come from the network center or information technology center. This causes bad user experience and is not what we have expected either.

© IFIP International Federation for Information Processing 2019
Published by Springer Nature Switzerland AG 2019
X. Tang et al. (Eds.): NPC 2019, LNCS 11783, pp. 225–234, 2019.
https://doi.org/10.1007/978-3-030-30709-7_18

In this paper, we introduce a generic solution suitable for both Web-Based and Non Web-Based federate accesses. We put the point on the coupling degree of federate access and service provider. The lower degree, the smaller differences for experience. If the authentication solution is completely transparent to the service provider, the latter cannot perceive the existence of federation authentication for there is no difference between Web-Based and Non-Web-Based service and the authentication experience will be consistent.

Taking eduroma as a model, users can add @domain as a suffix after the username, i.e. username@example.org, to roam to their home organization LDAP [5] and finish the LDAP authentication process, which is transparent to the service provider. The roaming LDAP federation can be regarded as a virtual LDAP which is shown in Fig. 1. The service provider does not need to participate in the details of the federate access; it only needs to support the LDAP protocol. In view of the extensive support of LDAP protocol, this solution will be very friendly to the service provider and will be easy to integrate.

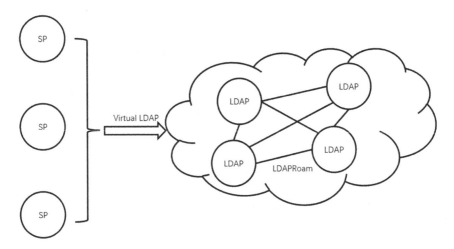

Fig. 1. LDAPRoam as a virtual LDAP

It looks similar to the radius proxy structure of eduroam. Considering the particularity protocol of LDAP, the coincident attribute in the federation, and privacy protection of user information, we choose a net structure similar to SAML federation instead of the tree structure of eduroam federation. We also use the asymmetric encryption technology to form the trust relationship. It allows encrypting passwords by using the public key, which can help to prevent service providers from obtaining any plaintext of the password. We construct an independent service, LDAPRoam, as a proxy for the actual LDAP at the back, which can forward authentication requests of roaming.

The contributions of this paper are:

- The concept of LDAP roaming. This Identity federation is transparent to the service provider and supports identity roaming for internet services anywhere.
- The architecture of LDAP roaming. This solves the difficulties of LDAP protocol application in federate access.
- Evaluated LDAPRoam from the perspective of usability and easy integration, and compared it with the existing federate access solutions.

2 Related Works

There are many Web-Based identity federation technologies, such as SAML, OAuth [6], and OpenID [7]. Most of the time, they all rely on browsers. In Shibboleth project (SAML-Based), ECP [8] (Enhanced Client or Proxy) is proposed to work in a non-browser environment. However, due to the ECP needs supported by client modification, there is little "real world" support other than Shibboleth.

In some specific Non-Web-Based scenarios, such as roaming of wireless networks, the AAA-Based technology is an option. Eduroam is a case that has been widely used in educational and scientific research institutions. Users can roam the authentication back to their own organization when accessing the eduroam network. Eduroam uses hierarchical Radius architecture. The roaming authentication request are protected by EAP methods, such as PEAP/EAP-MS-CHAPv2. Although the accounting request can also carry attributes [9], the standard of Radius attributes is designed for network accounting, which cannot meet the authorization requirements in several general scenarios. So, the federate access solution coming from AAA-Based cannot be applied in scenarios other than network authentication.

ABFAB [10] is the outcome of a project named Moonshot. The project is created to serve the programs built on Non-Web-Based services. The solution of this project is to extend eduroam to support SAML assertion. So ABFAB requires that application clients must support GSS-API [11]. Although many protocols are already supporting GSS-API (e.g., via ssh, nfs, ftp), the application client is still asked for modifications that intrude too deep, in order to turn on the function. The goal of Moonshot is to push all these changes into a standard and require updating at the clientend, but this is obviously unrealistic in the short run. This means that there is little possibility for ABFAB to be implemented at present.

FedKERB [12] and ABFAB have similar structures. FedKERB adds a KDC component in order to support Kerberos, which makes it difficult to change things on the client. Therefore, this solution is also very hard to be promoted.

Jens Köhler's [13] work is similar to ours. They also proposed an LDAP-based solution to keep it transparent to the service provider. However, in its solution, the attribute is acquired through SAML ECP, which also requires that the IdP and application client must support ECP first. Therefore, there are still difficulties in popularizing and implementing this solution.

Our solution does not require any modification on the application client, and all identity privacy and passwords can be well protected through external plugins, thus providing the maximum possibility for promote solution.

3   Challenges

Because the authentication and authorization mode of LDAP is very different from Radius, there are many challenges in LDAP-based roaming.

1. It must allow users to input plaintext passwords indirectly. It also cannot make extra modifications on application clients.
2. A trust relationship must be established between LDAP when providing roaming services, to avoid any possible hijacking in the process.
3. LDAP authentication is usually divided into two steps: Search and Bind. This may cause the loss of roaming domain name information in the second Bind step.
4. Since Bind must occur after Search, there is an unauthorized Search behavior. If there is no restriction, it may result in a leakage of user information on Search step.
5. There may be different attribute categories among different LDAP nodes, which need to be standardized using some methods.

We will describe in detail how to overcome these challenges in Sect. 4.

4   Solution

As mentioned in Sect. 3, the first challenge is the security of transmission. In the radius proxy structure of eduroam, there are intermediate forwarding nodes. Though passwords are well protected by client-side encryption, attribute information (e.g., username) can still be obtained by intermediate nodes, into which additional attributes can even be inserted. In fact, we have no choice but to believe that the intermediate forwarding node can be trusted. We also need to prevent possible cheating from middleman, which requires encryption. Since eduroam architecture itself does not support encryption, it needs another way to assist, such as connecting the radius nodes through the GRE tunnel.

Therefore, we draw on the experience from the SAML federation structure, a mesh point-to-point interconnection structure. The whole federation maintains a main metadata, which contains the basic information and certificate information of each LDAPRoam node.

The certificate is issued by LDAPRoam through sending the private key. Figure 2 shows the differences between LDAPRoam and eduroam architectures.

Through the asymmetric encryption system of the certificate and private key within the mesh point-to-point structure, the security of the whole transmission is well guaranteed. Firstly, there is no intermediate nodes in the point-to-point structure and the data transmission path has been minimized. Secondly, the private key-certificate system based on asymmetric encryption can not only prevent the request message from

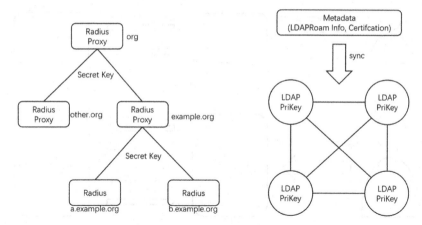

Fig. 2. Different structures with eduroam and LDAPRoam

being hijacked by the middleman through verifying the certificate, but also ensure that the request message comes from the trusted initiator by verifying the signature.

The second challenge is password protection. The password must be directly input from the service provider's client because there are no invasive changes taken by the service provider. When we are roaming in eduroam, we can directly type the password on the trusted operating systems (OS), because the OS usually uses mschapv2 encryption method to ensure that the password is securely encrypted at the beginning. But we cannot give the same trust to these third-party service provider clients. But in our mesh point-to-point structure, this problem is very easy to solve. We only need to encrypt the password through the public key of the other node. Then this password can only be decrypted after roaming to the destination. This process does not require any modification of the service provider's client. In fact, even if modified, it cannot be trusted as well. Using an external plugin is helpful to encrypt the password forms and keep user experience unchanged. The automatic filling of plugin forms has been verified by many password managers (e.g., via 1Password).

The third challenge is the authentication mode of LDAP. Unlike the AAA mode, which sends the user name and challenge message directly, LDAP uses DN (distinguished name) in Bind operation, while the username is an LDAP attribute, such as uid or sAMAccountName. Therefore, for an LDAP application, the standard practice is usually divided into two steps:

- First, search the DN of items by taking that username as the query condition of the attribute filter
- Second, bind with the DN and the password to verify whether the authentication is successful.

Figure 3 shows the difference on authentication between AAA and LDAP.

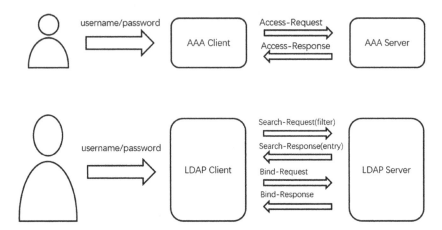

Fig. 3. Different authentication process with AAA and LDAP

In the second step of LDAP authentication, the DN information no longer contains the roaming domain name of the user. In order to forward the Bind-Request correctly to the appropriate node, we need maintain not only the correspondence between DN and LDAPRoam service but between domain and LDAPRoam service. For example, the roaming domain name of alice@a.example.org is example.org, and the DN record in LDAP of this account is cn=alice, dc=a, dc=example, dc=org. Then we need to map dc=a, dc=example, dc=org and a.example.org to the same LDAPRoam node. These information will be published by LDAPRoam to Metadata for other nodes to query.

The fourth challenge is the unauthorized LDAP Search. Due to the two-step nature of LDAP authentication, Bind must occur after Search. We cannot evade unauthorized LDAP Search behaviors. However, it is obviously inappropriate to return unauthorized user attributes to roaming LDAP. Our solution is to introduce the concept of authorization validity into LDAPRoam, and generate a cache record with valid authorization for users to have a successful Bind. The roaming party is allowed to query the user attribute within the validity period of authorization; otherwise only DN will be returned. This may lead to some abnormality of LDAP Client which does not meet the standards, but it has no influence on clients implemented according to the standards.

The fifth challenge is the standardization of attributes. The attribute standard among different LDAP may be completely different. Although there are a series of RFC standards for LDAP attributes, the understanding of the attribute fields may still be inconsistent. At least between OpenLDAP and Active Directory, the default field for the username is totally different. OpenLDAP usually uses uid, while Active Directory uses sAMAccountName. Moreover, since the details of LDAPRoam are transparent to the application, service providers have no way to adapt this attribute relationship to specific LDAPRoam nodes. In SAML federation, IdP can map attributes into specific attribute names and oid strings when it queries them, which makes the attribute exchange within SAML follow the fixed standards. LDAPRoam also uses the same method of mapping attribute relationships by LDAPRoam nodes to shield attribute differences among different LDAP. Same as SAML2, LDAPRoam nodes protect the identity privacy of users and reduce unnecessary provision by releasing attributes when roaming in various nodes.

Figure 4 shows the overall architecture of LDAPRoam. The LDAPRoam node provides an LDAP-style interface for application clients, which makes authentication roaming transparent to the service provider. LDAPRoam uses RESTful API interface to simplify roaming message processing between each other. All LDAPRoam nodes synchronize Metadata information at a regular time to obtain basic information and certificate information of each node in the federation. When LDAPRoam queries the attributes from Backend-LDAP, the authoritative data source, it will make an attribute-map according to the attribute standard of the federation.

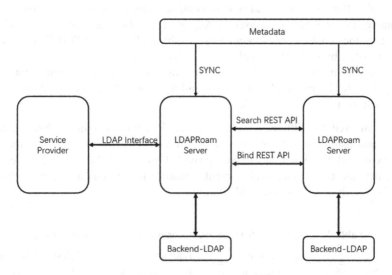

Fig. 4. LDAPRoam structure

Table 1 shows the LDAPRoam field information provided in metadata.

Table 1. Metadata field for LDAPRoam.

Field	Value	Comment
domain_name	ldap.b.example.org	Server domain name
served_domain	["b.example.org"]	Served domains
base_dn	"dc=b, dc=example, dc=org	The basedn
certification	"nLmIuZXhhbXB...."	The certification
bind_endpoint	"https://ldap.b.example.org/api/v1/bind"	Bind API endpoint
search_endpoint	"https://ldap.b.example.org/api/v1/search"	Search API endpoint

5   Evaluations

Whether the solution can be effectively promoted depends on the simplicity of deployment. Our solution does not require any changes from the service provider. For identity providers, they only need to deploy a LDAPRoam service.

For service provider, we have two modes, full trust mode and limit trust mode.

- In the Full trust mode, it is safe to users to directly enter the username and password to service providers that are usually government agencies. A typical application scenario is that a fully trusted service provider encapsulates the interface again, such as oAuth2, in order to support other service providers to interface. This mode can well solve the federate access requirements in some cookie limited scenarios. For example, the embedded browser in Alipay cannot be used normally in SAML2, because the cookies are completely disabled. But the LDAPRoam solution encapsulated by oAuth2 can solve this problem well.
- In Limit trust mode, in order to protect our password security. We must first encrypt the password through a plugin or tool, and then submit the encrypted password to the client of the service provider.

We compared the differences between LDAPRoam and other federation access solution, and listed the differences in several dimensions, mainly based on the seven requirement and the level of support provided by Alejandro Pérez-Méndez [14]. Table 2 lists the comparison with current mainstream solutions and Table 3 lists the comparison with other experimental solution.

Table 2. Comparison between mainstream federate access solutions.

Topic	LDAPRoam	Eduroam	SAML2 (shibboleth)
R1 – Authentication in the IdP	Roam back to IdP	EAP and AAA proxy back to IdP	Redirect to IdP Portal
R2 – High level authorization	LDAP attribute search	Limit support with radius attribute	SAML2 assertion from the IdP
R3 – Data transport security	Asymmetrical encryption, point to point	EAP Tunnel, with intermediate node	Asymmetrical encryption, point to point
R4 – Single Sign On	Not support	Not support	Support
R5 – Re-use of instructures and standards	Based on LDAP, TLS and RSA	Based on EAP and AAA	Based on SAML2
R6 – Usability	Username/password	Username/password	Username/password
R7 – Identity Privacy	Attribute map control	EAP Tunnel and pseudonyms	Attribute map control
Web-Based	Support	Not support	Support
Non Web-Based APP (Full trusted)	Send password directly	Password encrypt, such as mschapv2	Must modify the client
Non Web-Based APP (limit trusted)	Send password encrypt	Password encrypt, such as mschapv2	Must modify the client

Table 3. Comparison between experimental federate access solutions.

Topic	LDAPRoam	ABFAB	FedKERB
R1 – Authentication in the IdP	Roam back to IdP	EAP, AAA and GSS-API	EAP, AAA and GSS-API
R2 – High level authorization	LDAP attribute search	SAML2 assertion from the IdP	SAML2 assertion from the IdP
R3 – Data transport security	Asymmetrical encryption, point to point	EAP Tunnel and RadSec	EAP Tunnel and RadSec
R4 – Single Sign On	Not support	Not support	Based on Kerberos
R5 – Re-use of instructures and standards	Based on LDAP, TLS and RSA	Based on EAP, AAA and GSS-API	Based on EAP, AAA, GSS-API and Kerberos
R6 – Usability	Username/password	Username/password	Username/password
R7 – Identity Privacy	Attribute map control	EAP Tunnel and pseudonyms	EAP Tunnel and pseudonyms
Web-Based	Support	Support	Support
Non Web-Based APP (Full trusted)	Send password directly	Must modify the client	Must modify the client
Non Web-Based APP (limit trusted)	Send password encrypt	Must modify the client	Must modify the client

LDAPRoam can provide good support in all other dimensions except SSO. Compared with the current mainstream solutions, LDAPRoam combines the advantages of eduroam and SAML2, and can support both web-based and Non Web-Based applications with the same user experience. Compared with other experimental solutions, LDAPRoam does not require service provider clients to make any modifications, which is very helpful for practical application and promotion.

Taking HPC as an example, it is usually necessary to provide federation access on web-sites and give support for users' console access (i.e. ssh). As an HPC service with a long history, it is obviously unrealistic to require all users to upgrade their SSH clients in order to provide federate access. Now the solution of LDAPRoam can be operated and implemented easily without damaging the users' privacy interests.

6 Conclusion

In this paper, we introduce the concept of LDAPRoam, which combines the advantages of SAML2 alliance and eduroam alliance. We discuss many challenges that the model faces, and give solutions by designing and adapting the particularity of the LDAP protocol. After evaluation, the roaming solution of LDAPRoam is highly suitable for deployment and promotion. We have already implemented the experimental roaming verification between East China Normal University and the Information Center of

Shanghai Municipal Education Commission through LADPRoam. It is planned to carry out the promotion step by step. In a wide range of situations, the application still needs further observation and verification.

References

1. Torres, J., Nogueira, M., Pujolle, G.: A survey on identity management for the future network. IEEE Commun. Surv. Tutor. **15**(2), 787–802 (2013)
2. Cantor, S., Kemp, J., Philpott, R., Eve, M.: Assertions and Protocols for the OASIS Security Assertion Markup Language (SAML) v2.0, OASIS Standard, March 2005
3. Wierenga, K., et al.: Deliverable DJ5.1.4: Inter-NREN Roaming Architecture. Description and Development Items, GN2 JRA5. GEANT2, September 2006
4. Ylonen, T., Lonvick, C.: The Secure Shell (SSH) protocol architecture, IETF RFC 4251, January 2006
5. Sermersheim, J.: Lightweight Directory Access Protocol (LDAP): The Protocol, IETF RFC 4511, June 2006
6. Hardt, D.: The OAuth 2.0 Authorization Framework, IETF RFC 6749, October 2012
7. OpenID Connect Core 1.0 incorporating errata set 1. https://openid.net/specs/openid-connect-core-1_0.html. Accessed 08 Nov 2014
8. ECP-Shibboleth Concepts. https://wiki.shibboleth.net/confluence/display/CONCEPT/ECP. Accessed 05 Apr 2016
9. Rigney, C.: RADIUS Accounting, IETF RFC 2866, June 2000
10. Application Bridging for Federated Access Beyond Web (ABFAB) IETF Working Group. https://datatracker.ietf.org/wg/abfab/charter/. Accessed 30 Sept 2016
11. Linn, J.: Generic Security Service Application Program Interface Version 2, Update 1, IETF RFC 2743, January 2000
12. Pereniguez, F., Marin-Lopez, R., Kambourakis, G., et al.: PrivaKERB: a user privacy framework for Kerberos. Comput. Secur. **30**(6/7), 446–463 (2011)
13. Köhler, J., Simon, M., Nussbaumer, M., Hartenstein, H.: Federating HPC access via SAML: towards a plug-and-play solution. In: Kunkel, J.M., Ludwig, T., Meuer, H.W. (eds.) ISC 2013. LNCS, vol. 7905, pp. 462–473. Springer, Heidelberg (2013). https://doi.org/10.1007/978-3-642-38750-0_35
14. Perez-Mendez, A., Pereniguez-Garcia, F., Marin-Lopez, R., et al.: Identity federations beyond the web: a survey. IEEE Commun. Surv. Tutor. **16**(4), 2125–2141 (2014)

Characterizing Perception Module Performance and Robustness in Production-Scale Autonomous Driving System

Alessandro Toschi[1], Mustafa Sanic[1], Jingwen Leng[1(✉)], Quan Chen[1], Chunlin Wang[2], and Minyi Guo[1]

[1] Department of Computer Science and Engineering,
Shanghai Jiao Tong University, Shanghai, China
{aleto_95,leng-jw,chen-quan,guo-my}@sjtu.edu.cn, mtsanic@gmail.com
[2] Chuxiong Normal University, Chuxiong City, China
wcl@cxtc.edu.cn

Abstract. Autonomous driving is a field that gathers many interests in academics and industry and represents one of the most important challenges of next years. Although individual algorithms of autonomous driving have been studied and well understood, there is still a lack of study for those tasks in a production-scale system. In this work, we profile and analyze the perception module of the open-source autonomous driving system Apollo, developed by Baidu, in terms of response time and robustness against sensor errors. The perception module is fundamental to the proper functioning and safety of autonomous driving, which relies on several sensors, such as LIDARs and cameras, for detecting obstacles and perceiving the surrounding environment. We identify the computation characteristics and potential bottlenecks in the perception module. Furthermore, we design multiple noise models for the camera frames and LIDAR cloud points to test the robustness of the whole module in terms of accuracy drop against a noise-free baseline. Our insights are useful for future performance and robustness optimization of autonomous driving system.

Keywords: Autonomous driving · Robustness analysis · Performance profiling · Deep neural networks

1 Introduction

Autonomous driving is becoming one of the most important applications. We must understand the key characteristics of autonomous driving to build proper architectures and systems for it. There are prior efforts in that direction, but they mainly focus on individual tasks/kernels and rely on the usage of an autonomous

© IFIP International Federation for Information Processing 2019
Published by Springer Nature Switzerland AG 2019
X. Tang et al. (Eds.): NPC 2019, LNCS 11783, pp. 235–247, 2019.
https://doi.org/10.1007/978-3-030-30709-7_19

driving simulator, such as CARLA [1] or OpenPilot [2]. In other words, detailed knowledge about a production-scale autonomous driving software system is still a missing piece to the puzzle of understanding autonomous driving.

Motivated by that challenge, this work studies Apollo (version 3.5) [3], which is an open-sourced production-scale autonomous driving software, developed by Baidu. It has many complex and realistic modules, each of which targets a high-level feature of autonomous driving, such as perception, prediction, planning, as shown in Fig. 1. Modules are described by their input/output (I/O) relationship to other modules, modelled as stages of a pipeline, and not to be intended as monolithic pieces of software, so they can be further decomposed as a set of inner components, following the same architecture and design philosophy. The communication, among modules and components, is data-driven and is enabled by a runtime framework, named Cyber [4], that implements the publisher & subscriber architecture. Each component can write and read on multiple channels, and the messages are serialized using Google Protocol Buffer [5].

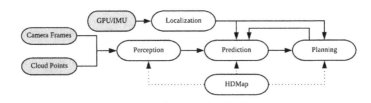

Fig. 1. Apollo software architecture

Among the many modules in Apollo, the perception module, which is built to perceive the environment, is the entry point to all the following modules. The module uses multiple sensors, including Full-HD cameras and LIDARs, and is also very computation-intensive as it relies on multiple deep neural networks (DNNs). The whole system depends on the accuracy of such algorithms and detectors to ensure responsiveness and safety. Thus, the perception module needs to be trustable. This paper focuses on the perception module through performance and robustness analysis.

We study the response time of different components in the perception module because it is critical for the predictability and accuracy of the entire system. The response time of each module has been set to 100 ms, which has been adopted as the standard maximum response time since it should ensure a proper and safe reaction to any possible situation. Besides that, response time is also crucial for many dynamic processing routines that use time-deltas to perform online corrections and discard past data. Exceeding these time requirements can lead to a potential loss of useful information that may affect the accuracy of the system. Different from prior efforts on individual sensors [6], our studies focus on the separate and concurrent processing in the presence of multiple sensors, which Apollo uses for safe and reliable output.

The robustness of perception module is also crucial for the autonomous driving system as failures in the module would cause disastrous consequences [7]. Meanwhile, DNN models are model-driven, and therefore, not all the possible scenarios are predictable in the training phase of machine learning algorithms, especially [8]. As such, we extend the robustness analysis methodology from recent efforts [9,10] to create noise models for both the camera and LIDAR sensors. We use them to test the robustness of DNN models deployed in the perception module.

Our performance analysis complies with prior work on that DNN models take the majority of the average response time, and therefore, they are great candidates for architectural specialization. However, prior work fails to recognize the importance of the CPU owing to the pre- and post-processing that only exists in a production-scale system. The robustness analysis tested the accuracy of the perception against camera and LIDAR noise, separately. The loss of accuracy has been evaluated on obstacles, lanes and the outcome of the experiments highlighted that even if detectors deteriorates, the whole module mitigates the noise thanks to the presence of the fusion component, which, combining the data coming from multiple sensors, can alleviate the effects.

The paper is organized as follows: in Sect. 2, a comprehensive view of algorithms within the perception module are presented. Section 3 discusses each component according to its response time. In Sect. 4, noise models and robustness experiments are introduced moreover, later in Sect. 5, conclusions are provided.

2   Perception Module Description

The perception module, as presented in Fig. 2, is composed of several components. The fusion camera detection is in charge of detecting obstacles, lanes and tracking them through past frames. The LIDAR segmentation identifies obstacles from cloud points, which are further classified and tracked by LIDAR recognition. The output of the LIDAR recognition and the fusion camera detection is now homogenous to be fused into a single coherent detection, taking into account all the obstacles, by the fusion component, which represents the last component of the module.

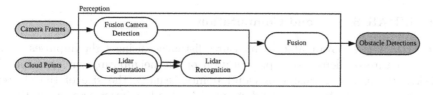

Fig. 2. Perception software architecture

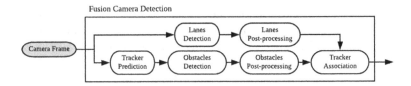

Fig. 3. Obstacle camera pipeline

2.1 Camera Sensor and Computation

The objective of the fusion camera detection is to detect obstacles within camera frames, such as vehicles, pedestrians, lanes, and keep the reference of previous obstacles to track them over frames. Detection is achieved by applying camera frames to an obstacle camera pipeline, that is shown in Fig. 3.

Lanes detection is performed using the state-of-art convolutional neural network (CNN) Spatial CNN [11], which outperformed other lanes detector. The main feature of the network is that exploits spatial relationship among pixels, being able to identify straight shaped objects, such as lanes, even if obstructed. Lanes post-processing is necessary to extract, from the raw output of the neural network, the coefficients of the polynomial that approximate each lane together with lane edge points. The points and coefficients belong to the image plane, so another post-processing operation is to project them first upon the car and ground planes and then to combine the two projections, obtaining the three-dimensional representation, which refers to world coordinates.

Obstacles detection is achieved using a CNN based on YOLO [12,13] that has been enhanced to identify the obstacles bounding boxes in three dimensions. Similar to lanes detection, each bounding box is further projected onto world coordinates by the obstacle post-processor.

Obstacles tracking consists of two main steps: prediction, in which previous obstacles positions and bounding boxes are updated according to time-delta, between the current iteration and the past one, and the car pose through an Adaptive Kalman Filter; association, in which the tracker tries to associate past obstacles to the new obstacles found by the detector, using the frame similarities as a metric, and eventually discard old obstacles.

2.2 LIDAR Sensor and Computation

The LIDAR is a sensor able to measure distances, using light impulses, generating a three-dimensional representation of the neighboring environment. A LIDAR message is a cluster of cloud points; usually, the size of the cluster is around one hundred thousand points. A cloud point represents a three-dimensional point, expressed in LIDAR coordinates, and a light intensity value, which corresponds to the object reflectance [14].

The computation regarding the LIDAR sensor is divided between two components: LIDAR segmentation and LIDAR recognition, as shown in Fig. 2.

LIDAR Segmentation. The LIDAR segmentation [15] is the process that detects obstacles, within the surrounding environment, given the cloud points coming from the LIDAR. The pre-processing filters illegal values and discards points that are outside the Region of Interest (RoI) respect the car position. The segmentation is performed using a CNN, so the filtered cloud points need to be converted into feature maps manageable by the neural network. The conversion into feature maps projects the (x, y) coordinates of cloud points over a quantized two-dimensional grid. The CNN is composed of a custom-design convolutional auto-encoder, which selects only the most semantic information for the segmentation, and then a classifier detects the obstacles attributes such as center position, height and class probability. Finally, a bounding box is built for each obstacle, by finding a 6-edge polygon of cloud points, which completely wraps the obstacle.

LIDAR Recognition. The LIDAR recognition [15] is in charge of tracking LIDAR obstacles over time. Similar to, the tracking of camera obstacles, each LIDAR obstacle tries to match to an existing obstacle by constructing a bipartite graph, in which to each obstacle is associated with its distance from existing tracks, and then the assignment problem is solved using the Hungarian algorithm [16]. Later, the class assigned to each obstacle is further assigned, taking into consideration past matched obstacles to reducing the class switch during the whole observation.

2.3 Fusion

The fusion component associates and merges obstacles' bounding boxes, coming from LIDARs and cameras and then updates the motion state of each obstacle. The bounding box [17], at time step k, is represented as a vector $\boldsymbol{x}(k) = [x(k), y(k), \theta(k), v(k), \omega(k), a(k), w(k), l(k), h(k)]^T$, where (x, y) is the center position, θ is the heading angle, v is the linear velocity, ω is the angular velocity, a is the acceleration and w, l, h define the width, length and height of the 3D box. The association among bounding boxes is achieved by minimizing the Euclidean distance of the center positions, using the Hungarian algorithm [16]. The motion state is estimated through an Adaptive Kalman Filter with constant acceleration model using the velocity and position provided by bounding boxes [15]. The fusion algorithm implemented is not sequential; observations from sensors are not treated equally since the component defines a main fusion sensor, which triggers the fusion action. The main fusion sensor is configurable and observations dispatched from it cause the execution, determining the fusion frequency. Measurements from other sensors are cached in an ordered queue, according to timestamps. When a new main fusion sensor's message arrives, the component assesses the reliability of cached measurements by checking the time-deltas, between the new message and their timestamps, discarding those who are above a threshold.

3   Performance Analysis

The objective of the performance analysis is to characterize the perception module, using the response time as a metric, to understand the computational effort required by each task. The study is a useful guide to figure out how a real autonomous driving system is designed and its response to real-world scenarios (Table 1).

Table 1. Neural network parameters - a summary of the neural networks present in the perception module.

Network	Input size	Layers	Parameters
SCNN [11]	$640 \times 480 \times 3$	143 convolutional 3 deconvolutional	13.68 M
YOLO [12,13]	$1440 \times 800 \times 3$	34 convolutional	6.54 M
Segmentation Sect. 2.2 Custom design	$864 \times 864 \times 4$	15 convolutional 10 deconvolutional	2.97 M

Experimental Setup. In our simulation, we configure the perception module to use one camera sensor and one LIDAR sensor. The hardware platform we study includes an Intel i7 8700 CPU and an NVIDIA GTX 1080TI GPU, which aligns with Apollo's officially recommended system Nuvo-6108GC [18]. The recommended system uses an Intel i7 6700 CPU and an NVIDIA GTX 1080 GPU. We compile the code with optimization enabled and GPU support. For the input data to the perception module, we use various scenarios taken from the Kitti dataset [19]. The dataset includes both camera frames and cloud points.

Figures 4 and 5 show our performance analysis result. Figure 4(a) shows the execution time breakdown of the fusion camera detector component, where the DNN computation (for lane detector and obstacle detector) accounts for the 97.26% of the whole component response time. In contrast, Fig. 4(b) shows the execution time breakdown of the LIDAR segmentation component, where the DNN computation (for segmentation) only accounts for the 50.55% of the whole component response time. The pre-processing of clout points (one-by-one filtering of cloud points) on the CPU takes about the other half.

Figure 4(c) and (d) show the execution time for the LIDAR recognition and fusion component, respectively. Those components run on the CPU and their execution time is much less than the previous two components. However, unlike the constant processing time of the previous two components, the execution time of LIDAR recognition and fusion component highly depend on the number of objects in the frame, which may make their real time processing more challenging. Figure 5 shows the averaged response time with standard deviation for all computation tasks in the perception module.

In summary, we make the following observations from the results:

O1 The GPU is an crucial architecture for the autonomous driving system since it enables the acceleration of DNNs. It is impossible to use only the CPU to meet the 100 ms because of the heavy computation requirement of DNNs.

O2 The CPU tasks (pre-processing in Fig. 4(b), LIDAR recognition in (c), and fusion in (d)) still take a large portion of the response time. Specific platform solutions, like NVIDIA Jetson, are built to fit the requirements of autonomous driving using ARM CPUs, which do not provide the same power as the high-end consumer CPU, i.e. the Intel i7 8700 [20]. As such, the CPU may become the bottleneck for the system.

O3 The DNN computation makes the response time highly predictable and can be assumed as constant due to the lower variance exposed by such tasks.

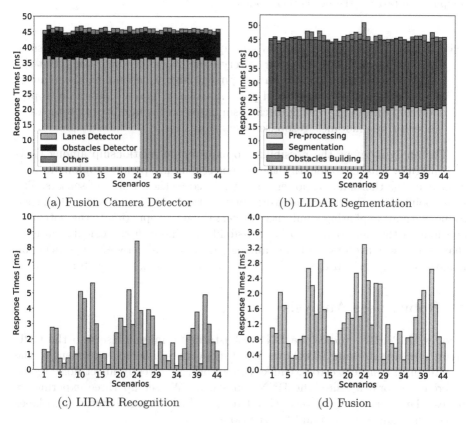

Fig. 4. These plots display the average response time, expressed in milliseconds, over the scenarios, highlighting the significant tasks where necessary.

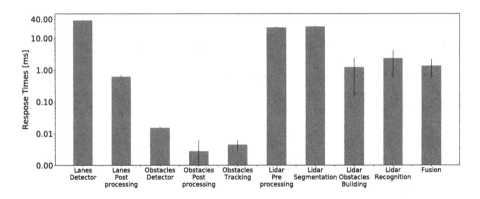

Fig. 5. Averaged response time (logarithmic scale) with standard deviation for different computation tasks in the perception module.

O4 The LIDAR pre-processing can still approximate as a constant task since the number of cloud points varies very little over different frames. But the LIDAR recognition and fusion component highly depend on the number of obstacles, either current and past. As such, it may be desirable to provide computation sprinting mechanism based on the number of obstacles to smooth those modules' response time.

The analysis opens the possibility to explore the relationship among GPU and multiple sensors to identify the proper scheduling policy to adopt in case of race, or the thermal and power impact related to the number of sensors. The study pointed out that neural network computation is statically executed for each frame, without exploiting similarities among them to approximate the outcomes and lighten the workload. Another possibility to diminish the calculation is to share the first convolutional layers among similar neural networks, operating on homogenous input, and then differentiate them according to their functions.

4   Robustness Analysis

Prior work [21,22] studies the DNN robustness by inject noise into the GPU architectural states and has shown DNN models are inherently resilient to the architectural transient errors. Unlike those work, we directly inject noise to the different sensors to study the DNN robustness. We set up three experiments to test the accuracy of the obstacles detection, LIDAR segmentation and lanes detection, against camera and LIDAR noise.

We conduct two experiments regard the camera noise, and one experiment regards the LIDAR noise, the trials, for obstacle detection and LIDAR segmentation, using scenarios from the Kitti dataset [19], while the lanes experiment use the TuSimple dataset [23]. The accuracy is evaluated using the F1 metric, which takes into account the presence of false positives and false negatives.

A true positive is a matching having an Intersection over Union (IoU) value, with the respects to the noise-free detection, higher than 0.5 for obstacles and 0.3 for the lanes experiment.

4.1 Noise Models

Our noise models between camera and LIDAR sensors are independent. In the real world, external factors like weather can impact the camera and LIDRA sensors simultaneously. Although camera noise models that simulate weather conditions [24] exist, LIDAR still lacks rigorous noise model definitions. Our work can be easily extended to use those joint models in the future.

The camera noise has been modelled using two classical image filters: brightness and contrast [25]. The filters can be expressed using the following mathematical function: $f(X) = \alpha X + \beta$, where α and β represent the contrast and the brightness factor, respectively.

The parameters, used in the camera noise experiment, are listed in Table 2, and an application example of such filters is shown in Figs. 6 and 7.

| $\beta = 30$ | $\beta = 60$ | $\beta = 90$ |

Fig. 6. Brightness

| $\alpha = 1.5$ | $\alpha = 2.0$ | $\alpha = 2.5$ |

Fig. 7. Contrast

Table 2. Camera noise model parameters

Filter	Values					
Brightness	30.0	60.0	90.0	0.0	0.0	0.0
Contrast	1.0	1.0	1.0	1.5	2.0	2.5

The LIDAR noise model is composed of two filters: drop filter and Gaussian noise. The drop filter simulates a cloud point loss by randomly discarding a fixed percentage of cloud points, 25% and 50% in this study. The Gaussian noise is applied to cloud points coordinates to simulate a measurement error, having zero mean and a standard deviation of 0.1, which represents a deviation of 10 cm.

$$p = [x, y, z] \qquad p' = p + \varepsilon \qquad \varepsilon \sim \mathcal{N}(0, 0.1^2 I_3) \tag{1}$$

4.2 Experiments

The first experiment involves obstacles detection; the camera noise is injected to camera frames to evaluate the accuracy of camera detection and fusion component. The main fusion sensor is set to be the camera, so, LIDAR cloud points are provided without alteration. Similar to the previous, the second experiment modifies cloud points according to LIDAR noise model. The accuracy evaluation accounts LIDAR segmentation together with the fusion component, having set LIDAR as main fusion sensor. The last experiment focuses on lane detection and concerns only the camera detection evaluation since the fusion component is not implicated in this task.

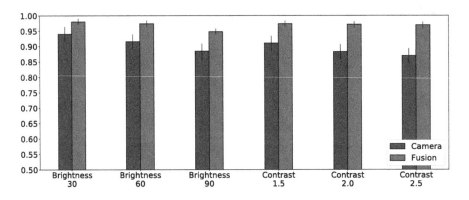

Fig. 8. Accuracies of camera and fusion components based on camera filters.

The experiments show similar behavior; however, there are several points they differ.

O6 The number of obstacles detected by LIDAR is higher than those of the camera, due to the narrower perspective and range of the camera view. The different perspective explains why LIDAR noise compromises more the fusion accuracy than the camera noise, proving that Apollo perception heavily relies on LIDAR.

O7 Figures 8 and 9 suggest that obstacles fusion makes the overall component more noise resistant. Moreover, this feature causes further modules to experience less noise by limiting propagation.

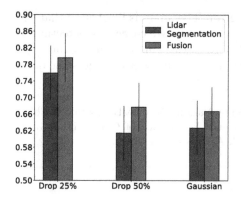

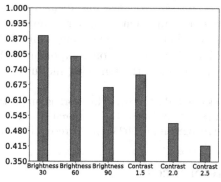

Fig. 9. Accuracies of LIDAR and fusion components based on LIDAR filters.

Fig. 10. Accuracies of lane detection based on camera filters.

O8 Figures 6 and 7 show that as brightness and contrast values increase, it gets harder to detect lanes, especially under the sunlight, due to the tendency of the image to become whiter. Figure 10 presents that contrast filter has a more significant impact on the accuracy of lane detection. The accuracy drops below 45% when the contrast value reaches 2.5.

The results show that the system is susceptible to vision disruptions. Further exploration should involve more complex simulation environments, populated by multiple sensors and validated using rigorous fault injection tools, like the recently proposed model [26], capable of generating faults covering different granularity, such as GPU bit flipping, obstacle obscuration, output alteration within ranges. Efforts should be spent to develop an enhanced version of the fusion component that dynamically changes the main fusion sensor according to the encountered scenario. Finally, it should be advantageous to study the effects of weather conditions on cloud points, to provide reproducible noise models to help the training phase; besides, data augmentation techniques should be adopted to counteract the effects of noise.

5 Conclusion

This work presents the performance and robustness analysis of the perception module belonging to a production-ready autonomous driving system. The performance analysis, focuses on the average response time, composed by neural networks and multiple sensors, which require acceleration via GPU to meet temporal constraints. This analysis demonstrated that neural networks take most of the calculation time and are fixed system, which cannot be dynamically adapted over similar inputs to reduce the inference time. The robustness analysis determines how accurate the camera and LIDAR are when they encounter challenging situations. The models proposed in this paper are similar to real events, such as brightness, contrast and measurement error. The analysis illustrated that the camera component is majorly affected by contrast, which causes detection of lanes and obstacles to be compromised. The robustness analysis also highlighted a new role for the fusion component within the module, which reduces the noise propagation to the following modules.

Acknowledgement. This work is supported by National Key R&D Program of China (2018YFB1305900); the National Natural Science Foundation of China under Grant (61702328 and 61602301); Microsoft Research Asia Research Grant.

References

1. Dosovitskiy, A., Ros, G., Codevilla, F., Lopez, A., Koltun, V.: CARLA: an open urban driving simulator. arXiv preprint arXiv:1711.03938 (2017)
2. CommaAI: Openpilot: Open source driving agent (2019). https://github.com/commaai/openPilot
3. Baidu: Apollo open platform (2019). https://github.com/ApolloAuto/apollo
4. Baidu: Apollo cyber (2019). https://github.com/ApolloAuto/apollo/tree/master/docs/cyber
5. Google: Protocol buffers (2008). https://developers.google.com/protocol-buffers/
6. Lin, S.-C., et al.: The architectural implications of autonomous driving: constraints and acceleration. ACM SIGPLAN Not. **53**, 751–766 (2018)
7. Jack Stewart, W.: People keep confusing their Teslas for self-driving cars (2018). https://www.wired.com/story/tesla-autopilot-crash-dui/
8. Tencent Keen Security Lab: Experimental security research of Tesla autopilot (2019). https://keenlab.tencent.com/en/whitepapers/Experimental_Security_Research_of_Tesla_Autopilot.pdf
9. Hendrycks, D., Dietterich, T.: Benchmarking neural network robustness to common corruptions and perturbations. arXiv preprint arXiv:1903.12261 (2019)
10. Pei, K., Cao, Y., Yang, J., Jana, S.: DeepXplore: automated whitebox testing of deep learning systems. In: Proceedings of the 26th Symposium on Operating Systems Principles, pp. 1–18. ACM (2017)
11. Pan, X., Shi, J., Luo, P., Wang, X., Tang, X.: Spatial as deep: spatial CNN for traffic scene understanding. In: Thirty-Second AAAI Conference on Artificial Intelligence (2018)
12. Redmon, J., Divvala, S., Girshick, R., Farhadi, A.: You only look once: unified, real-time object detection. In: Proceedings of the IEEE Conference on Computer Vision and Pattern Recognition, pp. 779–788 (2016)
13. Redmon, J., Farhadi, A.: Yolo9000: better, faster, stronger. In: Proceedings of the IEEE Conference on Computer Vision and Pattern Recognition, pp. 7263–7271 (2017)
14. National Ocean Service: What is lidar? (2018). https://oceanservice.noaa.gov/facts/lidar.html
15. Baidu: Apollo 3D obstacles perception (2019). https://github.com/ApolloAuto/apollo/blob/master/docs/specs/3d_obstacle_perception.md
16. Kuhn, H.W.: The hungarian method for the assignment problem. Nav. Res. Logist. Q. **2**(1–2), 83–97 (1955)

17. Cho, H., Seo, Y.-W., Kumar, B.V., Rajkumar, R.R.: A multi-sensor fusion system for moving object detection and tracking in urban driving environments. In: 2014 IEEE International Conference on Robotics and Automation (ICRA), pp. 1836–1843. IEEE (2014)
18. Neousys: Nuvo-6108gc series (2019). https://www.neousys-tech.com/en/product/application/rugged-embedded/nuvo-6108gc-gpu-computing
19. Geiger, A., Lenz, P., Stiller, C., Urtasun, R.: The kitti vision benchmark suite (2015). http://www.cvlibs.net/datasets/kitti
20. Michael Larabel, Phoronix: NVIDIA's Jetson AGX Xavier Carmel performance vs. low-power x86 processors (2019). https://www.phoronix.com/scan.php?page=article&item=nvidia-xavier-carmel&num=1
21. Jha, S., et al.: Kayotee: a fault injection-based system to assess the safety and reliability of autonomous vehicles to faults and errors. In: 3rd IEEE International Workshop on Automotive Reliability & Test (2018)
22. Jha, S., Banerjee, S.S., Cyriac, J., Kalbarczyk, Z.T., Iyer, R.K.: AVFI: fault injection for autonomous vehicles. In: 2018 48th Annual IEEE/IFIP International Conference on Dependable Systems and Networks Workshops (DSN-W), pp. 55–56. IEEE (2018)
23. TuSimple: Tusimple dataset (2019). https://github.com/TuSimple/tusimple-benchmark/issues/3
24. Garg, K., Nayar, S.K.: Photorealistic rendering of rain streaks. In: ACM Transactions on Graphics (TOG), vol. 25, pp. 996–1002. ACM (2006)
25. Rubaiyat, A.H.M., Qin, Y., Alemzadeh, H.: Experimental resilience assessment of an open-source driving agent. In: 2018 IEEE 23rd Pacific Rim International Symposium on Dependable Computing (PRDC), pp. 54–63. IEEE (2018)
26. Jha, S., et al.: ML-based fault injection for autonomous vehicles: a case for Bayesian fault injection, June 2019

Memory and File System

Spindle: A Write-Optimized NVM Cache for Journaling File System

Ge Yan, Kaixin Huang, and Linpeng Huang$^{(\boxtimes)}$

Department of Computer Science and Engineering, Shanghai Jiao Tong University,
Shanghai, China
{bueryg,kaixinhuang,lphuang}@sjtu.edu.cn

Abstract. Journaling techniques are widely employed in modern file systems to guarantee crash consistency. However, journaling usually leads to system performance decrease due to the frequent storage accesses it entails. Architects can utilize emerging non-volatile memory (NVM) as a persistent cache or journaling device to reduce the storage accesses of journaling file systems. Yet problems such as double writes, metadata write amplification and heavy transaction ordering overhead still exist in current solutions. Therefore, we propose Spindle, a write-optimized NVM cache to address these challenges. Spindle decouples data and metadata accesses by processing data in DRAM while pinning metadata in NVM. With redesigned metadata log and state switch mechanism, Spindle eliminates double writes and relieves metadata write amplification. Moreover, Spindle adopts a lightweight transaction scheme to guarantee crash consistency and reduce transaction ordering overhead. Experimental results reveal that Spindle achieves up to 47% throughput improvement compared with state-of-the-art design.

Keywords: File system · Non-volatile memory · Journaling ·
Data consistency

1 Introduction

Crash consistency is a significant feature that enables file systems to recover to a consistent state after unexpected system crashes or power failures. Journaling is a prevalent technique adopted by modern file systems, such as ext4 [11] and JFS [1], to maintain crash consistency. For example, redo journaling first writes the modified data to the journal area during the committing of a transaction. After the redo log is successfully committed, the modified data can be written to its desired location via checkpointing.

Although journaling can guarantee crash consistency, it significantly degrades system performance due to the heavy overheads entailed by frequent storage accesses. Precisely, journaling needs to write two blocks (i.e., the committed block and the checkpointed block) for every modified block and hence results in double disk I/Os. This problem is known as double writes of journaling.

© IFIP International Federation for Information Processing 2019
Published by Springer Nature Switzerland AG 2019
X. Tang et al. (Eds.): NPC 2019, LNCS 11783, pp. 251–263, 2019.
https://doi.org/10.1007/978-3-030-30709-7_20

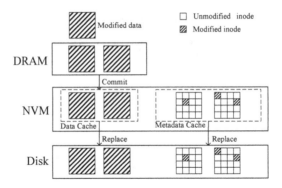

Fig. 1. Architecture of Spindle

Many in-memory file systems, such as BPFS [2], SCMFS [16], PMFS [4], NOVA [18], and HMVFS [20], have been proposed to utilize NVM as a replacement of disk. Since NVM provides persistency, byte-addressability and DRAM-like latency, these file systems avoid the bottleneck of disk I/O and provide high performance. However, two factors limit the utilization of these file systems in real environments. First, the price of NVM is predicated to be much higher than that of traditional storage (e.g., HDD and SSD). Second, current NVM technologies without broad application may be less stable than traditional storage.

To overcome the concerns discussed above, some researchers have investigated using NVM as a middle layer to accelerate disk-based file systems. For instance, UBJ [6] utilizes NVM to build a buffer cache united with the journaling layers; Tinca [15] proposes a transactional NVM cache with high performance and crash consistency. Although these systems utilize the low latency of NVM and the low cost of disk, write amplification and heavy transaction ordering overhead still exist due to their block-based updating strategy. Concretely, UBJ commits data in-place in NVM by freezing data and later checkpoints them to the disk. When updating a frozen block, UBJ can not overwrite it but must do it in a copy-on-write (COW) way, which incurs double writes on the critical path. Tinca does role switch to change the roles of modified blocks committed to the NVM cache. However, it does not differentiate metadata blocks from data blocks and metadata write amplification remains unsolved. Besides, it needs massive executions of cache line flush (e.g., clflush) and memory fence (e.g., mfence) to change the roles of modified blocks involved in the committing transaction.

With such observations, we propose Spindle, a write-optimized NVM cache for journaling file systems, as shown in Fig. 1. The key design of Spindle is decoupling data and metadata accesses. Data is processed in DRAM and committed to the NVM cache in the unit of block. Yet metadata is pinned in NVM and updated in the unit of inode. Such a design fully utilizes the byte-addressability of NVM and different updating granularities of data and metadata. Our main contributions can be summarized as follows.

- We propose a novel design to decouple data and metadata accesses by processing data in DRAM while pinning metadata in NVM at runtime. By redesigning the log and updating strategy of metadata, Spindle can efficiently reduce metadata writes to relieve metadata write amplification.
- We provide a lightweight transaction scheme to guarantee the consistency of data and metadata. Utilizing state switch and COW updating, Spindle can avoid double writes and reduce the transaction ordering overhead.
- We implement Spindle on ext4 file system and evaluate its performance with several benchmarks. Experimental results show that Spindle achieves up to 47% throughput improvement compared with state-of-the-art design.

The remainder of this paper is organized as follows. Section 2 introduces the background and motivation. Sections 3 and 4 describe the design and implementation of Spindle, respectively. Section 5 evaluates Spindle. We discuss related work in Sect. 6 and conclude this paper in Sect. 7.

2 Background and Motivation

2.1 NVM-based Systems

Emerging non-volatile memory technologies, such as STT-RAM [14], ReRAM [17], PCM [19], and 3D-XPoint [3], exhibit high storage density and low power consumption. Moreover, they can provide persistency, byte-addressability, DRAM-like latency and throughput. These properties enable architects to build fast and persistent systems with NVM.

Current NVM-based file systems can be classified into two categories. The first kind utilizes NVM as a replacement of disk, such as BPFS [2], SCMFS [16], PMFS [4], NOVA [18] and HMVFS [20]. Since the storage is substituted by NVM, these systems can exploit the properties of NVM to provide low latency and high throughput. The other one exploits NVM as a persistent cache or journaling device to improve the performance of disk-based file systems, such as UBJ [6] and Tinca [15]. In these systems, data is temporarily stored in NVM for fast accesses and eventually flushed to disk when data replacement is executed.

2.2 Crash Consistency

File systems should guarantee the consistency of data and metadata due to unpredictable power failures and system crashes. In-place updating is unfavored by file systems because a crash amid the updating process may cause inconsistency issue. Copy-on-write (COW) and journaling are two prevalent techniques for crash consistency. COW updates data out of place and substitutes the original data by changing respective metadata. Unlike COW, journaling does not change the metadata of a file but reserves backup copies of data for crash recovery.

The consistency of file system can be classified into metadata consistency, data consistency and version consistency. At the low level, metadata consistency only provides the consistency of metadata, while the consistency of data

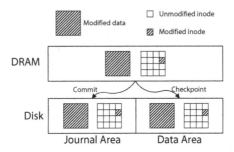

Fig. 2. Double writes and metadata write amplification

is ignored. At a higher level, data consistency guarantees that both data and metadata of a file system are consistent. Version consistency is an even higher level that ensures an old version of the file can be retrieved. This paper targets data consistency that can recover both data and metadata to a consistent state.

2.3   Motivation

Double Writes of Journaling. In journaling file systems, a committed block and corresponding checkpointed block have the same contents, but write operations are done twice. This problem is known as double writes of journaling. When NVM is provided as a persistent cache for disk-based file systems, double writes of journaling will lead to double memory copies and double executions of cache line flush and memory fence. Since data blocks in the NVM cache are flushed to the disk eventually, double writes of journaling also entail amplified disk I/O. Therefore, performance can get benefits if double writes can be eliminated.

Metadata Write Amplification. Metadata is organized in the unit of block according to the design of traditional journaling file systems. Partial-updated metadata results in whole-block write that seriously amplifies data I/O. For example, Tinca [15] writes data and metadata into NVM in the unit of block without differentiation. Suppose that an inode is 256B and a block is 4 KB, Tinca needs to write 4 KB data into NVM when only one inode is modified in the metadata block. Such a strategy leads to 16× execution time compared to solely updating an inode of 256B. Moreover, it squanders the byte-addressability of NVM. Therefore, it is desirable to reduce metadata write amplification to improve the overall performance. Figure 2 illustrates the issues of double writes and metadata write amplification in journaling file systems.

Transaction Ordering Overhead. Current CPU may reorder writes to the memory to optimize system performance [12]. A crash that happens amid reordered writes may cause inconsistency issue. For example, modification of file metadata should be performed after the updating of file data. If metadata is modified before the updating of file data and a crash happens, then the file will be inconsistent. To maintain the desired writing order, one widely-used method is to use regular store instructions (e.g., mov) followed by cache line flush and memory fence. Tinca adopts this method and entails massive executions of cache line flush and memory fence to change the roles of modified blocks.

3 Design

3.1 Cache Layout

As shown in Fig. 1, Spindle[1] consists of a data cache and a metadata cache.

Data Cache. Details of the data cache are depicted in Fig. 3(a). The data cache is made up of three components: the ring buffer, the data cache entries and cached data blocks. The ring buffer is used to coordinate a transaction and can be viewed as an array of 16-byte elements. Two pointers, Head and Tail, are provided to use the ring buffer in a round-robin way. Data cache entries with 16-byte elements are used for address mapping and crash recovery. The last part contains cached data blocks delivered by the file system.

Metadata Cache. Figure 3(b) shows the layout of metadata cache. The first part is a metadata log area with the granularity of inode size. After the metadata log area, there are multiple metadata cache entries that have the same structure with entries in the data cache. The ring buffer entries, data and metadata cache entries consist of an 8-byte disk block number and an 8-byte NVM block number. The major difference between these entries is that metadata cache entries are designed for metadata blocks while the others are used by data blocks. The last part is an area with cached metadata blocks.

A metadata block can be divided into many sub-blocks of inode size. Each running transaction in Spindle has a modified data block list and an original metadata sub-block list. When a metadata sub-block is going to be updated, we copy it to the log area and link it to the metadata sub-block list. Then we can perform in-place updates on the sub-block. When a transaction is successfully committed, related metadata sub-blocks in the log area can be removed.

3.2 Lightweight Transaction

Spindle provides a lightweight transaction scheme to ensure the consistency of file data. In particular, it exploits state switch and COW updating to avoid double writes on the critical path.

State Switch. In journaling file systems, a disk block has the state of being *committed* or *checkpointed*. When it is written to the journal, its state is *committed*. After it is flushed to its original location, its state becomes *checkpointed*. We utilize state switch to change the state of a cached data block. In our system, a data block with a ring buffer entry has the state of being *committed*,

---

[1] The memory hierarchy looks like a spindle with the data cache and the metadata cache interposed between DRAM and the disk.

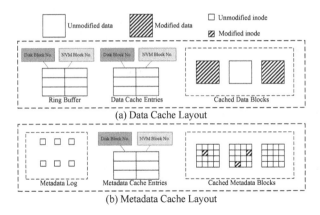

Fig. 3. Data and metadata cache layout of Spindle

which indicates it is involved in a committing transaction. After the transaction is successfully committed, corresponding ring buffer entry can be removed and the data block is switched to *checkpointed* state.

COW Updating. We adopt COW to write data into the data cache. Before writing a data block, we first search the data cache entries. If it has been cached, we copy its data cache entry to the ring buffer. Then we write the data to a new NVM block and update the data cache entry to point to the newly written block. If the block has not been cached, we create a new ring buffer entry with its disk block number and a special NON tag. Subsequently, we write the data to a new NVM block and create a new data cache entry. During the committing of a transaction, a cached data block may have two versions at the same time. Once the committing is successfully completed, the old one can be removed and the newly written one remains as persistent cached data.

Transaction Committing. We follow the routine of traditional journaling file systems to coalesce multiple blocks in a transaction. Since the NVM cache provides persistency that a file system desires, the committing of a transaction does not entail disk I/O when free data and metadata blocks can be found in the NVM cache. Before the committing of a transaction, Head and Tail point to the same entry in the ring buffer. Figure 4 illustrates the committing of a transaction, which can be summarized as follows.

(1) **Write ring buffer entry.** We search the block with the modified data block's disk block number. If it has been cached (i.e., cache hit), we copy its data cache entry to the ring buffer. Otherwise (i.e., cache miss), we create a new ring buffer entry with the disk block number and a special NON tag.
(2) **Write data block and data cache entry.** We write the data to a new NVM block and update corresponding data cache entry (cache hit) or create a new one (cache miss).
(3) **Update Tail pointer.** We move tail pointer forward by one.

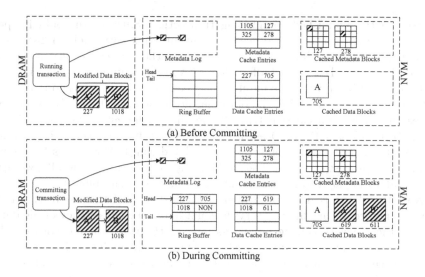

Fig. 4. Transaction committing of Spindle

(4) **Update Head pointer.** We repeat (1)–(3) until no block is left in the committing transaction. At last, we set `Head` to be identical to `Tail`.

When `Head` is equal to `Tail`, the committing is completed and the old data blocks can be recycled. Since metadata sub-blocks are first copied to the metadata log area and then in-place updated, we just need to remove the metadata sub-blocks in the log area.

3.3 Data Replacement

The NVM cache can not hold all the data of a file system due to its limited capacity. Therefore, we choose to write cold blocks back to disk. Data replacement is triggered when there is no free data or metadata block in the NVM cache. The replacement strategies for data and metadata are different.

(1) For data, the access unit and replacement unit is the same (i.e., block). We utilize the least-recently-used (LRU) algorithm to select the victim block. But there is an extra limit: data blocks involved in the running or committing transaction are not allowed to be swapped out.

(2) For metadata, the access unit is inode while the replacement unit is block. To select the victim, we maintain a counter for every cached metadata block. When any inode of a metadata block is accessed, the counter of that block is increased by one. If free metadata blocks can not be found in the cache, the block with the least counter value is chosen as the victim.

3.4 Crash Recovery

During crash recovery, we first scan the data and metadata cache entries to build an in-DRAM hash table. Then we scan the ring buffer and process cache entries involved in the unfinished committing transaction. Since crash could happen before the moving of `Tail`, we need to scan an extra entry behind `Tail`. For every ring buffer entry during the scan, we search the data cache entries with its disk block number. If the ring buffer entry's NVM block number is valid (i.e., not a `NON` tag), the block with this number has been cached and corresponding data cache entry exists. Otherwise, the block has not been cached. For the two cases, we search in the data cache entries. If the data cache entry exists, we revoke it (valid) or delete it (invalid) according to its ring buffer entry. If it does not exist, nothing needs to be done. After all related ring buffer entries have been scanned and processed, we set `Head` to be identical to `Tail`. Eventually, we scan the metadata log area to revoke the modified metadata sub-blocks.

Note that `Head` is not modified until the last step of the recovery. Therefore, any failure during the recovery does not affect the consistency of file data, because the recovery can be redone as long as the ring buffer entries can be retrieved. As for metadata, its consistency is maintained when it is successfully copied to the log area hence a crash during the recovery does not affect the consistency of metadata.

4   Implementation

We implement Spindle on `ext4` file system. The implementation mainly includes in-memory structures, data cache and metadata cache.

In-Memory Structures. A cuckoo hash table is used to accelerate the search with a disk block number for a data or metadata cache entry. Moreover, two LRU lists are managed in DRAM to select the victim to evict. Besides, bitmaps of data and metadata cache are also stored in DRAM. Since these structures can be reconstructed during system reboot, we do not keep them in NVM.

Data Cache. The data cache is implemented based on JBD2 [11]. We alter JBD2's interfaces to realize Spindle's characteristics. In particular, JBD2's descriptor block, revoke block and commit block are substituted by the ring buffer, data cache entries, `Head` and `Tail` pointers. Since the consistency of cached data and metadata is guaranteed when the transaction is successfully committed, the checkpointing function is removed from JBD2.

Metadata Cache. Implementation of the metadata cache primarily consists of three parts. First, metadata fetching and flushing is redesigned. Since metadata is not flushed to the disk unless under space pressure, we modify the implementation of `ext4-handle-dirty-metadata` to prevent metadata from being written back to disk. Second, the allocation and reclamation of metadata is renovated. Utilizing NVM as the metadata cache, we add a set of functions to allocate/free memory from/to the metadata cache. Third, the updating of metadata is substituted by our inode-based updating strategy.

5   Evaluation

5.1   Setup

We implement Spindle on Linux 4.18.1 and use `ext4` file system as our code base. All experiments are performed on an Intel Xeon E5 server with 98 GB DRAM and 1 TB HDD. The competitor we use to compare against Spindle is Tinca, which also utilizes NVM as a persistent cache to improve the performance of journaling file systems. Since Tinca is not open-source, we develop a prototype for it according to its implementation [15]. We use `ext4` file system to evaluate Tinca's performance and set the default cache mode as *writeback*. Since real NVM devices are not available to us yet, we add read/write latencies (50ns/180ns) to DRAM to simulate the NVM device. In the evaluation, 8 GB DRAM is used to simulate the NVM cache of Tinca and Spindle. For Spindle, we use 1 GB NVM as the metadata cache and the left 7 GB NVM as the data cache. Characteristics of the benchmarks in our evaluation are summarized in Table 1. All the results are averaged over five runs.

Table 1. Benchmark characteristics

Benchmark	Read/write ratio	Request size	Dataset size	Running time
Fio	0/10,3/7,5/5,7/3,10/0	4 KB	16 GB	20 min
Fileserver	1/2	16 KB	20 GB	30 min
Webproxy	5/1	16 KB	20 GB	30 min
Varmail	1/1	16 KB	20 GB	30 min

5.2   Microbenchmarks

Figure 5 shows the performance of Tinca and Spindle on `Fio` benchmark. As the write percentage declines from 100% (random write) to 0% (random read), the throughputs of Tinca and Spindle increase simultaneously. This is because writes will trigger data replacement to write cold data to disk. When the ratio of writes decreases, less disk writes are performed and the throughputs get promotion. Spindle outperforms Tinca with 1.13×–1.30× throughput under different read/write ratios. Concretely, Spindle achieves 30.0% throughput improvement for random writes and this is mainly due to the reduced cache line flush and memory fence. As for random reads, massive small metadata updates are executed and Spindle can effectively reduce metadata writes through its redesigned metadata updating strategy. The number of clflush per write operation is presented in Fig. 5(b). It can be observed that Spindle reduces up to 32.6% cache line flushes over Tinca. Although read operations only update the file metadata such as *access time*, Tinca needs to update extra cache entry and execute several clflush to change the role of a modified data block, which makes its transaction committing takes longer time than Spindle.

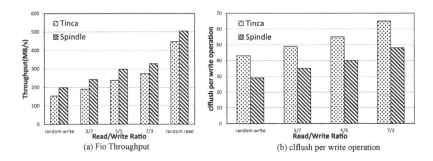

(a) Fio Throughput

(b) clflush per write operation

Fig. 5. Fio performance

The performance gaps are caused by two major differences between Tinca and Spindle: (1) Tinca writes metadata into NVM in the unit of block while Spindle updates metadata in the unit of inode. (2) Tinca needs to change the role of every modified block and entails extra cache line flush and memory fence.

5.3 Macrobenchmarks

We select fileserver, webproxy, and varmail as the macrobenchmarks to evaluate Spindle's performance. Figure 6 presents the throughputs of Tinca and Spindle under different file sizes.

Fileserver is a write-intensive workload that simulates a file server with operations like file creates, appends, writes, reads and deletes. We observe that for small files, the throughput improvement (40.7%) of Tinca is more striking. This is because metadata write amplification is more obvious for small writes. As the file size grows, the reduction of cache line flush, memory fence and metadata writes is counteracted by the writes of data.

Webproxy is a read-intensive workload that frequently reads the entire file. In such workload, Spindle can get few benefits from the clflush reduction as writes takes up only 16.7% in all read and write operations. However, webproxy entails massive small metadata updates and Spindle could benefit from its redesigned metadata updating strategy thus achieving up to 47% throughput improvement.

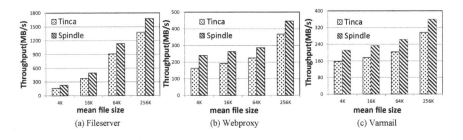

(a) Fileserver

(b) Webproxy

(c) Varmail

Fig. 6. Filebench performance

Varmail simulates the activity of an email server and performs a `fsync` after every file appending write. Both Spindle and Tinca do not need to write data to disk when there are free data and metadata blocks in the NVM cache. However, Tinca needs to write whole metadata blocks into NVM so its throughputs are 20.9%–33.5% lower than that of Spindle.

In conclusion, for read-intensive workloads like webproxy, Spindle benefits from its inode-based metadata updating strategy as it effectively reduces metadata write amplification; for write-intensive workloads like fileserver and varmail, the lightweight transaction scheme of Spindle plays an important role. Since writes to the NVM demands strict ordering, Spindle can effectively reduce the executions of clflush and mfence to promote system performance.

6   Related Work

Using flash-based SSD to develop devices with transactional supports or accelerate HDD-based systems has been explored. Nitro [9] utilizes data compression and deduplication into SSD cache to accelerate primary storage. LightTx [10] tries to reduce the transactional cost while providing better performance. TxFlash [13] utilizes the COW characteristic of NAND flash to provide transaction interface. However, these works focus on the flash-based SSD and do not take the emerging NVM into consideration.

Besides, many works utilize NVM to accelerate disk-based file systems. UBJ [6] is a buffer cache that commits in-place in NVM by freezing data and later checkpoints them to disk. Lee et al. [7] proposes to store modified data in the NVM cache and improve performance with space-efficient management techniques. Tinca [15] is a transactional NVM cache that utilizes the role switch mechanism to reduce write amplification of journaling file systems.

Managing NVM with in-memory file systems has been investigated as well. SCMFS [16] utilizes existing memory management to do the block management and keep space contiguous for each file. PMFS [4] avoids the overheads of block-based storage and enables direct persistent memory access with memory mapped I/O. Lee et al. [8] and Hwang et al. [5] propose tree structures that are optimized for non-volatile memory systems. These works indicate the promising potential of NVM in the future.

7   Conclusion

In this paper, we propose Spindle, a write-optimized NVM cache for journaling file systems. It decouples data and metadata accesses to update them in different granularities. Spindle adopts redesigned metadata log and metadata updating strategy to relieve metadata write amplification. Moreover, it utilizes state switch and COW updating to write data only once in the critical path. Besides, a novel committing protocol is proposed to reduce the transaction ordering overhead. Experiment results confirm that Spindle achieves up to 47% throughput improvement compared with state-of-the-art design.

Acknowledgement. This work is supported by National Key Research & Development Program of China (Grant No. 2018YFB10033002), the National Nature Science Foundation of China (Grant No. 61472241).

References

1. Best, S.: JFS overview (2000)
2. Condit, J., et al.: Better I/O through byte-addressable, persistent memory. In: Proceedings of the ACM SIGOPS 22nd Symposium on Operating Systems Principles, pp. 133–146. ACM (2009)
3. Dadmal, U.D., Vinkare, R.S., Kaushik, P., Mishra, S.: 3D X point technology. Int. J. Electron. Commun. Soft Comput. Sci. Eng. (IJECSCSE) 13–17 (2017)
4. Dulloor, S.R., et al.: System software for persistent memory. In: Proceedings of the Ninth European Conference on Computer Systems, p. 15. ACM (2014)
5. Hwang, D., Kim, W.H., Won, Y., Nam, B.: Endurable transient inconsistency in byte-addressable persistent B+-TREE. In: FAST, pp. 187–200 (2018)
6. Lee, E., Bahn, H., Noh, S.H.: Unioning of the buffer cache and journaling layers with non-volatile memory. In: FAST, pp. 73–80 (2013)
7. Lee, E., Kang, H., Bahn, H., Shin, K.G.: Eliminating periodic flush overhead of file I/O with non-volatile buffer cache. IEEE Trans. Comput. **65**(4), 1145–1157 (2016)
8. Lee, S.K., Lim, K.H., Song, H., Nam, B., Noh, S.H.: WORT: write optimal radix tree for persistent memory storage systems. In: FAST, pp. 257–270 (2017)
9. Li, C., Shilane, P., Douglis, F., Shim, H., Smaldone, S., Wallace, G.: Nitro: a capacity-optimized SSD cache for primary storage. In: ATC, pp. 501–512 (2014)
10. Lu, Y., Shu, J., Guo, J., Li, S., Mutlu, O.: LightTX: a lightweight transactional design in flash-based SSDs to support flexible transactions. In: 2013 IEEE 31st International Conference on Computer Design (ICCD), pp. 115–122. IEEE (2013)
11. Mathur, A., Cao, M., Bhattacharya, S., Dilger, A., Tomas, A., Vivier, L.: The new ext4 filesystem: current status and future plans. In: Proceedings of the Linux symposium, vol. 2, pp. 21–33 (2007)
12. Pelley, S., Chen, P.M., Wenisch, T.F.: Memory persistency. In: ACM SIGARCH Computer Architecture News, vol. 42, pp. 265–276. IEEE Press (2014)
13. Prabhakaran, V., Rodeheffer, T.L., Zhou, L.: Transactional flash. In: OSDI, vol. 8 (2008)
14. Sun, Z., et al.: Multi retention level STT-RAM cache designs with a dynamic refresh scheme. In: Proceedings of the 44th Annual IEEE/ACM International Symposium on Microarchitecture, pp. 329–338. ACM (2011)
15. Wei, Q., Wang, C., Chen, C., Yang, Y., Yang, J., Xue, M.: Transactional NVM cache with high performance and crash consistency. In: Proceedings of the International Conference for High Performance Computing, Networking, Storage and Analysis, p. 56. ACM (2017)
16. Wu, X., Reddy, A.: Scmfs: a file system for storage class memory. In: Proceedings of 2011 International Conference for High Performance Computing, Networking, Storage and Analysis, p. 39. ACM (2011)
17. Xu, C., Niu, D., Muralimanohar, N., Jouppi, N.P., Xie, Y.: Understanding the trade-offs in multi-level cell reram memory design. In: 2013 50th ACM/EDAC/IEEE Design Automation Conference (DAC), pp. 1–6. IEEE (2013)
18. Xu, J., Swanson, S.: NOVA: a log-structured file system for hybrid volatile/non-volatile main memories. In: FAST, pp. 323–338 (2016)

19. Yoon, D.H., Chang, J., Schreiber, R.S., Jouppi, N.P.: Practical nonvolatile multilevel-cell phase change memory. In: Proceedings of the International Conference on High Performance Computing, Networking, Storage and Analysis, p. 21. ACM (2013)
20. Zheng, S., Huang, L., Liu, H., Wu, L., Zha, J.: HMVFS: a hybrid memory versioning file system. In: 2016 32nd Symposium on Mass Storage Systems and Technologies (MSST), pp. 1–14. IEEE (2016)

Two-Erasure Codes from 3-Plexes

Liping Yi[✉], Rebecca J. Stones[✉], and Gang Wang[✉]

College of Computer Science, Nankai University, Tianjin, China
{yiliping,becky,wgzwp}@nbjl.nankai.edu.cn

Abstract. We present a family of parity array codes called 3-PLEX for tolerating two disk failures in storage systems. It only uses exclusive-or operations to compute parity symbols. We give two data/parity layouts for 3-PLEX: (a) When the number of disks in array is at most 6, we use a horizontal layout which is similar to EVENODD codes, (b) otherwise we choose hybrid layout like HoVer codes. The major advantage of 3-PLEX is that it has optimal encoding/decoding/updating complexity in theory and the number of disks in a 3-PLEX disk array is less constrained than other array codes, which enables greater parameter flexibility for trade-offs in storage efficiency and performances.

Keywords: Latin squares · Array codes · Storage efficiency · Computational complexity · Data/parity layout

1 Introduction

As a fault-tolerant technology, erasure codes have been widely used to ensure the reliability of storage systems. According to whether encoding/decoding is based on exclusive-or operations, erasure codes can be roughly divided into two categories: Reed-Solomon codes and parity array codes. The common property of these codes is that they tolerate two simultaneous disk failures, but they also have their own trade-offs in terms of storage efficiency and computational complexity.

Traditional Reed-Solomon code [12] is based on finite field operations which results in high computational complexity. Quite a few studies have tried to improve the computational performance of RS codes by using special hardware or dedicated algorithms. For example, the modified Cauchy RS code [13] replaces the finite field operation with exclusive-or operation, but it still has higher computational complexity than array codes. Although RS codes have the above shortcoming, it is the only erasure code that can achieve arbitrary fault tolerance, and it is an MDS code [2] which means it has optimal storage efficiency. Another advantage of RS codes is that there is no assumption on the number of disks which enables it to be applied in more scenarios.

Parity array codes arrange data/parity blocks according to the structure of arrays [5]. Compared with RS codes, since array codes are completely based on

© IFIP International Federation for Information Processing 2019
Published by Springer Nature Switzerland AG 2019
X. Tang et al. (Eds.): NPC 2019, LNCS 11783, pp. 264–276, 2019.
https://doi.org/10.1007/978-3-030-30709-7_21

exclusive-or operations, its encoding and decoding algorithms are relatively simple and easy to implement. According to the data/parity layout, array codes can be divided into three types: (a) horizontal codes (such as EVENODD [1], RDP [3], etc.), (b) vertical codes (X-Code [16], WEAVER [8], etc.), and (c) hybrid codes (HoVer [9]). All of these array codes tolerate two faults.

EVENODD code [1] as the original horizontal array code, provides an efficient means to tolerate double disk failures in RAID architecture. Figure 1-a shows an example of EVENODD with five data disks and two parity disks. EVENODD code only uses two individual parity disks (stores horizontal parities and secondary diagonal parities respectively) to achieve two fault tolerance and the implementation of EVENODD code does not require any hardware modification of the standard RAID-5 controller [1]. It's also a MDS code with optimal storage efficiency. Another advantage of EVENODD is that the number of disks of EVENODD disk array is adjustable. The array size of EVENODD is $(n-1) \times (n+2)$, where n is the number of data disks. Although its encoding definition requires prime n, it can also achieve arbitrary number of disks by horizontal shortening. Since EVENODD code defines a special check sum S (the XOR-sum of "$D4$"s in Fig. 1-a), its computational complexity is higher than other array codes.

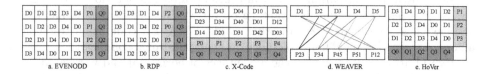

Fig. 1. Array codes of 5 data blocks

RDP (Row-Diagonal Parity) code [3] is another horizontal array code whose array size is $(p-1) \times (p+1)$ (where p is a controlling parameter which must be a prime number greater than 2). RDP code replaces the special check sum S of EVENODD with a missing diagonal and uses a "parity dependent" architecture (where row parities also contribute to the calculation of diagonal parities), which reduces the computational complexity compared with EVENODD. Figure 1-b illustrates a RDP disk array with four data disks and two parity disks. Corbett et al. [3] proved that both encoding/decoding and updating complexity of RDP is almost optimal.

Updating complexity is an important metric of array codes; we generally use the number of blocks required to be updated when a data block is updated to measure updating complexity. In a storage system with frequent updating operations, higher updating complexity will lead to worse performance. In order to alleviate this problem, vertical array codes have been proposed.

X-Code [16] is the first vertical array code which tolerates two disk failures. The array size of X-Code is $n \times n$ (where n is a prime); the first $n-2$ rows are filled with data blocks, and the last two rows store parity blocks, i.e., each disk stores both data and parity blocks. X-Code computes the XOR sum of all data blocks on the diagonals of slope 1 and −1 to generate parity blocks, hence the name X-Code. Figure 1-c depicts an example of X-Code with 5 disks. This special structure enables X-Code to achieve optimal computational complexity. X-Code is also an MDS code, and thus it has optimal storage efficiency. An important limitation of X-Code is that the number of disks in X-Code array must be a prime and it cannot admit arbitrary array sizes using horizontal shortening like EVENODD. If we impose simple horizontal shortening on a X-Code to break through its constraint of array size, the original relationship between data blocks and corresponding parity blocks will be destroyed, thereby losing the ability to tolerate two erasures.

WEAVER code [8] is another typical vertical array code. There are four types of WEAVER codes proposed in [8], and only WEAVER(n, k, t) when $k = t = 2$ can tolerate two disk failures. The array size of WEAVER$(n, 2, 2)$ is $2 \times n$, the first row only stores data blocks and the last row is filled with parity blocks. Figure 1-d shows an example of WEAVER$(n, 2, 2)$ with five disks. WEAVER code computes parity blocks by calculating XOR sum of two adjacent data blocks like weaving, hence named WEAVER. The definition of encoding enables WEAVER to have no constraint in array size, which makes up for the deficiency of X-Code. Besides, a key advantage of WEAVER is that it can achieve up to twelve fault tolerance by adjusting parameters. It also has optimal computational complexity. Nonetheless, the main disadvantage of WEAVER code is that it has low storage efficiency (up to 50%). In other words, WEAVER code has a trade-off among storage efficiency, fault tolerance and computational complexity.

Hybrid array codes combine the advantages of horizontal and vertical array codes. HoVer code [9] with $(v + r) \times (h + n)$ array size is a representative hybrid array code, hence named HoVer (Horizontal and Vertical). Figure 1-e shows an example of HoVer code with 5×8 array size. The advantages of HoVer code is that it's a near-MDS code, which means it has high but not optimal storage efficiency, no limitation of array size, parameter flexibility enables it to achieve higher fault tolerance, and its updating complexity is optimal. However, the encoding and decoding complexity become higher as the number of disks increases, so HoVer code also has a trade-off between storage efficiency and encoding/decoding speed.

Hafner [8] pointed out that there is no perfect array code, and each code has a trade-off among fault tolerance, storage efficiency and computational complexity. In this paper, we aim to find an array code with a better trade-off.

We follow ideas in e.g. [4–7] describing methods of using Latin squares to construct array codes. Motivated by these methods, this paper proposes a new double-erasure array code named 3-PLEX based on Latin squares. The advantages of 3-PLEX are that it has optimal computational complexity and no constraint of array size. The storage efficiency of 3-PLEX is always 60%. In other words, 3-PLEX also has a trade-off between storage efficiency and performances.

This paper is organized as follows: we give the relationship between Latin squares and k-plex in Sect. 2. Then, we describe the encoding procedure of 3-PLEX with mathematical definition in Sect. 3. In Sect. 4, we present the decoding procedure with proof for two fault tolerance. In Sect. 5, we compare 3-PLEX and other array codes. In Sect. 6, we describe an implementation and performance tests of 3-PLEX based on NCFS. Section 7 summarizes the correspondence and presents some future works.

2 Latin Squares and k-plexes

A k-plex of order n is an $n \times n$ matrix whose symbols belong to a set of size n, with the following properties:

– each row contains k distinct symbols and $n - k$ empty positions,
– each column contains k distinct symbols and $n - k$ empty positions, and
– each symbol occurs exactly k times.

Latin squares are k-plexes of order n when $k = n$, and thus k-plexes generalize Latin squares [14]. A 1-plex embedded in a Latin square is called a transversal. The union of 3 disjoint transversals forms a 3-plex, and we use this method to construct 3-plexes. Figure 2-a is a Latin square of order 5 containing the 3-plex of order 5 in Fig. 2-b. Figure 2-c indicates to obtain a 3-PLEX code.

$$
\begin{bmatrix}
① & 2 & 3 & ④ & 5 \\
2 & ③ & 4 & 5 & ① \\
③ & 4 & ⑤ & 1 & 2 \\
4 & ⑤ & 1 & ② & 3 \\
5 & 1 & ② & 3 & ④
\end{bmatrix}
\quad
\begin{bmatrix}
\cdot & 2 & 3 & \cdot & 5 \\
2 & \cdot & 4 & 5 & \cdot \\
\cdot & 4 & \cdot & 1 & 2 \\
4 & \cdot & 1 & \cdot & 3 \\
5 & 1 & \cdot & 3 & \cdot
\end{bmatrix}
\quad
\begin{bmatrix}
\cdot & d_2 & d_3 & \cdot & d_5 \\
d_2 & \cdot & d_4 & d_5 & \cdot \\
\cdot & d_4 & \cdot & d_1 & d_2 \\
d_4 & \cdot & d_1 & \cdot & d_3 \\
d_5 & d_1 & \cdot & d_3 & \cdot
\end{bmatrix}
$$

a. DCLS of order 5   b. 3-plex of order 5   c. 3-PLEX code with 5 data disks

Fig. 2. The relationship among DCLS, 3-plex and 3-PLEX code

In this paper, we construct 3-plexes from diagonally cyclic Latin squares (DCLSs), which we define as a Latin square having forward broken diagonals which have a constant increment from top to bottom (this is slightly different from the standard definition [15]). Each broken diagonal in the Latin square in Fig. 2-a increases by 2 (we highlight two forward broken diagonals). In a DCLS, the union of any three forward broken diagonals is a 3-plex. For simplicity, we use back-circulant Latin squares (i.e., constant broken antidiagonals) as DCLSs. DCLSs only exist when n is odd, although 3-plexes exist for all orders $n \geq 3$.

3 Encoding

Assuming that there are n data disks, in Sect. 3.1 we discuss how to arrange the parity blocks. We also assume each block occupied the space of 1 bit. In actual storage, each block may occupy 1 byte or more of disk space, which will not affect the encoding procedure of 3-PLEX. Similar to other array codes [1,3,9,16], the parity blocks in 3-PLEX code can also be computed by calculating the XOR sum of data blocks on the diagonal with slope 1 or -1. Figure 2 shows how to map a 3-plex of order 5 to a 3-PLEX code with five data disks. Data blocks in a 3-PLEX array correspond to symbols in the 3-plex marked with the same colors. The data blocks placed in the same column are stored in the same disk in the 3-PLEX disk array. Note that data blocks are placed consecutively on each disk rather than arranged in line with the data layout of 3-PLEX code, i.e., the empty positions denoted '.' do not actually exist. To guarantee 3-PLEX with two-fault tolerance, we encode data blocks in 3-PLEX array to row parity blocks and diagonal parity blocks respectively like EVENODD code.

3.1 Data/Parity Layout

Before giving the definition of the encoding procedure, we need to choose one appropriate data/parity layout of 3-PLEX. There are three layouts introduced previously: horizontal, vertical, and hybrid layout. The storage efficiency and encoding/updating complexity of the three layouts of 3-PLEX are same, and only decoding complexity will be meaningfully affected by data/parity layout, therefore in order to attain the lowest decoding complexity, we need to adapt a proper data/parity layout.

Let's take a 3-PLEX code of order 5 as a running example. Figure 3 plots three data/parity layouts of 3-PLEX. The row parity blocks are shaded light gray, and the diagonal parity blocks are shaded dark gray and represented by the same number as their data blocks. In real storage systems, typically blocks in each layout compose a *stripe*, and blocks in the same column form a *stripe unit* as the basic storage unit. Stripe units in a stripe are stored on different disks. A storage system contains many stripes. In addition, the empty positions in each array layout do not exist actually and blocks in the same column are consecutively stored practically. Thus, data blocks and parity blocks have different sizes in horizontal layout and hybrid layout. This will not cause unbalance in storage because data and parity blocks from different stripes typically are evenly distributed over all the disks in a modern distributed storage system. Moreover, encoding/decoding/updating computation will be performed correctly as long as we keep track of the relationship between data blocks and parity blocks.

We compute the average number of exclusive-or operations required to reconstruct two failed disks to measure the decoding complexity of 3-PLEX with different data/parity layouts separately. For horizontal layout, we calculate the decoding complexity from threes cases: (a) when two data disks fail, the decoding complexity is 12, (b) when one data disk and one parity disk fail: $6 + 2n$ (where n is the number of data disks), and (c) when two parity disks fail: $4n$. Hence, the mean decoding complexity of reconstructing two disk failure is $6 + 2n$. In terms of vertical layout, we find it can not support two fault tolerance after theoretical analysis. As for hybrid layout, we compute the decoding complexity in two ways: (a) two hybrid disks fail: 14, and (b) one hybrid disk and row parity disk fail: $24 + 2n$. Thus, the horizontal layout has lower decoding complexity than

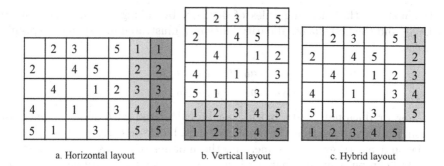

a. Horizontal layout   b. Vertical layout   c. Hybrid layout

Fig. 3. Three alternative data/parity layouts of 3-PLEX

the hybrid layout when $n < 6$, otherwise the hybrid layout has lower decoding complexity. For simplicity, we focus on 3-PLEX with horizontal layout.

3.2 Encoding Procedure

Let $a_{i,j}$ be the data/parity block at the ith row and jth column ($i \in \{0, 1, \ldots, n-1\}$ and $j \in \{0, 1, \ldots, n+1\}$), then the parity blocks of 3-PLEX code with horizontal layout can be calculated according to the following rules:

$$a_{i,n} = \bigoplus_{j=0}^{n-1} a_{i,j}, \tag{1}$$

$$a_{i,n+1} = \bigoplus_{j=0}^{n-1} a_{\langle i-j \rangle_n, j}. \tag{2}$$

where n is the number of data disks and $\langle x \rangle_n = x \bmod n$. For simplicity, we choose a 3-plex from three non-consecutive forward diagonals so that the difference between the indices of two of the three is indivisible by 3.

In order to make the encoding rules defined by (1) and (2) easy to understand, we give an encoding example of 3-PLEX with horizontal layout and order five showed in Fig. 4-a.

a. Two consecutive failed data disks   b. Two inconsecutive failed data disks

Fig. 4. Two sub-cases of two failed data disks

To illustrate the encoding rules of 3-PLEX by taking the first row parity block $a_{0,5}$ and the first diagonal parity block $a_{0,6}$ illustrated in Fig. 4-a as specific examples:

$$a_{0,5} = a_{0,1} \oplus a_{0,2} \oplus a_{0,4}$$

$$a_{0,6} = a_{4,1} \oplus a_{3,2} \oplus a_{2,2}$$

Other row or diagonal parity blocks can be calculated with same manner. In addition, it's easy to see that the number of data blocks in each row parity group and diagonal parity group is same, i.e., the number of exclusive-or operations required to compute one row parity block and one diagonal parity block is equal. This feature explains why the encoding/updating complexity of horizontal and hybrid layouts are identical.

4   Decoding and Proof for Two Fault Tolerance

When a single disk of failure occurs in 3-PLEX, lost data blocks of failed disk can be directly reconstructed through row parity or diagonal parity. In this Section, a corresponding decoding algorithm is given to prove that 3-PLEX code has the ability to correct two disk failures.

Assuming columns c_1 (disk c_1) and c_2 (disk c_2) fail in the 3-PLEX array, we describe the decoding algorithm split into four cases:

Case: $c_1 = n$; $c_2 = n+1$. I.e., both of the parity disks fail. This case is equivalent to re-encoding all data blocks, hence the decoding algorithm is the same as the encoding procedure.

Case: $c_1 < n$; $c_2 = n$. I.e., one data disk and the row parity disk fail simultaneously. In this case, the invalid data disk should be first recovered through diagonal parity blocks and surviving data blocks, and then the failed row parity disk can be reconstructed according to the encoding rules of row parity. Hence, the decoding algorithm can be derived on the basis of (1) and (2):

$$a_{i,c_1} = \left(\bigoplus_{j=0}^{n-1} a_{<i-j>_n,j} \right) \oplus a_{i,n+1}, \quad j \neq c_1,$$

$$a_{i,n} = a_{i,c_2} = \bigoplus_{j=0}^{n-1} a_{i,j}.$$

Case: $c_1 < n$; $c_2 = n + 1$. I.e., one data disk and the diagonal parity disk fail simultaneously. This situation is similar to the previous case. we can directly reconstruct failed data disk using row parity blocks and survival data blocks, then failed diagonal disk can be recovered according to (2). Therefore, the decoding algorithm of this case can be derived from (1) and (2):

$$a_{i,c_1} = \left(\bigoplus_{j=0}^{n-1} a_{i,j} \right) \oplus a_{i,n}, \quad j \neq c_1,$$

$$a_{i,n+1} = a_{i,c_2} = \bigoplus_{j=0}^{n-1} a_{<i-j>_n,j}.$$

Case: $c_1 < c_2 < n$. I.e., two data disks fail simultaneously. We divide this case into two sub-cases which are illustrated in Fig. 4: the two failed data disks are (a) consecutive, (b) not consecutive. For case (a), since any consecutive two data disks only have at most one common row filled with data blocks, we recover the two data blocks in this special row (if it exists) according to diagonal parity; we mark these two data blocks blue in Fig. 4. The remaining data blocks are only one failed data blocks in their respective rows, and are thus simply recovered via row parity, marked in green in Fig. 4. As for case (b), we recover the failed data blocks marked green on the basis of row parity, and we also recover data blocks marked blue according to diagonal parity. After the blue data blocks are recovered, then we use row parity blocks and other related survival data blocks to recover the last unrepaired data blocks marked red. Hence, the decoding algorithm of situation (b) can be derived according to (1) and (2):

$$a_{i,c_1} = \Big(\bigoplus_{j=0}^{n-1} a_{<i-j>_n,j} \Big) \oplus a_{i,n+1}, \quad j \neq c_1,$$

$$a_{i,c_2} = \Big(\bigoplus_{j=0}^{n-1} a_{i,j} \Big) \oplus a_{i,c_1} \oplus a_{i,n}.$$

Since the decoding algorithms above are all derived from encoding rules defined by (1) and (2), we conclude that 3-PLEX code has two fault tolerance.

5   Comparison with Existing Schemes

In this section, we compare 3-PLEX code with array codes proposed in [1, 3, 8, 9, 16] in terms of fault tolerance, storage efficiency, encoding/decoding/updating complexity and constraint of array size.

Fault Tolerance: fault tolerance is a basic indicator for measuring erasure codes. Table 1 lists the fault tolerance and requirements for the number of data disks of multiple erasure codes. RS code is the only code applied in practice that can support arbitrary fault tolerance, but it has higher computational complexity. Although HoVer code also can achieve arbitrary fault tolerance, its high

Table 1. Comparison of multiple erasure codes in fault tolerance

Erasure code	Fault tolerance	The number of data disks
RS	Arbitrary	Depends on the length of code words
EVENODD	2	Prime
RDP	2	Prime−1
X-Code	2	Prime
WEAVER	≤ 12	No constraints
HoVer	Arbitrary	No constraints
3-PLEX	2	No constraints

fault tolerance will significantly influence storage efficiency and computational complexity. EVENODD, RDP and X-Code can only tolerant two simultaneous disks failures, and they also have array size limitation. WEAVER code can support up to 12 fault tolerance due to its special definition, however there is no systematic construction method for WEAVER codes and most of them need to determine the encoding rules using special search technology, hence WEAVER code has worse scalability. 3-PLEX code proposed in this paper has been proved that it has two fault tolerance.

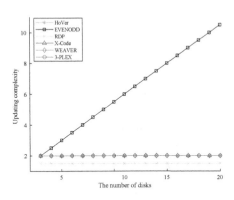

Fig. 5. Storage efficiency

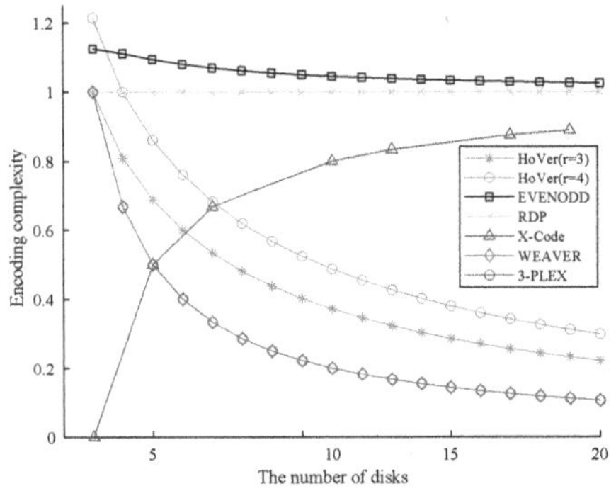

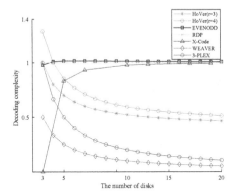

Fig. 6. Encoding complexity

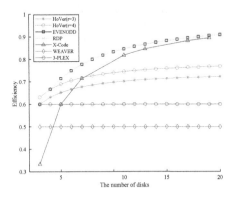

Fig. 7. Updating complexity

Fig. 8. Decoding complexity

Storage Efficiency: Storage efficiency is the proportion of data blocks among all blocks, and is another important criterion for evaluating erasure codes. In Fig. 5, we see that WEAVER$(n, 2, 2)$ code has the lowest storage efficiency (50%) which is a trade-off between storage efficiency with other indicators. The storage efficiency of 3-PLEX code is always 60% which is higher than WEAVER$(n, 2, 2)$

code. Since EVENODD, RDP, and X-Code are MDS codes, and HoVer is a near-MDS code, they have (near) optimal storage efficiency. Therefore, 3-PLEX also has a trade-off between storage efficiency and the other performance metrics.

Encoding Complexity: Encoding overhead of erasure codes determines practicability, and hence is an important indicator for evaluation. We generally use the number of exclusive-or operations required to generate one single parity block during encoding procedure to measure the encoding complexity. Figure 6 plots the encoding complexity vs. the number of disks. In order to facilitate the observation of the relationship between practical encoding complexity and theoretical optimal encoding complexity (the generation of a single parity block requires $n - 1$ exclusive-or operations in theory), we use the ratio of practical and theoretical optimal encoding complexity as the ordinate, i.e., integer 1 in ordinate indicates that the encoding complexity is theoretically optimal. It can be seen from Fig. 6 that the encoding complexity of WEAVER code and 3-PLEX code is equal and lowest compared to other array codes. HoVer, X-Code and RDP codes reach the optimal encoding complexity in theory. Since the encoding procedure of EVENODD code involves a special checksum S, its encoding complexity approaches but is not theoretical optimal as the number of disks increases.

Updating Complexity: A large number of small IO operations increases the updating overhead of a storage system, hence updating complexity is an important parameter for judging the performance of erasure codes. Updating complexity is usually represented by the average number of parity blocks needing recalculation after updating one data block. Figure 7 illustrates the variation in updating complexity of the various array codes as the number of data disks increases. Except for EVENODD code, the updating complexity of other array codes is 2 which implies these array codes have optimal updating complexity. Since EVENODD requires the special checksum S due to the calculation of diagonal parity blocks, when one data block on this special diagonal is updated, S also will be updated and all diagonal parity blocks need to be updated, this is the reason why EVENODD's updating complexity is higher than other array codes.

Decoding Complexity: The decoding complexity of erasure codes directly affects the availability and reliability of a storage system. Therefore, decoding complexity is another key indicator for evaluating erasure codes. We usually use the number of exclusive-or operations required to recover a single failed block during decoding procedure to measure decoding complexity. We use the ratio of practical and theoretical optimal decoding complexity as ordinate of Fig. 8 and integer 1 indicates that the decoding complexity of corresponding array code is optimal in theory. We see in Fig. 8 that 3-PLEX, WEAVER$(n, 2, 2)$, X-Code and RDP code all reach or approach the theoretical optimal decoding complexity. Also due to the existence of S, EVENODD's decoding complexity is higher than other array codes but also close to be optimal in theory.

To summarize, 3-PLEX code exchanges lower computational complexity with some storage space, i.e., there is a trade-off between storage efficiency and the other performance metrics.

6   Implementation and Performance

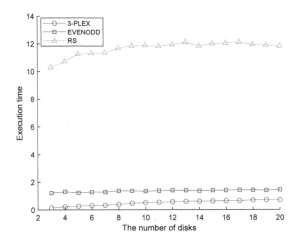

Fig. 9. Single fault reconstruction time of 3-PLEX, EVENODD and RS code

In this section, we implement the encoding and decoding algorithm of 3-PLEX code on Network-Coding-Based Distributed File System (NCFS) [10]. We focus on measuring the decoding complexity of repairing after single disk failure of each erasure code on NCFS [11]. We deploy NCFS on the virtualized Linux platform with 8 cores (2 GHz) of CPU, 16G of memory, 200G of HDD, and use actual execution time of reconstructing one failed disk to represent the decoding complexity of each erasure codes. For each erasure code, we respectively measure its execution time of decoding 100 times when the total number of disks in this code ranges from 3 to 20 and calculate the average of the results of 100 times to record the decoding time. Figure 9 plots the test results for RS, EVENODD and 3-PLEX code.

We observe that 3-PLEX code has the lowest decoding complexity, EVEN-ODD code follows, and RS code has the highest decoding complexity. As we assume that the capacity of each disk is 100 MB, it seems that 3-PLEX code has a marginal improvement over EVENODD code illustrated in Fig. 9. If each disk stores far greater than 100 MB of data, the advantage of 3-PLEX code will be more obvious. Another conclusion we draw is that the decoding complexity of the three erasure codes becomes gradually stable as the number of disks increases to 15. Both of the two conclusions are consistent with the performance analysis in theory. In addition, we observe that performances of each code in the simulated storage system are in agreement with the theoretical analysis in Sect. 5, which indicates that the theoretical performance comparison analysis of the codes in Sect. 5 is reliable.

7 Conclusions

In this paper, we present a novel erasure code called 3-PLEX code which can tolerate two disk failures. It has optimal computational complexity and no constrains of array size. It exchanges low computational complexity with storage efficiency, i.e., there is a trade-off between storage efficiency and performances. In addition, since the idea for 3-PLEX code generated from Latin squares, we can continue to research the following aspects: (a) study 3-plex and orthogonal 3-plexes to extend 3-PLEX code with higher fault tolerance, (b) adjust the values of k and n to find a k-PLEX code that can achieve a better trade-off in fault tolerance, storage efficiency and performances.

References

1. Blaum, M., Brady, J., Bruck, J., Menon, J.: EVENODD: an efficient scheme for tolerating double disk failures in RAID architectures. IEEE Trans. Comput. **44**(2), 192–202 (1995)
2. Blaum, M., Roth, R.M.: On lowest density MDS codes. Trans. Inform. Theory **45**(1), 46–59 (1999)
3. Corbett, P., et al.: Row-diagonal parity for double disk failure correction. In: Proceedings of the FAST, pp. 1–14 (2004)
4. Gang, W., Sheng, L., Xiaoguang, L., Jing, L.: Representing X-Code using latin squares. In: Proceedings of the PRDC, pp. 177–182 (2009)
5. Gang, W., Xiaoguang, L., Sheng, L., Guangjun, X., Jing, L.: Constructing double-erasure HoVer codes using Latin squares. In: Proceedings of the ICPADS, pp. 533–540 (2008)
6. Gang, W., Xiaoguang, L., Sheng, L., Guangjun, X., Jing, L.: Constructing liberation codes using Latin squares. In: Proceedings of the PRDC, pp. 73–80 (2008)
7. Gang, W., Xiaoguang, L., Sheng, L., Guangjun, X., Jing, L.: Generalizing RDP codes using the combinatorial method. In: Proceedings of the NCA, pp. 93–100 (2008)
8. Hafner, J.L.: WEAVER codes: Highly fault tolerant erasure codes for storage systems. In: Proceedings of the FAST, vol. 5 (2005)
9. Hafner, J.L.: HoVer erasure codes for disk arrays. In: Proceedings of the DSN, pp. 217–226 (2006)
10. Hu, Y., Yu, C.M., Li, Y.K., Lee, P.P., Lui, J.C.: NCFS: on the practicality and extensibility of a network-coding-based distributed file system. In: Proceedings of the NetCod, pp. 1–6. IEEE (2011)
11. Huang, C., Xu, L.: Star: an efficient coding scheme for correcting triple storage node failures. Trans. Comput. **57**(7), 889–901 (2008)
12. MacWilliams, F.J., Sloane, N.J.A.: The theory of error-correcting codes, vol. 16. North Holland (1977)

13. Plank, J.S.: Optimizing Cauchy Reed-Solomon codes for fault-tolerant storage applications. University of Tennessee, Technical Report CS-05-569 (2005)
14. Wanless, I.M.: A generalisation of transversals for Latin squares. Electron. J. Combin. $9(1)$, r12 (2002)
15. Wanless, I.M.: Diagonally cyclic Latin squares. Euro. J. Combin. 25, 393–413 (2004)
16. Xu, L., Bruck, J.: X-Code: MDS array codes with optimal encoding. IEEE Trans. Inform. Theory 45, 272–276 (1999)

Deep Fusion: A Software Scheduling Method for Memory Access Optimization

Yimin Zhuang[1,2]([✉]), Shaohui Peng[1,2], Xiaobing Chen[1,2], Shengyuan Zhou[1,2], Tian Zhi[1,2], Wei Li[1,3], and Shaoli Liu[1,3]

[1] SKL of Computer Architecture, Institute of Computing Technology, CAS, Beijing, China
{zhuangyimin,pengshaohui18z,chenxiaobing,zhousy,zhitian, liwei2017,liushaoli}@ict.ac.cn
[2] University of Chinese Academy of Sciences, Beijing, China
[3] Cambricon Tech. Ltd, Shanghai, China

Abstract. Deep neural networks (DNNs) have been considered to be the state-of-the-art artificial intelligence methods in a very broad range of applications. However, DNNs are compute intensive and memory intensive which are difficult to be employed in practical scenarios. Due to their favorable parallel computing ability, a series of DNN accelerators have been proposed. However, the improvement of on-chip computing capacity and the increasing number of parameters in the neural networks make access to memory a bottleneck. In this paper, we analyze the existing DNN algorithms. We observe that the special structure of neural networks makes it have two useful characteristics, which are unilateral directivity and local independence. Based on these characteristics, we propose a general software scheduling method to reduce memory access cost. Based on the experimental results, our method can reduce 32% memory access cost and achieve a speedup of 1.6x in average on our experiment platform and the best result is in ResNet-50, which is up to 56% and 2.62x.

Keywords: Fusion · Reuse · On-chip Memory

1 Introduction

Deep neural networks (DNNs) are ubiquitous in a very broad range of applications, such as speech recognition [1], object detection [2,3], semantic segmentation [4] and so on. With the continuous development of DNNs both the number of neurons and synapsis increases exponentially. As a result, the operations of computing and memory accessing will grow far beyond the hardware processing capability especially for the embedded systems. A large number of solutions have been proposed by the researchers to address this limitation, such as pruning [5],

© IFIP International Federation for Information Processing 2019
Published by Springer Nature Switzerland AG 2019
X. Tang et al. (Eds.): NPC 2019, LNCS 11783, pp. 277–288, 2019.
https://doi.org/10.1007/978-3-030-30709-7_22

data compressing [6], low-precision quantization [7], etc. However, the existing general processor platforms (such as CPU, FPGA, DSP, etc.) are still difficult to fully meet the requirements of practical applications.

Some researchers considered the general characteristics of DNN algorithms and designed neural network accelerators [11–13]. DianNao [8] is a dedicated accelerator which makes advantages of the data locality and computational properties of DNNs. DaDianNao [9] adopts time-division multiplexing of neurons to acquire high performance. EIE [10] utilizes sparse data to speed up the process of computation. Generally, DNN accelerators prefer to add private on-chip memory for performance improvement. Data is loaded from DRAM to on-chip memory and then the results are stored back to DRAM after computation. However, for most of the neural network accelerators, a large increase in the computational resources will aggravates the shortage of memory bandwidth and resource contention of on-chip network. The data transmission latency between internal and external storage will make up a large portion in the program execution time.

In this work, we propose a general software scheduling method to optimize the memory access by making advantages of both unidirectional data transportation and local data independence. Besides, we propose an on-chip memory reuse method to expand the on-chip memory size.

The paper is organized as follows. In Sect. 2, we show the bottleneck that we face of memory access and the optimization potential of DNNs. In Sect. 3, we introduce the details of our method. The experimental methodology and experimental results are presented in Sect. 4. Section 5 makes a conclusion at last.

2 Motivation

2.1 Memory Access Bottleneck

Most DNN algorithms are computational and memory intensive. A number of accelerators which can offer high compute capability have been proposed to solve the computationally intensive problem. As a matter fact, the current mainstream neural network accelerators have TFLOPS-level operation capability which is far beyond the bandwidth of the current external memory. However, most of these work assume away the question of memory access. To illustrate this problem, we analyze the amount of computation and memory access for all layers in ResNet-18 [21].

As shown in Fig. 1, the ratio of computation to memory access for each layer is different in ResNet-18 which need different requirement for bandwidth and compute capability. Taking the element-wise layer as example, we need a bandwidth of 12 GB/s if our compute capability is 1 GFLOPS. Meanwhile, the requirement of bandwidth is only 10 MB/s for convolution layers with the same compute capability of 1 GFLOPS. Although the hardware architecture of neural network accelerators is well-designed to make a balance between memory bandwidth and computation capability, they will never reach their full potential without software optimization. We further statistics the proportion of computation and memory access for each layer in the whole ResNet-18. As shown in Table 1, more than 95% of data transmission account is in certain layers including convolution layers, BatchNorm layers, scale layers, ReLU layers and eltwise layers. However, the computation amount of these layers is small except convolution layers, which is less than 1% in the whole network. Thus, the bottleneck of memory access is serious in these layers.

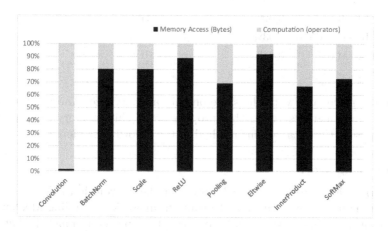

Fig. 1. Compute-to-global-memory-access ratio for each layer in ResNet-18

2.2 Potential of Optimization

To solve this bottleneck, we further analyze the characteristics of DNN algorithms. We can neutralize the inhomogeneity between memory access and computation of different layers by fusing multiple layers together because of the important characteristic of unilateral directivity in DNN algorithms. Since the dataflow of DNNs is unilateral, multiple layers can be computed together. Neurons are stored on-chip and are considered as input neurons for one hidden layer. The output neurons are still stored on-chip and are used as input neurons for next hidden layers. Besides, we can obviate a large number of data transmission from DRAM to on-chip memory or on-chip memory to DRAM. We observe that more than 99.6% data transmission can be reduced in ResNet-18 if all layers can be fused together.

However, it is almost unavailable to fuse all layers in real neural networks. Among the reasons for this state of affairs, one may cite the mismatch between the size of on-chip memory and neurons, the small on-chip memory size limited by the hardware overhead and the vast on-chip memory size needed to cache the intermediate data of the fused layers. Another important characteristic of DNN algorithms to relieve this problem is local independence. Each point of output neurons in one layer only depends on a defined region of input neurons. Hence, the neurons can be tiled into pieces and we can compute each piece separately. Thus,

Table 1. Proportion of computation and memory access for each layer in ResNet-18

Layer	Computation amount	Memory accessing amount
Convolution	99.55%	45.59%
BatchNorm	0.15%	15.27%
Scale	0.15%	15.24%
ReLU	0.06%	13.10%
Pooling	0.05%	2.92%
Eltwise	0.02%	6.41%
InnerProduct	0.03%	1.46%
SoftMax	0.00%	0.01%

we can fuse more layer with the same on-chip memory size. And combining the aforementioned features, we further propose an on-chip memory reuse method. We will present the detail of our method in the following section.

2.3 Existing Works

Some works which fuse the active layers after the convolution layers have been done in some mainstream machine learning frameworks, including MXNet [14] and Tensorflow [15,16]. However, the compute capacity is much higher than bandwidth in most accelerators. The ratio in GPU V100 is 10x and it is much greater in other accelerators. Thus, it is meaningful to fuse more layers. Manoj Alwani et al. [17] proposed a method to fuse multiple convolution layers, but it aims at hardware implementation. As a result, it is not general and flexible enough. Thus, a general software scheduling method with deeper fusion is important to solve the bottleneck of memory access.

3   Optimization Method

In this section, we propose a software scheduling method on neural network accelerators. The method consists of two mainly parts. One is layer fusion by software which can greatly reduce the demand of memory access. The other is on-chip memory reuse method, which can solve the large memory space required by layer fusion and the limitation of on-chip memory size in the accelerators. We will tile the data of each layer into pieces, then for each calculation, we get a piece of output from corresponding pieces of input, as shown in Fig. 2. We will describe our method in detail in this section.

3.1   Layer Fusion

At first, we show the details of the software scheduling method. To make the program of software more flexible, we decoupled the fusion process into two phases. One phase is operational-related shape deduction (SD) and the other phase is operational-independent shape transfer (ST).

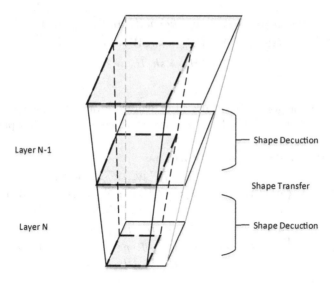

Fig. 2. An example for layer fusion process

SD. Shape deduction is to get the coordinate relationship between input data and output data. Most of layers in DNNs have four dimensions including batch, height, width and channel. In this subsection, for simplicity, we use the shape deduction between height dimension and width dimension as an example. For other dimension or dimensions greater than four, the method is almost the same. Besides, we prefer to infer shape from output shape to input shape, because sometimes there is redundant data for input data which will affect the shape deduction.

We use $Range(W)$ to represent an interval and $W \in [w_b, w_e)$. Similarly, we can use $Range(X, Y)$ to represent a range on a two-dimensional plane. Hence, we can use the following expressions to represent the process of shape deduction.

$$Range(W_i, H_i) = kernel(W_o, H_o)$$

Where $kernel(\cdot)$ is the function of shape deduction, $Range(X_i, Y_i)$ is coordinate range of input data and $Range(X_o, Y_o)$ is the coordinate range of output data. The shape deduction is related to the operators of each layer. For different layers, the deduction formulas are different. We take some typical layers as examples and we use xxx_kernel to distinct different kernel functions. Here xxx is usually an abbreviation of layer name.

Convolution/Pool. Convolution and pool are the most typical layers in DNNs. There are some basic parameters in these operators, such as kernel size, stride, etc. We use kh, kw, sh and sw as abbreviations.

$$Range(W_i, H_i) = cvpl\_kernel(Range(W_o, H_o)) :$$
$$(W_{i_b}, W_{i_e}) = (W_{o_b} * sw, W_{o_e} * sw + kw)$$
$$(H_{i_b}, H_{i_e}) = (H_{o_b} * sh, H_{o_e} * sh + kh)$$

Pad. Pad operator generally occurs in convolution or pooling layers. However, pad operator will change the shape of data, thus we make it as a separate layer.

$$Range(W_i, H_i) = pad\_kernel(Range(W_o, H_o)) :$$

$$W_{i_b}(W_{i_e}) = \begin{cases} 0 & if \quad W_{o_b}(W_{o_e}) < pad\_left \\ W_{o_b}(W_{o_e}) - pad\_left & if \quad pad\_left \leq W_{o_b}(W_{o_e}) \leq W + pad\_left \\ W & if \quad W_{o_b}(W_{o_e}) > W + pad\_left \end{cases}$$

$$H_{i_b}(H_{i_e}) = \begin{cases} 0 & if \quad H_{o_b}(H_{o_e}) < pad\_up \\ H_{o_b}(H_{o_e}) - pad\_up & if \quad pdf\_up \leq H_{o_b}(H_{o_e}) \leq H + pad\_up \\ H & if \quad H_{o_b}(H_{o_e}) > H + pad\_up \end{cases}$$

BatchNorm/Scale/Active. For these layers, they will not change the shape of data, thus it makes shape deduction directly.

$$Range(W_i, H_i) = elt\_kernel(Range(W_o, H_o)) :$$
$$(W_{i_b}, W_{i_e}) = (W_{o_b}, W_{o_e})$$
$$(H_{i_b}, H_{i_e}) = (H_{o_b}, H_{o_e})$$

ST. In shape deduction phase, each layer only focuses on the coordinate of output data and returns the coordinate of the input data. In shape transfer phase, it will call the kernel function defined in shape deduction phase. The coordinate of output data will be set as input to the kernel function of current layer and the result of kernel function will be passed to the kernel function of the previous layer. Thus, we will get all coordinate information of all layer be fused after we go through all these layers. The pseudocode is shown in Fig. 3.

```
ShapeTransfer (Range (Xo, Yo)):
Coordinates [FusionLayerNum] = Range (xo, Yo)
For (LayerIndex=FusionLayerNum-1; LayerIndex>=0; LayerIndex--):
    Coordinates [LayerIndex]=kernel (Coordinates [LayerIndex+1])
Return coordinates;
```

Fig. 3. Pseudocode for shape transfer

3.2   On-Chip Memory Reuse

Although layer fusion can greatly reduce the requirements of memory access, it is limited by the size of on-chip memory. In this part, we analyze the characteristics of DNN at first, and then introduce the on-chip memory reuse method which makes use of the characteristics to break the memory limitation.

Base on the characteristic of unilateral directivity for DNNs, once the input data of one layer has been used and this data does not need by other layers, the memory space of this data can be reused. Besides, the shape of data can be gotten in advance in most of inference phase, which makes data reused on-chip is available.

The process of data distribution can be represented in a simplified sequence. Here, we only care about the point when the memory usage status changes, and we define the equivalent life time of each data from the allocate point to the free point. To illustrate this process more clearly, an example sequence is shown in left side of Fig. 4.

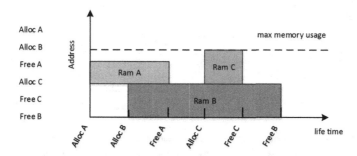

Fig. 4. The left side shows an example of memory distribution sequence. The right side shows a possible memory distribution result.

Figure 4 also shows an intuitive memory reuse method. We consider two data dependent if there is overlap between the life times of these two data. Otherwise, the memory space can be shared by these two data. As what we show in Fig. 4, ram A and ram B are dependent, and ram B and ram C are dependent too. But for ram A and ram C, they are independent, thus they share the same memory space.

The intuitive memory reuse method can save a large number of space, but it is limited by the data size. Once the size of input data or the size of output data is larger than on-chip memory size, we cannot fuse more layers.

To make these cases can be fused, we propose a deep memory reuse method. Base on the characteristics of local independence, when a local part of input data has been used to get a local part of output data, the memory space of this local part can reused. Thus, even if the life time of input data and output data has overlap, the output data can reuse part of memory space of the input space. The most special case is some element wise layers, such as add, BatchNorm, scale, etc. The input data and output data of these layers have the same size and can share the same memory space.

To illustrate the point more clearly for general cases, we take a concrete example of convolution layer which is shown in Fig. 5.

The horizontal axis in Fig. 5 is the growth of output data in H and W dimension. Here, we consider the multiplication of height and width as one dimension. For the calculation of each point, we get the address of first point in the piece of data we need and the address of current output point. Then, we join these points into two lines, as shown in Fig. 5. For each point, the memory space for those input data whose address is below the input address line can be reused, because these data have been used to calculate the output data before current point.

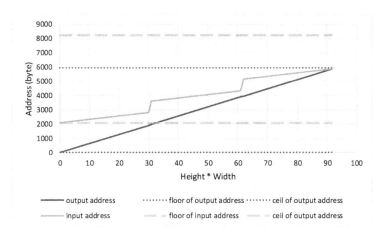

Fig. 5. The height, width and channel of input data and output data for this convolution is [8, 64, 3] and [3, 31, 16] and the sizes of kernel and stride are 4×4 and 2×2.

Even the life time of output data and input data is overlapped, they can still share a part of memory space. As shown in Fig. 5, the address range of input data is from 2096 (bytes) to 8240 (byte) and that of output data is from 0 (byte) to 5952 (byte). Thus, we reduce 31% memory usage in this case.

3.3   Fusion Method

Combining the methods above, we will show the implementation of our fusion method in this part. Figure 6 is an executive flow chart of our method. For each fusion, we first tile output data of the last layer into pieces. Then we do shape transfer for each piece and in the phase of shape transfer, it will call the kernel functions defined by each fused layer to do shape deduction. We then allocate memory space for all data. If the memory distribution is well-done, we can try to fuse the next layer. Otherwise, we shrink the tiled size and try to do shape transfer and memory distribution again. If the tiled size is the smallest tiled size, it means the current layer cannot be fused and returns the already fused layer list.

This process can be done before networks execution and we can get the coordinate and memory address for each piece of data. According to these information, we can execute the entire network through the specific instruction set or opcodes provided by the accelerators. The pseudocode is shown in Fig. 7. The in and out are the first input data and last output data for current fused list. The compute function is defined by each layer according to their own algorithms. The input data is loaded from DRAM and all middle data is stored on-chip. The output data is stored back to DRAM when we get a piece of final results.

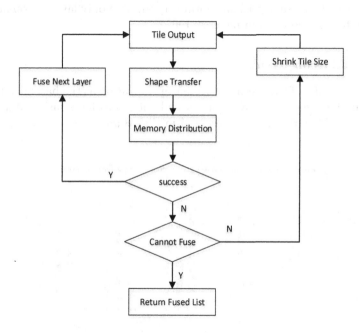

Fig. 6. Flowchart of layer fusion method

```
Execution ( in , out ):
    For ( i = 0; i < PiecesNum; i++):
        mid_data = Load( in [ i ])
        For ( j = 0; j < FusionLayerNum; j++):
            mid_data = compute<i >(mid_data)
            out [ i ] = mid_data
        store ( out [ i ])
```

Fig. 7. Pseudocode for execution

4   Experiment

4.1   Experiment Methodology

We design a prototype accelerator as our experiment platform. The structure of the prototype refers to the design of DaDianNao [9]. In our experiment platform, we limit the bandwidth between DRAM and on-chip memory to 1.5 GB/s. the compute capability of the prototype accelerator is 200 GFLOPS and we set 768 KB size of on-chip memory.

We choose five typical NN models as the benchmarks to evaluate our method, i.e. VGG-19 [18], GoogLeNet [19], InceptionV3 [20], ResNet-18 [21] and ResNet-50 [21]. Besides, we evaluate our optimization in the prototype accelerator, and compare the result of the memory access reduction and execution time improvement between the method without optimization, with only layer fusion and with both layer fusion and on-chip memory reuse.

4.2   Layer Fusion Result

We take the results of the method without optimization as the baseline. Then we test two methods with only layer fusion and with both layer fusion and on-chip memory reuse. The results are presented in Figs. 8 and 9.

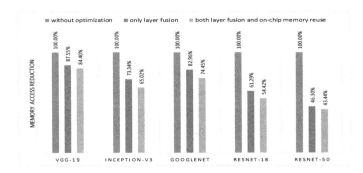

Fig. 8. Ratio of memory access reduction w.r.t the baseline.

Figure 8 shows the result of memory access reduction rate compared with baseline. We get more than 15% reduction of memory access in our benchmarks, especially for ResNet-50 which acquires 56% reduction. The performance improvement of execution time has similar tendency as shown in Fig. 9. We get at least 1.26x performance improvement in VGG-19 and at most 2.62x performance improvement in ResNet-50. Besides, we can get a better effect which is more than 5% improvement in average for both memory access reduction and performance by using on-chip memory reuse in addition.

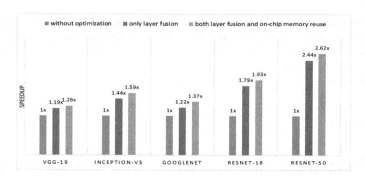

Fig. 9. Speedup w.r.t. the baseline (execution time).

We find that the result in VGG-19 is not much better than other networks. Then we further analyze the fusion status in VGG-19. We observe that the most layers in VGG-19 is convolution layers and the kernel size of all these layers is 3x3 which result in a large synapse data size. However, synapse is shared by all input data in convolution layers. Thus, if we tile input data into pieces, each piece of data need the same synapse data and these synapse data will take up an independent memory space. The memory space is run out of soon if we fuse more convolution layers. If we have much more on-chip memory space, we can layer out more synapse data or we can reuse the space of synapse data if the input data is not tiled into pieces. However, it is a tradeoff between performance and the area of accelerators.

5 Conclusion

The development of neural network accelerators makes DNNs run faster and faster, but the slowly development of bandwidth for DRAM makes accelerators stuck in a memory access bottleneck. It is very important to solve this problem to utilize accelerators more effective.

In this paper, we propose a new software scheduling method to optimize memory access. It mainly consists of two parts, one is layer fusion and the other is on-chip memory reuse. They utilize the properties of DNNs. That is unilateral directivity and local independence. Based on the experimental results, our method achieves 32% memory access reduction and 1.6x speedup in average. To get a better performance, we can expand the amount of space on chip, however, it is a tradeoff between performance and the area of accelerators.

Acknowledgment. This work is partially supported by the National Key Research and Development Program of China (under Grant 2017YFB1003104), the NSF of China (under Grants 61432016, 61532016, 61672491, 61602441, 61602446, 61732002, 61702478, 61732007 and 61732020), Beijing Natural Science Foundation (JQ18013), the 973 Program of China (under Grant 2015CB358800), National Science and Technology Major Project (2018ZX01031102), the Transformation and Transfer of Scientific and Technological Achievements of Chinese Academy of Sciences (KFJ-HGZX-013), Key Research Projects in Frontier Science of Chinese Academy of Sciences (QYZDB-SSW-JSC001) , Strategic Priority Research Program of Chinese Academy of Science (XDB32050200, XDC01020000) and Standardization Research Project of Chinese Academy of Sciences (BZ201800001).

References

1. Xiong, W., et al.: Achieving human parity in conversational speech recognition. In: IEEE/ACM Transactions on Audio, Speech, and Language Processing, p. 99 (2016)
2. Ren, S., et al.: Faster R-CNN: towards real-time object detection with region proposal networks. IEEE Trans. Pattern Anal. Mach. Intell. **39**(6), 1137–1149 (2017)
3. Redmon, J., Farhadi, A.: YOLOv3: An Incremental Improvement (2018)
4. Noh, H., Hong, S., Han, B.: Learning Deconvolution Network for Semantic Segmentation (2015)
5. Han, S., et al.: Learning both Weights and Connections for Efficient Neural Networks (2015)
6. Han, S., Mao, H., Dally, W.J.: Deep compression: compressing deep neural networks with pruning, trained quantization and huffman coding. Fiber **56**(4), 3–7 (2015)
7. Jacob, B., et al.: Quantization and Training of Neural Networks for Efficient Integer-Arithmetic-Only Inference (2017)
8. Chen, T., et al.: DianNao: a small-footprint high-throughput accelerator for ubiquitous machine-learning. ACM Sigplan Not. **49**(4), 269–284 (2014)
9. Chen, Y., et al.: DaDianNao: A Machine-Learning Supercomputer (2014)
10. Han, S., et al.: EIE: efficient inference engine on compressed deep neural network. ACM Sigarch Comput. Archit. News **44**(3), 243–254 (2016)
11. Shen, Y., Ferdman, M., Milder, P.: Escher: a CNN accelerator with flexible buffering to minimize off-chip transfer. In: 2017 IEEE 25th Annual International Symposium on Field-Programmable Custom Computing Machines (FCCM) IEEE Computer Society (2017)
12. Chen, Y.-H., et al.: Eyeriss: an energy-efficient reconfigurable accelerator for deep convolutional neural networks. IEEE J. Solid-State Circuits **52**(1), 127–138 (2017)
13. Liu, S., et al.: Cambricon: an instruction set architecture for neural networks. In: ACM/IEEE International Symposium on Computer Architecture (2016)
14. Chen, T., et al.: MXNet: A Flexible and Efficient Machine Learning Library for Heterogeneous Distributed Systems. Statistics (2015)
15. Abadi, M., et al.: TensorFlow: a system for large-scale machine learning (2016)
16. Abadi, M., et al.: TensorFlow: Large-Scale Machine Learning on Heterogeneous Distributed Systems (2016)
17. Alwani, M., et al.: Fused-Layer CNN Accelerators. In: IEEE/ACM International Symposium on Microarchitecture (2016)
18. Simonyan, K., Andrew Z.: Very deep convolutional networks for large-scale image recognition. arXiv preprint arXiv:1409.1556 (2014)
19. Szegedy, C., et al.: Going Deeper with Convolutions (2014)
20. Xia, X., Cui, X., Bing, N.: Inception-v3 for flower classification. In: International Conference on Image (2017)
21. He, K., et al.: Deep Residual Learning for Image Recognition (2015)

Optimizing Data Placement on Hierarchical Storage Architecture via Machine Learning

Peng Cheng[1,2], Yutong Lu[3(✉)], Yunfei Du[3], Zhiguang Chen[3], and Yang Liu[4]

[1] College of Computer, National University of Defense Technology, Changsha, China
peng.cheng@nscc-gz.cn
[2] State Key Laboratory of High Performance Computing, Changsha, China
[3] National Supercomputer Center in Guangzhou, School of Data
and Computer Science, Sun Yat-Sen University, Guangzhou, China
{yutong.lu,yunfei.du,zhiguang.chen}@nscc-gz.cn
[4] Department of Computer Science and Technology,
Tsinghua University, Beijing, China
liuyang2011@tsinghua.edu.cn

Abstract. As storage hierarchies are getting deeper on modern high-performance computing systems, intelligent data placement strategies that can choose the optimal storage tier dynamically is the key to realize the potential of hierarchical storage architecture. However, providing a general solution that can be applied in different storage architectures and diverse applications is challenging. In this paper, we propose adaptive storage learner (ASL), which explores the idea of using machine learning techniques to mine the relationship between data placement strategies and I/O performance under varied workflow characteristics and system status, and uses the learned model to choose the optimal storage tier intelligently. We implement a prototype and integrate it into an existing data management system. Empirical comparison based on real scientific workflows tests shows that ASL is capable of combining workflow characteristics and real-time system status to make optimal data placement decisions.

Keywords: Storage optimization · Machine learning ·
Hierarchical storage · Data placement

1 Introduction

With the converging of High-Performance Computing (HPC) and big data, massive datasets are produced and analyzed by HPC systems. For example, the large N-body simulation that evolved more than a trillion particles on the BG/Q Mira system generates approximately 5PB of raw outputs [1]. The exascale deep learning on the Summit system analyzes 3.5TB climate data [2] to detect

© IFIP International Federation for Information Processing 2019
Published by Springer Nature Switzerland AG 2019
X. Tang et al. (Eds.): NPC 2019, LNCS 11783, pp. 289–302, 2019.
https://doi.org/10.1007/978-3-030-30709-7_23

extreme weather. As scientific workflows become more complex and more data-intensive, supporting these workflows on HPC systems present serious challenges in I/O performance [3,4].

To improve the I/O performance, many modern HPC systems use middleware or heterogeneous storage devices to expand the storage subsystem in a hierarchical manner. Memory-based staging solutions, such as DataSpaces [5], use memory space of compute nodes to stage intermediate data. Shared burst buffer strategy, such as Cori system [6], uses SSDs near I/O nodes and Cray DataWarp software [7] to implement a shared staging area for coupling applications. Local burst buffer strategy, such as Summit system [8], equips per-node attached SSD for each compute node to buffer intermediate data locally. Meanwhile, current research on storage class memories (SCM) is expected to be able to add additional layers to the memory and storage hierarchy in future HPC systems [9].

Due to the largely different latency, bandwidth and capacity of heterogeneous storage devices, different data placement strategies could lead to wildly varying I/O performance. However, existing storage systems usually apply one-layout-fits-all strategy or leave the burden of making data placement decisions to users [10]. Such fixed and manual data placement might result in the inefficient use of hierarchical storage architecture because of load imbalance [11] and resource contention [12]. Previous works [13,14] provide workflow-aware data placement mechanism but depend on user-provided hints, which is infeasible for complex workflows. While Stacker [15] leverages hierarchical n-grams model to predict upcoming read request and guide data prefetching, smart data placement decisions across different storage tiers remain to be researched.

In this paper, we propose adaptive storage learner (ASL), which explores the idea of leveraging machine learning techniques to mine the relationship between data placement strategies and I/O performance under varied workflow characteristics and system status, and uses the learned model to choose the optimal storage tier dynamically during the workflow execution. Compared with previous works, ASL focus on scientific workflows and enables intelligent data placement strategy to make the most benefit of hierarchical storage architecture without any user-provided hints. We provide a prototype implementation of ASL and integrate it with Alluxio [16] data management system. Our evaluations of two scientific workflows validate the effectiveness of ASL in combining workflow characteristics and real-time system status to make optimal data placement decisions. While this paper focuses on optimizing data placement across different storage tiers, the idea of training a classification model to guide storage optimization can be applied in different scenarios. Our contributions in this paper can be summarized as follows.

- A workflow simulator with I/O performance model that reflects the changing workloads, heterogeneous storage devices, and varying configurations.
- A classification model that leverages workflow characteristics and system status to make optimal data placement decisions.

- A prototype implementation that manages data on tiered storage architecture and enables intelligent data management decisions.
- An extensive evaluation with two scientific workflows on a typical HPC system.

The rest of this paper is organized as follows. Section 2 presents a brief background and the motivation of this paper. We explore the idea of training a classification model for storage optimization in Sect. 3 and present the design and implementation in Sect. 4. We validate the effectiveness of ASL in Sect. 5. Section 6 discusses some related studies currently existing in the literature. We conclude the paper and talk about future works in Sect. 7.

2   Background and Motivation

Scientific workflows: A scientific workflow is the assembly of complex sets of scientific data processing activities with data dependencies between them [17]. Due to the repetitive nature of scientific discovery, scientific workflow management system (SWfMS) like Pegasus [18] and Swift [19] are increasingly used in HPC environments to manage the complex simulations and analyses. Workflow description file contains the necessary information about the entire workflow, including the tasks to be executed and data-flow/control-flow dependencies between these tasks. SWfMSs take the description of the abstract workflow as input and coordinate and execute workflow tasks over available computing resources.

Motivation: Data placement strategies on hierarchical storage architecture can be divided into horizontal placement (how data are distributed inside a storage layer) and vertical placement (how data are distributed across different storage layers). In this paper, we focus on vertical data placement since heterogeneous storage devices show largely different latency, bandwidth, and capacity. While different data placement strategies could lead to wildly varying I/O performance, existing data management systems [15,16] often apply the one-layout-fits-all strategy that uses the top storage tier (e.g., memory tier) as the performance tier and uses lower storage tier (e.g., SSD tier or HDD tier) as capacity tier. However, such fixed data placement strategy might result in the inefficient use of hierarchical storage architecture. Firstly, serious **load imbalance** occurs since the fixed data placement strategy keeps staging data to a specific storage layer, while other storage layers keep unused [11]. Secondly, as the available space of that layer gets insufficient, backend data migration requests need to move data to lower tiers. The **resource contention** between regular write requests and backend data migration requests could even lead to almost 70% performance degradation [12].

3   Training Classification Model for Storage Optimization

The basic idea that guides our design is that both data access patterns and real-time system status should be taken into consideration to make the optimal data placement decision. For example, if the top storage layer has plenty of space to stage all the intermediate data during the execution of scientific workflows, choosing the top storage layer to serve every write request will provide superior I/O performance. Otherwise, only data will be accessed by subsequent task immediately can be written into the top storage layer, and other data should be written to the lower storage layer to keep load balance. While setting these rules manually may perform well for some applications, providing a general solution that can be applied in different storage architectures and diverse applications is challenging. In this paper, we propose Adaptive Storage Learner (ASL), which explore the idea of using machine learning techniques to solve this challenge.

3.1   Problem Definition

Selecting the optimal storage layer can be regarded as a multi-classification problem. We want to learn a model that takes parameters related to workflow characteristics and system status as input and predicts the optimal storage layer for each output file. We maintain three principles to train the prediction model:

- Both workflow characteristics and real-time system status are taken into consideration to make the optimal data placement decision.
- The optimization goal of the prediction model is not to minimize the I/O time of a single task, but to minimize the overall I/O time of the entire workflow by leveraging data access patterns and preventing resource contention. In other words, the optimal storage tier for an output file may not be the fastest storage tier, even if there is plenty of space currently.
- We set the granularity of a data placement decision to a file instead of each write request since dividing a file across a slow and a fast tier may lead to the problem that a slow tier becomes the bottleneck.

To train such a multi-classification model, we first identify parameters that affect I/O performance.

3.2   Parameters Affecting I/O Performance

Scientific workflows might demonstrate different I/O performance based on the workflow characteristics and system specifications. We summarize three sets of parameters that might affect the I/O performance and explain some of them because of space limitations.

Workflow Characteristics: These parameters include the scale of the workflow, the control-flow dependencies and data-flow dependencies between tasks. Specifically, data-flow dependencies contain data access pattern information of each output file. As previous works have demonstrated [14], data access patterns can be leveraged to improve I/O performance. For example, staging a file that will be accessed immediately to the top storage tier can reduce the data read time. Compared with previous works depend on user-provided hints to identify the data access patterns, we don't set any rules manually but provide all these information to the prediction model and let the model learns from it. All these parameters are statically determined before running a workflow and can be retrieved by parsing the workflow description file.

Runtime Storage Information: These parameters describe the storage information of generated intermediate files during the workflow execution, such as the size of each intermediate file and the storage tier that file resides. All these parameters are collected during the execution of the workflow.

Table 1. Variables contained in each I/O record

Variable	Description
V1	ID of the current output file
V2	Type of current task
V3	Number of tasks of the current type
V4	Number of input files of the current task
V5	Number of output files
V6	Number of tasks that is dependent with current output file
V7	Minimum distance between the output file and dependent tasks
V8	Total size of input files of the current task
V9	The remaining capacity of the memory storage tier
V10	The remaining capacity of the SSD storage tier
V11	The remaining capacity of the HDD storage tier
Prediction	The optimal storage tier

System Status: These parameters are specifications of the system where the workflow runs, including the deployment of the storage subsystem, the performance metrics of different storage tier, etc. All these parameters affect the I/O performance of a given workflow. While parameters like bandwidth and latency of each storage tier can be obtained statically, parameters like the remaining capacity of each storage tier require real-time monitoring.

3.3   Collecting I/O Records

After identifying parameters that affect the I/O performance, we are able to collect the I/O records during the workflow execution. We model these parameters into 11 variables listed in Table 1. Each I/O record represents a data placement decision for a given output file under the conditions described by 11 variables. Specifically, V1 and V2 are used to identify the data producer for each file create request. V3-V7 reflect the workflow characteristics and V8-V11 describe the real-time storage information and system status.

It's deserved to be mentioned that the size of each output file will definitely influence the I/O performance, but we abandon it to avoid the contradiction against the usage of the prediction model. The goal of the prediction model is to make data placement decisions before data are written to the target storage tier. However, the size of an output file can be calculated only after the write operation is finished. A compromise solution is to train another regression model to predict the size of the output file, but we do not implement it in our current work.

Collecting I/O records is complicated and cumbersome for two reasons. Firstly, all the I/O records need to be labeled since multi-classification problem requires supervised learning. In other words, each record must be labeled with the target storage tier explicitly before they can be used to train the prediction model. Secondly, recall the principle that the optimization goal of the prediction model is to minimize the overall I/O time of the entire workflow by leveraging data access patterns and preventing resource contention. This principle exacerbates the complexity of labeling records since inappropriate training data leads to the inaccurate prediction model.

To solve this challenge, we extend a workflow simulator [20] to simulate the I/O performance and label the I/O records automatically. In general, the extended workflow simulator has the following design considerations:

Tiered Storage Architecture: we add hybrid storage module consists of three storage tiers to model the I/O performance. Specifically, the specification of each storage tier is set based on real system tests. Detailed configurations are discussed in Sect. 5.

Simulation Rules: Several rules are set empirically to simulate the actual I/O performance. These include: the location of data contribute to the maximum read/write bandwidth, the bandwidth degradation once the available space of a storage tier exhausted. We do not list all the rules here because of space limits. To validate the effectiveness of the simulation, we compare the simulated I/O time and the actual I/O time of the Binary-Tree workflow [21]. Figure 1 illustrates the result of each type of task. When storage tier is set to memory, SSD, and HDD, the I/O time generated by workflow simulator are noted as *Sim-Mem*, *Sim-SSD*, and *Sim-HDD*, respectively. Similarly, the I/O time of running workflows on the real system is noted as *Real-Mem*, *Real-SSD*, and *Real-HDD*. Overall, the difference between simulation and real system tests is less than 10%.

Genetic Algorithm (GA): To search the optimal combinations of storage tiers for a given workflow, we implement a genetic algorithm in the workflow simulator. We treat a candidate storage tier combination as an individual, and the storage tier of each output file is represented as 2 genes (since each gene is a Boolean variable, at least two genes are needed to represent 3 storage tiers). The GA starts from a population of randomly generated individuals and evolves in an iterative process. The overall I/O time of a workflow is used to calculate the fitness of each individual. An individual is qualified to have the next generation only when its fitness is no less than the average fitness. The iterative crossover and mutation between qualified individuals improve the quality of the represented solution. Finally, the best individual is chosen to be the optimal combinations of storage tiers.

We randomly chose 58 workflows with varying scales and I/O characteristics from the synthetic workflow dataset [22]. We ran these workflows on top of the simulator and collect 3810 labeled I/O records.

3.4 Model Training

Gradient boosting algorithm with Classification and Regression Tree (CART) as base learners is used to train the prediction/classification model. While there are lots of machine learning algorithms, including logistic regression and support vector machines, we chose CART as the basic learner for two reasons. Firstly, CART is easy to understand and interpret. Secondly, the prediction overhead of CART is negligible. Since a single CART model might suffer from the poor generality, we use the gradient boosting technique, which averages over multiple CART classifiers and produces the final prediction, to enhance its generalization ability. The final prediction model can be treated as an ensemble of CART models.

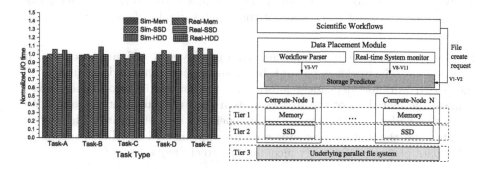

Fig. 1. Accuracy of simulation output   **Fig. 2.** ASL architecture overview

4   Design and Implementation

We design and implement a prototype of ASL that uses the prediction model presented in Sect. 3 to make optimal data placement decisions.

Figure 2 presents the architecture overview of ASL. ASL acts as a middleware integrated with an existing data management system that can manage data on tiered storage architecture. The key component of ASL includes workflow parser, real-time system monitor and storage predictor. The workflow parser extracts workflow characteristics form the workflow description file before running a given workflow. The real-time system monitor collects system status during the workflow execution. For each file create request, the storage predictor combines workflow characteristics and system status to make data placement decisions.

Parser and Monitor: To enable workflow-aware storage optimization, we implement a workflow parser that extracts workflow characteristics from the workflow description file. The extracted data are stored in multiple in-memory data structures and transform into variables V3-V7 as listed in Table 1. Although the workflow description file contains lots of valuable information, information like the size of a specific input file and the remaining capacity of each storage tier can only be collected during the workflow execution. The system monitor is used to collect such information dynamically. The dynamically collected information transformed into variables V8-V11 as listed in Table 1.

Prediction: The storage predictor uses the prediction model to make the data placement decision for every incoming file create request. Specifically, the decision-making process can be summarized into the following steps: 1. After receiving the file create request, the storage predictor verifies the type of the current task based on the name of the created file and the extracted workflow description info. 2. Retrieving workflow characteristics and system status from the workflow parser and the system monitor, respectively. 3. Constructing variables related to workflow characteristics and system status and feed these variables to the prediction model. 4. Predicting the optimal storage tier for the newly created file. As in the case of new workflows, the storage predictor can also predict the result since none of the input variables depend on historical information.

Implementation: We have implemented a prototype of ASL and integrated it with Alluxio. The workflow parser is implemented as a command line utility. All the parsed workflow dependent information are sent to Alluxio master, which manage the metadata of the storage system and serve metadata requests. We add extra modules to manage workflow dependent data structures. The system monitor and the storage predictor are also implemented in Alluxio master to guide data placement for every incoming file create request.

5   Evaluation

To demonstrate the effectiveness of ASL, we evaluate its performance on a typical HPC system with real scientific workflows.

5.1   Experimental Setup

Our testbed consists of 32 nodes configured in one rack on the on data analytics cluster of the Tianhe-2 system [23]. Each node is equipped with two 2.20 GHz Intel Xeon E5-2692-v2 processors (24 cores per node), 64 GB of RAM and one PCIe 1.5TB SSD. Alluxio is used to manage data on top of heterogeneous storage devices. Specifically, two nodes are used as the master nodes to manage the global metadata, and the other 30 nodes are used as the workers to stage data into local memories or per-node attached SSDs. We allocate 2–15 GB RAM of each node to constitute the memory storage tier. The per-node attached SSDs constitute the SSD storage tier, and the underlying Lustre file system acts as the HDD storage tier. Detailed specifications of each storage tier under the management of Alluxio are listed in Table 2.

We use Binary tree workflow [21] and GenBase workflow [24] to validate the effectiveness of ASL. These workflows are also used to train the prediction model, but none of the simulated workflow scales are used during the real-system evaluation. All intermediate data are staged into hierarchical storage architecture managed by Alluxio.

5.2 Decision-Making Under Varied Workflow Scales

Firstly, we validate the effectiveness of ASL in making optimal data placement decisions based on workflow characteristics. We set capacity of memory storage layer to 300 GB and vary the scales of workflows. For GenBase workflow with 80k*80k input data scales, 120 GB of raw data are processed and will generate more than 400 GB intermediate data. While the initial storage configuration of each run is fixed, we evaluate the performance of four data placement strategies. Staging all data into memory storage tier and SSD storage tier are noted as *Memory* and *SSD*, respectively. Selecting the memory or the SSD tier for each file randomly is noted as *Random*. Using the predictor and make data placement decision intelligently is noted as *ASL*. Since the SSD storage tier is sufficient to stage all the intermediate data during the evaluation, HDD tier is not used in consideration of I/O performance. Figure 3 shows the performance of the Binary-Tree and the GenBase workflow.

For the Binary-Tree workflow, when data size is smaller than 300 GB, memory storage layer has enough space to stage all the intermediate data. As a

Table 2. Storage configurations under the management of Alluxio

Tier	Allocated Capacity	Write BW	Read BW
Memory Tier	60–450 GB	950 MB/s	1100 MB/s
SSD Tier	1200 GB	800 MB/s	850 MB/s
HDD Tier	1200 GB	500 MB/s	550 MB/s

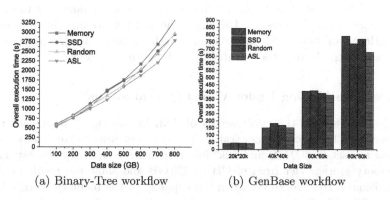

(a) Binary-Tree workflow   (b) GenBase workflow

Fig. 3. Performance of varied scales workflows

result, *Memory* strategy performs better than *SSD* and *Random* strategy. As data size keeps increasing, *Memory* strategy migrates data from the memory tier to the SSD tier to make room for the newly created file. The resource contention between regular write request and backend data migration request leads to performance degradation. Since SSD tier has enough space to stage all intermediate data during our evaluations, resource contention problem does not occur in *SSD* strategy. The *Random* strategy alleviates the load imbalance problem to some extent, but data access patterns of workflows are not taken into consideration.

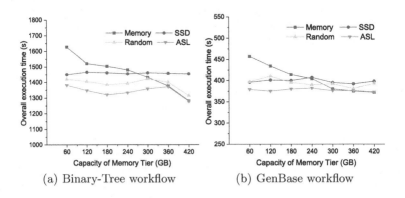

(a) Binary-Tree workflow   (b) GenBase workflow

Fig. 4. Performance of varied memory capacity

Contrast to these strategies, *ASL* combines information including available space of each storage layer, input size of the current task, and distance of dependent task to choose the optimal storage tier for each intermediate file. When the memory storage tier has plenty of space to stage all the intermediate data, *ASL* choose the memory tier as the primary storage tier to minimize the data access time. As data size increases and the memory storage tier gets exhausted, *ASL* leverages the hidden data access pattern info and only stage data that will be accessed by subsequent tasks intermediately to the memory tier to prevent resource contention. As a result, *ASL* shows the best performance in all cases. For the GenBase workflow, *ASL* shows similar performance with *Memory* strategy at first and outperforms other strategies as data size increases.

5.3 Decision-Making Under Varied System Status

Secondly, we validate the effectiveness of ASL in making optimal data placement decisions based on system status. The scales of Binary-Tree and GenBase workflow are set to 400 GB and 60k*60k, respectively. We vary the capacity of the memory storage tier from 60 GB to 420 GB and show the result in Fig. 4. For the Binary-Tree workflow, when the capacity of the memory storage tier is less than 240 GB, *Memory* strategy performs the worst because of the serious load imbalance and resource contention. As available capacity keeps increasing,

Memory strategy shows better performance. While *SSD* strategy shows stable performance as expected, *Random* strategy performs better as memory capacity increases. In comparison, *ASL* performs the best by making optimal data placement decision based on workflow characteristics and system status. For the GenBase workflow, since 60k*60k input data scale generates almost 250 GB intermediate data, *ASL* performs the best at first and shows similar performance with *Memory* strategy as the capacity of memory tier is larger than 300 GB.

In summary, our evaluations validate the effectiveness of ASL in combining workflow characteristics and real-time system status to make intelligent data placement decisions.

6   Related Work

As storage hierarchies are getting deeper on HPC systems, managing data on tiered storage architecture are getting increased attention. Data Elevator [12] enables asynchronously data flushing from burst buffer to the PFS, but different storage layers are managed separately. Heterogeneity-Aware Tiered Storage [25] and OctopusFS [11] extend HDFS to support tiered data storage architecture. They propose data placement and retrieval policies based on I/O throughput and capacity of storage devices, however, data access patterns are not used to make data management decisions. While Alluxio [16] and UniStor [26] provide a unified view across different storage layers, both of them lack the ability to choose the optimal storage tier dynamically. Multi-tiered data staging framework [13], Hermes [27] and TDMS [14] provide application-aware data placement mechanism but depend on user-provided hints to identify the data access patterns. Compared with these works, we treat selecting the optimal storage tier as a multi-classification problem and use machine learning techniques to make data placement strategies intelligently.

Many efforts have been made to enable adaptive and intelligent storage optimization. Stacker [15] chooses hierarchical n-grams model to predict upcoming read request and guide data prefetching on hierarchical storage architecture. Since stripe size and the distribution of correlated blocks dominate the aggregate bandwidth of parallel file systems, Dong Dai et al. [28] explored the idea of using word embedding technique to mine the block correlations. Erica Tomes et al. [29] combined graph coloring, bin packing, and network flow techniques to distribute correlated data blocks to different storage servers adaptively. Compared with these works, we focus on data placement and load balance across different storage hierarchies. Ziggurat [30] profiles the application's access stream online to predict the behavior of individual writes and chooses non-volatile main memory or disks to serve the write request. While the classification criteria in Ziggurat is set empirically, ASL does not depend on any manual rules and combine both workflow characteristics and real-time system status to choose the optimal storage tier.

7   Conclusion

Due to the largely different performance characteristics of hierarchical storage layers and the variety of scientific workflows, providing a general solution that can make intelligent data placement decisions is challenging. In this paper, we explore the idea of using machine learning techniques to solve this challenge. We propose that selecting the optimal storage layer under varied workflow characteristics and system status can be regarded as a multi-classification problem. We implement a workflow simulator to collect labeled I/O records automatically and use the gradient boosting algorithm with CART as base learners to train the classification model. We implement a prototype and integrate it into Alluxio system. Our evaluations on two scientific workflows validate the effectiveness of using machine techniques to optimize I/O performance. In our current implementation, the prediction model is not modified once it is deployed. In future work, we plan to collect histories records of workflows and update the model dynamically.

Acknowledgment. This work was supported by the National Key R&D Program of China under Grant No. 2017YFB0202204 and No. 2017YFB0202201, the National Science Foundation of China under Grant NO.U1811464, and the Program for Guangdong Introducing Innovative and Entrepreneurial Teams under Grant NO. 2016ZT06D211.

References

1. Habib, S., et al.: Hacc: simulating sky surveys on state-of-the-art supercomputing architectures. New Astron. **42**, 49–65 (2016)
2. Kurth, T., et al.: Exascale deep learning for climate analytics. In: Proceedings of the International Conference for High Performance Computing, Networking, Storage, and Analysis, SC 2018, Dallas, TX, USA, 11–16 November 2018, pp. 51:1–51:12 (2018)
3. Miyoshi, T., et al.: Big data assimilation toward post-petascale severe weather prediction: an overview and progress. Proc. IEEE **104**(11), 2155–2179 (2016)
4. Liu, N., Cope, J., Carns, P.H., Carothers, C.D., Ross, R.B., et al.: On the role of burst buffers in leadership-class storage systems. In: IEEE 28th Symposium on Mass Storage Systems and Technologies, MSST 2012, 16–20 April 2012, Asilomar Conference Grounds, pp. 1–11. Pacific Grove, CA, USA (2012)
5. Docan, C., Parashar, M., Klasky, S.: Dataspaces: an interaction and coordination framework for coupled simulation workflows. Cluster Comput. **15**(2), 163–181 (2012)
6. Bhimji, W., Bard, D., Romanus, M.: Accelerating science with the nersc burst buffer early user program. In: LBNL LBNL-1005736, May 2016
7. Cray. Datawarp user guide s-2558-5204, June 2016. http://docs.cray.com/books/S-2558-5204/S-2558-5204.pdf
8. Oak Ridge National Laboratories. Summit user guide, May 2019. https://www.olcf.ornl.gov/for-users/system-user-guides/summit
9. Swami, S., Mohanram, K.: Reliable non-volatile memories: techniques and measures. IEEE Des. Test **99**, 1 (2017)

10. Li, H., Ghodsi, A., Zaharia, M., Shenker, S., Stoica, I.: Tachyon: reliable, memory speed storage for cluster computing frameworks. In: Proceedings of the ACM Symposium on Cloud Computing, pp. 6:1–6:15. Seattle, WA, USA (2014)

11. Kakoulli, E., Herodotou, H.: Octopusfs: a distributed file system with tiered storage management. In: Proceedings of the 2017 ACM International Conference on Management of Data, SIGMOD 2017, pp. 65–78 (2017)

12. Dong, B., Byna, S., Wu, K.P., Johansen, H., Johnson, J.N., Keen, N.: Data elevator: low-contention data movement in hierarchical storage system. In: 23rd IEEE International Conference on High Performance Computing (HiPC 2016), pp. 152–161. Hyderabad, India (2016)

13. Jin, T., et al.: Exploring data staging across deep memory hierarchies for coupled data intensive simulation workflows. In: 2015 IEEE International Parallel and Distributed Processing Symposium, IPDPS 2015, pp. 1033–1042 (2015)

14. Cheng, P., Lu, Y., Du, Y., Chen, Z.: Accelerating scientific workflows with tiered data management system. In: IEEE International Conference on High Performance Computing and Communications (2018)

15. Subedi, P., Davis, P.E., Duan, S., Klasky, S., Kolla, H., Parashar, M.: Stacker: an autonomic data movement engine for extreme-scale data staging-based in-situ workflows. In: Proceedings of the International Conference for High Performance Computing, Networking, Storage, and Analysis (SC 2018), pp. 73:1–73:11 (2018)

16. Alluxio Inc., Alluxio overview, May 2019. https://docs.alluxio.io/os/user/stable/en/Overview.html

17. Deelman, E., Gannon, D., Shields, M.S., Taylor, I.J.: Workflows and e-science: an overview of workflow system features and capabilities. Future Gener. Comp. Syst. **25**(5), 528–540 (2009)

18. Deelman, E., et al.: Pegasus, a workflow management system for science automation. Future Gener. Comput. Syst. **46**, 17–35 (2015)

19. Wilde, M., Hategan, M., Wozniak, J.M., Clifford, B., Katz, D.S., Foster, I.: Swift: a language for distributed parallel scripting. Parallel Comput. **37**(9), 633–652 (2011)

20. Chen, W., Deelman, E.: Workflowsim: a toolkit for simulating scientific workflows in distributed environments. In: 8th IEEE International Conference on E-Science, pp. 1–8 (2012)

21. Hazekamp, N., et al.: Combining static and dynamic storage management for data intensive scientific workflows. IEEE Trans. Parallel and Distrib. Syst. **99**, 1 (2018)

22. Pegasus. Pegausus syntheticworkflows, February 2019. https://download.pegasus.isi.edu/misc/SyntheticWorkflows.tar.gz

23. Liao, X., Xiao, L., Yang, C., Yutong, L.: Milkyway-2 supercomputer: system and application. Front. Comput. Sci. **8**(3), 345–356 (2014)

24. Taft, R., Vartak, M., Satish, N.R., Sundaram, N., Madden, S., Stonebraker, M.:. Genbase: a complex analytics genomics benchmark. In: Proceedings of the 2014 ACM SIGMOD International Conference on Management of Data (SGIMOD 2014). ACM (2014)

25. Krish, K.R., Anwar, A., Butt, A.R.: hats: a heterogeneity-aware tiered storage for hadoop. In: IEEE/ACM International Symposium on Cluster, Cloud and Grid Computing, pp. 502–511 (2014)

26. Wang, T., Byna, S., Dong, B., Tang, H.: Univistor: integrated hierarchical and distributed storage for HPC. In: IEEE International Conference on Cluster Computing, CLUSTER 2018, Belfast, UK, 10–13 September 2018, pp. 134–144 (2018)

27. Kougkas, A., Devarajan, H., Sun, X.H.: Hermes: a heterogeneous-aware multi-tiered distributed I/O buffering system. In: Proceedings of the 27th International Symposium on High-Performance Parallel and Distributed Computing (HPDC 2018), pp. 219–230 (2018)
28. Dai, D., Bao, F.S., Zhou, J., Shi, X., Chen, Y.: Vectorizing disks blocks for efficient storage system via deep learning. Parallel Comput. **82**, 75–90 (2019)
29. Tomes, E., Rush, E.N., Altiparmak, N.: Towards adaptive parallel storage systems. IEEE Trans. Comput. **67**(12), 1840–1848 (2018)
30. Zheng, S., Hoseinzadeh, M., Swanson, S.: Ziggurat: a tiered file system for non-volatile main memories and disks. In: 17th USENIX Conference on File and Storage Technologies, FAST 2019, Boston, MA, 25–28 February 2019, pp. 207–219 (2019)

Short Papers

I/O Optimizations Based on Workload Characteristics for Parallel File Systems

Bing Wei[1,2], Limin Xiao[1,2(✉)], Bingyu Zhou[1,2], Guangjun Qin[1,2(✉)],
Baicheng Yan[1,2], and Zhisheng Huo[1,2]

[1] State Key Laboratory of Software Development Environment,
Beihang University, Beijing, China
{weibing,xiaolm,qingj}@buaa.edu.cn
[2] School of Computer Science and Engineering, Beihang University, Beijing, China

Abstract. Parallel file systems usually provide a unified storage solution, which fails to meet specific application needs. In this paper, we propose an extended file handle scheme to address this problem. It allows the file systems to specify optimizations for individual file or directory based on workload characteristics. One case study shows that our proposed approach improves the aggregate throughput of large files and small files by up to 5% and 30%, respectively. To further improve the access performance of small files in parallel file systems, we also propose a new metadata-based small file optimization method. The experimental results show that the aggregate throughput of small files can be effectively improved through our method.

Keywords: Parallel file systems · Workload characteristics ·
Extended file handle · Small file optimizations

1 Introduction

Applications with different workload characteristics usually have different access requirements for storage resources. The unified storage solution of parallel file systems fails to meet specific application needs. Many approaches [2–4] have been proposed to address this issue. However, these approaches cannot meet the following three requirements at the same time: (1) flexible management of I/O optimizations; (2) dynamical selection of I/O optimizations; (3) adaptive adjustment of I/O optimizations at runtime. In this paper, we propose an extended file handle (EFH) scheme to meet the above-mentioned requirements. The serving process of an I/O request can be customized with the EFH; hence, the corresponding optimization information can be achieved. To further improve the access performance of small files, we describe performance trade-off between small file load and metadata load based on the metadata-based method [5]. The steady trade-off model and the burst load trade-off model are established to determine the small file threshold. Small files are migrated across file system

© IFIP International Federation for Information Processing 2019
Published by Springer Nature Switzerland AG 2019
X. Tang et al. (Eds.): NPC 2019, LNCS 11783, pp. 305–310, 2019.
https://doi.org/10.1007/978-3-030-30709-7_24

servers based on load condition, thereby improving the access performance of small files while avoiding overload on metadata servers.

The rest of this paper is organized as follows: Sect. 2 describes the design of extended file handle. Section 3 presents the small file optimization method. Section 4 presents the experimental results and discussions. Section 5 presents the conclusions.

2   Design of Extended File Handle

We describe the definition of the EFH model in this section. An example of extended file handle structure is shown in Fig. 1, an EFH consists of five elements, including logical file handle, real file handle, version, optimization indices, and handle types. The logical file handle is used to uniquely identify a file. It is assigned by using a simple random distribution method when creating a file. The real file handle is the unique identifier of a file in the file system.

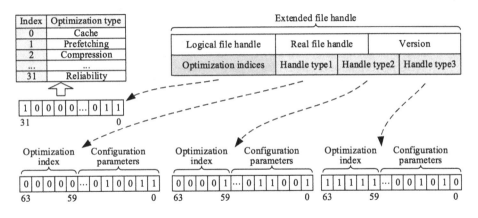

Fig. 1. An example of extended file handle structure.

The EFH version number is used for consistency maintenance. The 32-bit optimization index element indicates which optimization type is enabled. Each bit corresponds to an optimization type. If the bit is set to 1, then the corresponding optimization type is enabled; otherwise, it is not enabled. As a result, the I/O optimizations can be managed in fine grain. The handle types are used to record the customized configuration parameters for corresponding optimization type. The high 5-bit of a handle type records the index of optimization type that is corresponding to the handle type and the low 59-bit of a handle type records the corresponding configuration parameters. The EFH is stored in the directory entry that is stored on the metadata servers. Multi-type optimization information is managed with small memory overhead.

We abstract the processing of an I/O request across file system servers as a file I/O path. A proper file I/O path is selected based on the extended file handle.

The process of selecting I/O path consists of four modules: EFH buffer, EFH parser, decision maker, and I/O path set. The recently used EFHs are cached in the EFH buffer. The main job of the EFH parser is to parse the EFH and transfer the parsed information to decision maker. The decision maker selects the proper I/O path based on the EFH parsed information to serve the I/O request. The I/O path set contains all the available I/O paths on the server. When changing the optimizations, enabling or updating a handle type may involve updating data between client-side and server-side. The version number of an EFH is incremented by 1 if updating successfully.

3   Small File Optimization Method

The steady trade-off model determines the small file threshold based on the long-term running status of system. The information of unused space capacity and the load of the metadata server is periodically collected to calculate the global threshold (Gl_t), which is used to determine the threshold for a specific file and can be calculated by the following equation:

$$Gl_t = \begin{cases} Gl_{pre-t} - xGl_{pre-t} & Ca_{unused} \geq Ca_t \\ Max(Gl_{max}, Gl_{pre-t} + yGl_{pre-t}) & Ba_{io} < Ba_{low-t}, Ca_{unused} < Ca_t \\ Gl_{pre-t} - zGl_{pre-t} & Ba_{io} \geq Ba_{low-t}, Ca_{unused} < Ca_t \end{cases} \quad (1)$$

Ca_{unused} is the ratio of unused space capacity to the total space capacity. Ca_t is the threshold of unused space capacity. Gl_{pre-t} is the global threshold of the previous moment. Parameters x, y, and z are empirical adjustment parameters. Ba_{io} is the ratio of the current I/O bandwidth to the maximum I/O bandwidth. Ba_{high-t} and Ba_{low-t} are the high and low load threshold, respectively. Gl_{max} is the given maximum global threshold.

The migration frequency of a file is used to avoid frequent migrations of small files. The target threshold $(F_{target-t})$ for a file is the larger one between Gl_{max} and the fine-adjusted threshold. It can be calculated by the following equation:

$$F_{target-t} = Max(Gl_{max}, \frac{(\theta + Fre_m)Gl_t}{\theta}) \quad (2)$$

Fre_m is the migration frequency and θ is the empirical adjustment parameter. Once receiving the access request of a small file that is stored on a metadata sever, the target threshold for the file is calculated by Eqs. 1 and 2. If the file size exceeds the target threshold, the file will be migrated to other servers. Reversely, if a file stored on a data server is truncated to a size below the target threshold, the file will be migrated to a metadata server.

The burst trade-off model determines the small file threshold in the burst load situation. The exponential smoothing method (ESM) calculates prediction value by the following equation:

$$E(t) = \lambda V(t-1) + (1-\lambda)E(t-1) \quad (3)$$

$E(t)$ and $E(t\text{-}1)$ are the prediction values for the moment t and $t\text{-}1$, respectively. λ is the smoothing parameter. $V(t\text{-}1)$ is the observed value for the moment $t\text{-}1$. The prediction load can be easily calculated by Eq. 3. However, the prediction accuracy is low because of lacking of the consideration of the current I/O request status. A burst load sensing model (BLS-ESM) based on ESM is proposed to improve the prediction accuracy.

The I/O scheduler in the metadata server is used to determine the execution order of the I/O requests that are sent from the clients, and the requests that cannot be served at the current moment are blocked in the queue. $S_{t-2,t-1}$ is the amount of requested data that is served in the queue between moment $t\text{-}2$ and $t\text{-}1$. $S'_{t-2,t-1}$ is the total amount of data that is blocked in the queue between moment $t\text{-}2$ and $t\text{-}1$. The probability of burst load at the moment t can be calculated by the following equation:

$$R_{i-1} = \frac{S'_{t-2,t-1}}{S_{t-2,t-1}} \tag{4}$$

The larger the R_{i-1}, the greater the possibility of a burst load, and vice versa. Therefore, the predicted value at the moment t can be calculated by the following equation:

$$Gl_t = \begin{cases} (R_{i-1} - 1 + \lambda)V(t-1) + (1-\lambda)E(t-1) & R_{i-1} \notin (\mu, \nu) \\ \lambda V(t-1) + (1-\lambda)E(t-1) & R_{i-1} \in (\mu, \nu) \end{cases} \tag{5}$$

In the above equation, μ represents the low threshold of the burst load and ν represents the high threshold of the burst load. BLS-ESM is used to calculate the small file load prediction value at next moment for the metadata server.

4   Evaluation

Our experiments were conducted on a 5-node cluster of machines. Each machine was configured with two 20-core 2.2 GHz Intel Xeon 4114 CPUs, 128 GB of memory, two 7.2 K RPM 4 TB disks, and the Centos7 operating system. Each machine was configured with 5 virtual machines, which had the same configuration. The network was 1-Gigabit Ethernet. Our proposed approaches were conducted in PVFS [1].

4.1   Case Study: Directory Hint Optimization

We used traces *pweb* [6] and *pgrep* [6] to test data I/O performance for the three approaches, including default PVFS, PVFS-EFH (EFH), and directory hint (DH) [7]. Figure 2 shows the aggregate throughput of the three above-mentioned approaches when replaying the two traces. EFH improves the aggregate throughput over PVFS in terms of small files for the two trace by up to 11% and 30%, respectively. Meanwhile, EFH improves the aggregate throughput over PVFS in terms of large files by up to 5% for *pweb* and has no significant impact on large files for *pgrep*.

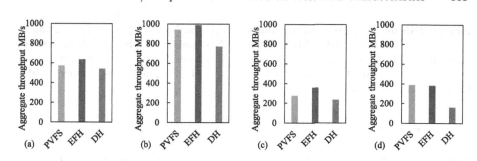

Fig. 2. The aggregate throughput of data I/O: (a) small files of pweb; (b) large files of pweb; (c) small files of pgrep; (d) large files of pgrep.

4.2 Testing Small File Optimization Methods

We used IOR [8] benchmark to test the performance of small file optimization methods. Figure 3 shows the aggregate throughput of the original metadata-based method (OMB) [5] and our method under single metadata server. When increasing the number of client processes from 2 to 20, the metadata performance degradations for OMB and our method are 62% and 11%, respectively; the small file performance improvement for OMB and our method are 150% and 196%, respectively.

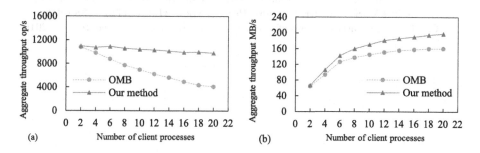

Fig. 3. The aggregate throughput: (a) metadata; (b) small files.

5 Conclusion

To meet the various requirements of multiple applications on storage resources, we propose an extended file handle scheme, which allows parallel file systems to specify customized optimizations for each file or directory based on workload characteristics. Our approach enables fine-grained management of selecting I/O optimizations for serving multiple workloads. We propose an adaptive optimization method to further improve small file performance. Performance trade-off between small file load and metadata load is achieved by our proposed method.

Acknowledgment. This work was supported by the National key R&D Program of China under Grant NO. 2017YFB1010000, the National Natural Science Foundation of China under Grant No. 61772053, the Science Challenge Project, No. TZ2016002, and the fund of the State Key Laboratory of Software Development Environment under Grant No. SKLSDE-2017ZX-10.

References

1. Ross, R.B., Thakur, R.: PVFS: a parallel file system for Linux clusters. In: Proceedings of the 4th Annual Linux Showcase and Conference, pp. 391–430 (2000)
2. Isaila, F.: Collective I/O tuning using analytical and machine learning models. In: 2015 IEEE International Conference on Cluster Computing. pp. 128–137. IEEE (2015)
3. Zhang, S., Catanese, H.: The composite-file file system: decoupling the one-to-one mapping of files and metadata for better performance. In: 14th USENIX Conference on File and Storage Technologies. pp. 15–22 (2016)
4. Byna, S., Chen, Y.: Parallel I/O prefetching using MPI file caching and I/O signatures. In: Proceedings of the 2008 ACM/IEEE Conference on Supercomputing. pp. 44. IEEE (2008)
5. Carns, P., Lang, S.: Small-le access in parallel le systems. IEEE IPDPS **2009**, 1–11 (2009)
6. Uysal, M., Acharya, A.: Requirements of I/O systems for parallel machines: An application-driven study (1998)
7. Kuhn, M., Kunkel, J.M.: Dynamic le system semantics to enable metadata optimizations in PVFS. Concurr. Comput. Pract. Exper. **21**(14), 1775–1788 (2009)
8. LNCS Homepage. https://sourceforge.net/projects/ior-sio. Accessed 16 May 2019

Energy Consumption of IT System in Cloud Data Center: Architecture, Factors and Prediction

Haowei Lin[1], Xiaolong Xu[1(✉)], and Xinheng Wang[2]

[1] Jiangsu Key Laboratory of Big Data Security and Intelligent Processing,
Nanjing University of Posts and Telecommunications, Nanjing 210023, China
{1218043318,xuxl}@njupt.edu.cn
[2] University of West London, London W55RF, UK
Henry.Wang@uwl.ac.uk

Abstract. In recent years, as cloud data center has grown constantly in size and quantity, the energy consumption of cloud data center has increased dramatically. Therefore, it is of great significance to study the energy-saving issues of cloud data centers in depth. Therefore, this paper analyzes the architecture of energy consumption of IT system in cloud data centers and proposes a new framework for collecting energy consumption. Based on this framework, the factors affecting energy consumption are studied, and various parameters closely related to energy consumption are selected. Finally, the RBF neural network is used to model and predict the energy consumption of the cloud data centers, which is aim to prove the accuracy of the framework for collecting energy consumption and influencing factors. The experimental results show that these parameters under the framework for collecting energy consumption have better accuracy and adaptability to the prediction of energy consumption in cloud data centers than the previous model of energy consumption prediction.

Keywords: Cloud computing · Cloud data center ·
Energy consumption · Prediction · Architecture

1 Introduction

In recent years, cloud data centers are facing more and more traffic demands, resulting in the continuous formation and expansion of cloud data centers around the world [1]. Although their economic profits are increasing, huge energy consumption has also received more and more attention. Since cloud computing can

This work was jointly supported by National Key Research and Development Program of China under Grant 2018YFB1003702, Jiangsu Key Laboratory of Big Data Security & Intelligent Processing, and the Talent Project in Six Fields of Jiangsu Province under Grant 2015-JNHB-012.

realize the flexibility and scalability of computing resources [2], and the sheer size of its scale, the problem about energy consumption of cloud data centers has changed from decentralized way in the past to the current centralized approach [3,4]. In order to optimize the use of energy consumption in cloud data centers, it is necessary to establish a high-precision prediction model of energy consumption for cloud data centers.

In view of the above situation, the RBF neural network is used to model and predict the energy consumption of the cloud data centers, which is aim to prove the accuracy of the framework for collecting energy consumption and influencing factors.

2 Architecture

The framework for collecting energy consumption of IT system proposed in this paper is extended on the basis of the framework for collecting energy consumption of IT system proposed by Zhao, Z. [5] in 2016, and is improved in the details. Figure 1 is the framework for collecting energy consumption of IT system.

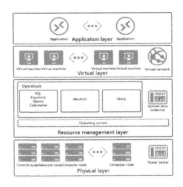

Fig. 1. The framework for collecting energy consumption of IT system

The framework can be divided into physical layer, resource management layer, virtual layer and application layer. The physical layer is located at the bottom of the framework of cloud environment and is divided into physical resources for building cloud environments and power tester. The physical resources used to build the cloud environment is the infrastructure built by the cloud environment, which are the fundamental source of energy consumption of the IT system, and the main target of cloud data center energy consumption prediction. The resource management layer is located above the physical layer and is divided into the operating system, cloud computing resource management service and system data collector. The virtual layer is the third layer of the framework of cloud environment and is to provide users with virtualization platforms and virtualized resources. The application layer is located above the virtual layer and is to run the required application according to different requirements on the virtual machine that the users applies for.

3   Factors Affecting Energy Consumption of IT System

The key innovations presented in this paper are: By decomposing the energy consumption of IT system of the cloud data center and selecting better factors affecting energy consumption to train the RBF neural network, the prediction ability of the RBF neural network model is better than the previous model of energy consumption of IT system, which is aim to prove that these factors play a crucial role in the energy consumption of IT systems.

Modeling energy consumption based on system usage is a usual method of energy modeling. According to the related research [6], the top six system parameters in power consumption of virtual machine have a significant nonlinear relationship with the energy consumption of virtual machine, which are user mode runs using the percentage of total CPU time, core state runs using the percentage of total CPU time, memory utilization, the total amount of I/O transfer per physical device per second, number of pages missing per second of system, and the physical machine load of each physical device. Moreover, because there is a massive scale of the data in the cloud data center, this paper will eliminate the number of pages missing per second of system, which affects the model of energy consumption minimally.

Through experiments, it is found that if the system information is collected for only this part of the equipment, not only can the burden of equipment which collect system information be reduced, but also a predictive model with higher fitting degree can be obtained.

As the experiments showed, we found that the energy consumption of IT systems is extremely sensitive to changes in the energy consumption in a few seconds nearby. Therefore, in the case of maintaining the energy object model with strong objectivity and robustness, and improving the accuracy of the energy consumption model, this paper keeps the modeling method of randomly disturbing the training samples, and adds the energy consumption in the time of last unit to the data set of the training model.

4   Application and Experiments

This paper compared the data from the above experiments from two aspects. On the one hand, we observed the result about whether the training data of the RBF neural network included the compute nodes that enter the virtual layer. On the other hand, we observed the result about whether the training data of the RBF neural network included energy consumption data in the time of last unit.

The data of this experiment were all analyzed using RBF neural network. The same part of the training data included the system information data of the control nodes, the network nodes, and the compute nodes that entered the virtual layer and the energy consumption in the time of last unit. The different parts were that the first training data did not include the compute nodes that did not enter the virtual layer, and the second training data included the compute

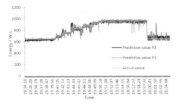 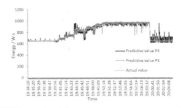

(a) With and without the compute nodes that entered the virtual layer.   (b) With and without energy consumption data in the time of last unit.

Fig. 2. The training results of the RBF neural network

Table 1. Error comparison in two cases

Model	P1	P2	P3
The sum of squared errors	38.415	43.827	63.181
The relative error	4.2%	4.8%	6.9%

nodes that did not enter the virtual layer. In Fig. 2(a), the predicted value P1 was the training situation of the training data that did not include the compute nodes that did not enter the virtual layer; the predicted value P2 was the training situation of the training data that included the compute nodes that did not enter the virtual layer.

Although the prediction result of P1 at a few special points was extremely extreme, the predicted value P1 had a better fit with the actual value in comparison with the predicted value P2.

As shown in Table 1, we found that there was better predicted effect of the RBF neural network model, which was trained by the training data that did not include the compute nodes that did not enter the virtual layer. From this experiment, we found that when predicting energy consumption in a cloud environment, it was correct to use the system data of the control nodes, the network nodes, and the compute nodes that entered the virtual layer as the training data.

The data of this experiment were all analyzed using RBF neural network. The same part of the training data included the system information data of the control nodes, the network nodes, and the compute nodes that entered the virtual layer. The different parts were that the first training data included the energy consumption in the time of last unit, and the second training data did not include the energy consumption in the time of last unit. In Fig. 2(b), the predicted value P1 was the training situation of the training data that included the energy consumption in the time of last unit; the predicted value P3 was the training situation of the training data that did not include the energy consumption in the time of last unit.

Although the prediction result of P1 at a few special points was extremely extreme, the predicted value P1 had a better fit with the actual value in comparison with the predicted value P3.

As shown in Table 1, we found that there was better predicted effect of the RBF neural network model, which was trained by the training data that included the energy consumption in the time of last unit. From this experiment, we found that when predicting energy consumption in a cloud environment, it was correct to use the energy consumption in the time of last unit as the part of training data.

5   Conclusion

This paper analyzes the architecture of energy consumption of IT system in cloud data centers and proposes a new framework for collecting energy consumption. Based on this framework, the factors affecting energy consumption are studied, and various parameters closely related to energy consumption are selected.

Acknowledgement. This work was jointly supported by National Key Research and Development Program of China under Grant 2018YFB1003702, Jiangsu Key Laboratory of Big Data Security & Intelligent Processing, and the Talent Project in Six Fields of Jiangsu Province under Grant 2015-JNHB-012.

References

1. Yeganeh, H., Salahi, A., Pourmina, M.A.: A novel cost optimization method for mobile cloud computing by capacity planning of green data center with dynamic pricing. Can. J. Electric. Comput. Eng. **42**(1), 41–51 (2019)
2. Wang, L., Gelenbe, E.: Adaptive dispatching of tasks in the cloud. IEEE Trans. Cloud Comput. **6**(1), 33–45 (2018)
3. Li, C., Ruijin, Z., Li T.: Enabling distributed generation powered sustainable high-performance data center. In: IEEE 19th International Symposium on High Performance Computer Architecture, pp. 35–46 (2013)
4. Valentini, G.L., Lassonde, W., Khan, S.U.: An overview of energy efficiency techniques in cluster computing systems. Cluster Comput. **16**(1), 3–15 (2013)
5. Zhou Z.: Energy Consumption Acquisition and Prediction Method for Cloud Computing Services. PhD thesis, East China Normal University, Shanghai, China (2016)
6. Hao X.: Research on Neural Network Based Virtual Machine's Power Prediction Mode. PhD thesis, Beijing University of Posts And Telecommunications, Beijing, China (2015)

Efficient Processing of Convolutional Neural Networks on SW26010

Yi Zhang[1(✉)], Bing Shu[2], Yan Yin[1], Yawei Zhou[1], Shaodi Li[1], and Junmin Wu[1]

[1] University of Science and Technology of China, Hefei, Anhui, China
`colezhan@mail.ustc.edu.cn`
[2] Jiangnan Institute of Computing Technology, Wuxi, Jiangsu, China

Abstract. Artificial intelligence has developed rapidly in recent years. Deep neural networks are the basis of many artificial intelligence applications. How to accelerate the computational processing of deep neural networks is very important. To explor the potential for accelerating the process deep neural networks on various hardware platforms, we propose a convolutional neural network optimization method based on the Weight-Stationary for SW26010 processor. We re-circulate convolution loops and use hybrid DMA transmission mode to increase memory bandwidth and reduce memory access overhead. On top of those, further optimizations are done based on register communication, asynchronous DMA transfer double buffering, instruction scheduling and other schemes. Finally, we achieve a double-precision convolution performance over 2.4 Tflops, achieving 81% of the processor's peak performance. In multiple parameters, we achieve a proforamnce acceleration of $2.4 - 4.0\times$ speedup compared to the Tesla K80 GPU with cuDNNv7.

Keywords: SW26010 processor · Convolutional neural networks · Weight-stationary · Parallel model · Many-core architecture · Deep learning

1 Introduction

Deep neural networks (DNNs) are the foundation of many modern artificial intelligence (AI) applications. The high accuracy of DNNs is at the expense of high computational complexity and requires more computing power. In order to satisfy the computational requirements of DNNs, various acceleration hardware such as graphics processing units (GPU) [1] and FPGA [2] are applied to DNNs processing. Exploring DNNs calculation acceleration on various hardware platforms is a very valuable job. The SW26010 heterogeneous multicore processor is the processor chip of the Sunway TaihuLight supercomputer. In order to

Supported by the National Key Research and Development Program of China under Grant 2018YFB1003601.

explore the combination of DNNs and SW26010, accelerate the processing of DNNs on SW26010, we first optimize the computational processing of the convolutional neural network (CNN), a common form of DNNs on SW26010, and plan to expand to more forms in future. The main contributions of this work are as follows:

- We design a convolution algorithm based on the idea of Weight-Stationary and map it to SW26010.
- We analyze the DMA memory access characteristics of SW26010, use hybrid DMA transmission mode instead of bstride DMA transmission mode to increase memory bandwidth.
- We use an instruction scheduling method to reduce the idling time of computation units.
- We finally achieve over 2.4 Tflops double-precision convolution performance on SW26010, achieving 81% of the processor's peak performance.

2   Background and Related Work

2.1   Convolutional Neural Networks(CNN)

CNN is a common form of deep neural networks. In most CNNs, the operation of the convolutional layer occupies the largest portion of the total computation (more than 90%). The main operation of the convolutional layer is high-dimensional convolution. Let D be the input image, F be the convolutional filter, O be the output. Using the variable definitions in Fig. 1, the algorithm formula for convolution operations [1] can be expressed as follows:

$$O[n,k,p,q] = \sum_{c=0}^{C-1} \sum_{r=0}^{R-1} \sum_{s=0}^{S-1} F[k,c,r,s] \cdot D[n,c,p+r,q+s] \tag{1}$$

2.2   Related Work

SW26010 Heterogeneous Many-core Processor is manufactured by the Shanghai High Performance IC Design Center through independent technology(see Fig. 2). As an emerging hardware platform, SW26010 has less work on efficient processing of DNNs. The authors of swDNN [3] have developed deep learning framework swCaffe [4] and deep learning acceleration library swDNN [3] for SW26010. However, swDNN does not consider the balance between memory access and computation, their double buffering scheme cannot completely cover the cost of memory access and their instruction scheduling scheme is not the best. In order to address these problems, we design a new algorithm based on the Weight-Stationary to balance the memory access and calculation, achieve almost complete cover-up of the memory access cost, and we also draw on the efficient instruction scheduling provided by swDGEMM. Finally we get very good performance by these optimization.

3   Optimization of CNN on SW26010

3.1   Mapping CNN to SW26010

Considering the memory access characteristics of the SW26010 processor, the efficiency of accessing the main memory from the CPE using global read/store (gld/gst) is extremely low. The calculated data should be prefetched into the CPE's local memory (LDM) by means of DMA transfer when designing the algorithm. At the same time, considering the limited size of the LDM, it is necessary to keep the data reuse as much as possible when designing the algorithm.By analyzing the convolution algorithm, we can find that there are two forms of data reuse: one is the Output-Stationary, that is, output matrix is always kept in the LDM. When the matrix multiplication and addition operation associated with output matrix is completed, output matrix will be written back to the main memory. Another form of data reuse is the Weight-Stationary, which converts the loop's order, always keeps weight matrix in the LDM, reads in and reads out the output matrix in stages to complete the convolution calculation. We finally choose to design the algorithm based on the Weight-Stationary, it can reduce bandwidth requirements and adapt to the SW26010's limited bandwidth

---

Algorithm 1. CNN on SW26010 with double buffering

1: Load O to the CPEs, O are bcp start matrixs(from$(0,0)$to$(bcp,0)$) of size $n \times k$
2: load D to the CPEs, D is $n \times c$ start matrix in$(0,0)$
3: **for** $cr = 0 : bcr : R$ **do**
4:     **for** $cs = 0 : S$ **do**
5:         Load F to the CPEs,F are bcr matrixs(from$(cr,cs)$to$(cr+bcr,cs)$) of size $k \times c$

6:         **for** $cp = 0 : bcp : P$ **do**
7:             **for** $cq = 0 : Q$ **do**
8:                 Load next O to the CPEs for next computation, O are bcp matrixs(from$(cp,cq)$to$(cp + 2bcp, cq)$) of size $n \times k$
9:                 **for** $ch = cp : cp + bcp + bcr - 1$ **do**
10:                    Load next D to the CPEs for next computation, D is $n \times c$ matrix in$(ch + 1, cq + cs)$
11:                    **if** $ch >= cp + cr$ and $ch < cp + cr + R$ **then**
12:                        $O_{cp,cq} + = D_{ch,cq+S} \times F_{cr,cs}$
13:                    **end if**
14:                    sync
15:                **end for**
16:                Store current O to the CPEs, O are bcp matrixs(from$(cp,cq)$to$(cp + bcp, cq)$) of size $n \times k$
17:                sync
18:            **end for**
19:        **end for**
20:    **end for**
21: **end for**

---

and memory. And we also use double buffering for the data transmission. The optimized algorithm is as Algorithm 1.

3.2 DMA Transfer Optimization

To map GEMM onto the processor array, we use matrix block multiplication to block the input and output data in an 8x8 array structure, to transfer data from main memory to LDM through the DMA, it is necessary to perform DMA stride transmission, resulting in low DMA transmission rate. But by the analysis of Algorithm 1, we can find that the DMA stride transmission will be used to transfer the output matrix O back to the main memory only when the matrix multiplication and addition is completed, and the rest of the O's transmission can use a continuous DMA transmission mode. Therefore, we use a hybrid DMA transmission mode mixed with stride transmission and continuous transmission to improve the program's memory access bandwidth.

Para	Meaning
N	Number of images in mini-batch
C	Number of input feature maps
H	Height of input image
W	Width of input image
K	Number of output feature maps
R	Height of filter kernel
S	Width of filter kernel
P	Height of output feature maps
Q	Width of output feature maps

Fig. 1. Convolutional parameters

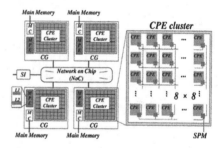

Fig. 2. The general architecture of the SW26010 processor

3.3 Instruction Scheduling

An important feature of the SW26010 is its dual instruction pipeline. Floating-point instructions and register communication instructions can be issued simultaneously by dual instruction pipeline. In some specific scenarios, great performance gains can be achieved by instruction scheduling, which was used in the design of swDNN [3] and swDGEMM [5]. After comparison, we selected the instruction scheduling provided by swDGEMM. Through the instruction scheduling method, the execution cycle of the instruction stream is reduced from 26 to 16, the overall operation efficiency of the program is greatly improved.

4   Results and Analysis

We use a number of sets of parameters to test our program performance and compare it to the K80 GPU using cuDNNv7. The final test results are shown in

Figs. 3, 4 and 5. Finally, we achieve a double-precision convolution performance over 2.4 Tflops, achieving 81% of the processor's peak performance, which is higher than swDNN [3]'s measured performance (only 54%) shown in their paper, and under different convolution filters, the performance is still stable(see Fig. 3). Compared to the K80 using cuDNNv7, we achieve 2.4 − 4.0× acceleration.

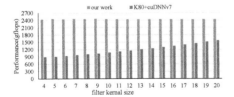

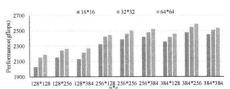

Fig. 3. Performance with different filter kernal (batchsize=128, output=64*64)

Fig. 4. Performance with different output size (batchsize=128, r=s=3)

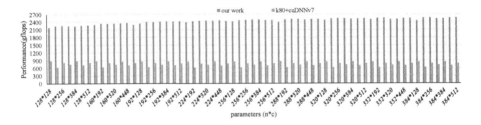

Fig. 5. Performance with different $n \times c$ (batchsize=128, output=64*64, r=s=3)

5   Conclusion

In this article, we present an efficient acceleration scheme for CNNs on SW26010 and finally achieve a double-precision convolution performance over 2.4 Tflops, achieving 81% of the processor's peak performance. In multiple parameters, we achieve a proforamnce acceleration of 2.4 − 4.0× speedup compared to the Tesla K80 GPU with cuDNNv7.

References

1. Chetlur, S., et al.: cudnn: Efficient primitives for deep learning. arXiv preprint arXiv:1410.0759 (2014)
2. Chen, Y., Chen, T., Xu, Z., Sun, N., Temam, O.: Diannao family: energy-efficient hardware accelerators for machine learning. Commun. ACM **59**(11), 105–112 (2016)

3. Fang, J., Fu, H., Zhao, W., Chen, B., Zheng, W., Yang, G.: swdnn: a library for accelerating deep learning applications on sunway taihulight. In: 2017 IEEE International Parallel and Distributed Processing Symposium (IPDPS), pp. 615–624. IEEE (2017)
4. Li, L., et al.: swcaffe: A parallel framework for accelerating deep learning applications on sunway taihulight. In: 2018 IEEE International Conference on Cluster Computing (CLUSTER), pp. 413–422. IEEE (2018)
5. Jiang, L., et al.: Towards highly efficient dgemm on the emerging sw26010 many-core processor. In: 2017 46th International Conference on Parallel Processing (ICPP), pp. 422–431. IEEE (2017)

ADMMLIB: A Library
of Communication-Efficient AD-ADMM
for Distributed Machine Learning

Jinyang Xie[✉] and Yongmei Lei[✉]

School of Computer Engineering and Science, Shanghai University, Shanghai, China
{jyxie,lei}@shu.edu.cn

Abstract. Alternating direction method of multipliers (ADMM) has recently been identified as a compelling approach for solving large-scale machine learning problems in the cluster setting. To reduce the synchronization overhead in a distributed environment, asynchronous distributed ADMM (AD-ADMM) was proposed. However, due to the high communication overhead in the master-slave architecture, AD-ADMM still cannot scale well. To address this challenge, this paper proposes the ADMMLIB, a library of AD-ADMM for distributed machine learning. We employ a set of network optimization techniques. First, hierarchical communication architecture is utilized. Second, we integrate ring-based allreduce and mixed precision training into ADMMLIB to further effectively reduce the inter-node communication cost. Evaluation with large dataset demonstrates that ADMMLIB can achieve significant speed up, up to 2x, compared to the original AD-ADMM implementation, and the overall communication cost is reduced by 83%.

Keywords: Asynchronous ADMM · Consensus optimization ·
Distributed machine learning

1 Introduction

With the ever-increasing sizes of datasets and models, machine learning model training often takes so long time. Due to the single machine's limited computing resources, it is reasonable to distribute large scale machine learning workloads across multiple computing nodes. Distributed machine learning (ML) jobs often involve solving a non-convex, decomposable and regularized optimization problem. Distributed optimization is becoming a prerequisite for solving large scale ML problems. The alternating direction method of multipliers (ADMM) [1] is an optimization technique by decomposing the original problem into subproblems for parallel iterations. Usually, ADMM was implemented in master-slave architecture, in which a master coordinates the computation of a set of distributed

Supported by the National Natural Science Foundation of China under grant No. U1811461.

workers. To reduce the synchronization overhead, recently, the synchronous distributed ADMM has been extended to the asynchronous setting [2,8]. Asynchronous updates would improve the computation efficiency of the distributed ADMM.

A major performance bottleneck of AD-ADMM is the high communication overhead due to the following factors. First, large-scale ML are trending to learn large models with tens or hundreds of millions of parameters, generating a large amount of network traffic for distributed training. Second, under the master-slave architecture, a single incoming link to the master is shared across multiple workers, causing network congestion. Hence, it is important to reduce the communication overhead when scaling AD-ADMM to large-scale clusters.

In our work, we focus directly on the problem of improving the performance and scalability of the AD-ADMM. We employ a set of network optimization techniques, such as hierarchical communication architecture, ring-based allreduce and mixed precision training, to achieve load balancing and reduce communication overhead. We build a library named ADMMLIB integrating our optimization techniques. ADMMLIB manages details of parallelism, synchronization and communication. It provides simple programming interfaces for users to implement scalable AD-ADMM.

2   Related Work

Because of the demand for faster training of ML model, several frameworks have been proposed, such as Petuum [7]. The standard distribution strategy in ML is data parallelism. To implement the data-parallel model training, there are two design choices: the parameter server (PS) [3] approach using master-slave architecture and the ring-based allreduce [4] approach with P2P architecture. In PS, a logical parameter server aggregates model updates from all workers and broadcasts to all workers. One bottleneck of the PS is the high communication cost on the central server. In the ring-based allreduce, all the nodes are organized as a logical ring, and each node communicates with two of its peers. Ring-based allreduce is an algorithm with constant communication cost. Recent literature [5] has shown clear benefits of the ring-based allreduce. Distributed ADMM has been widely studied as an alternative method for distributed stochastic gradient descent algorithms. Recently [1] proved that the ADMM is suitable for distributed optimization problems. [8] has considered a version of asynchronous ADMM to speed up the ADMM. [2] added a penalty term based on [8] to improve the convergence efficiency of non-convex problems. [6] uses hierarchical communication structure to improve the communication efficiency of distributed ADMM.

3   ADMMLIB: System Design and Optimization

3.1   Hierarchical Communication Architecture

Although master-slave architecture has been widely used in the ADMM, it is not quite suitable for large scale machine learning. As shown in Fig. 1, ADMMLIB adopts hierarchical communication architecture (HCA) to scale up to multicores on a single node, as well as scale out to multiple nodes in a cluster.

To scale up, ADMMLIB start a number of worker threads on each node and each worker thread is assigned to a dedicated CPU core. ADMMLIB also starts a coordinator thread on each node, and ADMMLIB will choose a coordinator as the master coordinator. Each worker only communicates with the coordinator on the same node. Coordinators communicate with each other to coordinate the computation of all workers in the cluster, therefore ADMMLIB can scale out to multiple nodes.

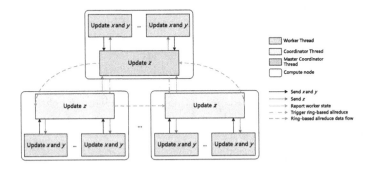

Fig. 1. Overview of ADMMLIB architecture.

Each worker i owns a train dataset partition and is responsible for updating x_i and y_i. x_i and y_i represent local model variable and dual variable [1], respectively. After sending up-to-date (x_i, y_i) to the coordinator on the same node, worker will block until they receive the updated z. z represents the global model variable [1]. Each coordinator takes charge of caching the latest (x_i, y_i) from workers. And each coordinator maintains its own copy of z. Replicas are kept consistent by exchanging data between coordinators. Specifically, when a coordinator receiving update from worker i, it reports to the master coordinator. Therefore, the master coordinator can know the status of all workers in the cluster. Once the partial barrier and bounded delay conditions [8] are satisfied, the master coordinator will inform all coordinators to perform an allreduce operation to update z.

Compared with the master-slave architecture, HCA can balance the load, because each coordinator (including the master coordinator) only needs to manage a small subset of workers.

3.2 Improvements on Internode Communication Strategies

The simple architecture of ADMMLIB also makes it easy to identify the limiting factors to scaling. Optimizing inter-node communication overhead is clearly the key to scaling.

Most of the inter-node communication overhead comes from the allreduce operation. Out of the possible allreduce implementation strategies, we choose the ring-based allreduce algorithm. If the input data is m bytes, ring-based allreduce equally partition data into N_n chunks, N_n is the number of coordinators. Each coordinator sends and receives m/N_n bytes of data $2(N_n - 1)$ times to complete an allreduce operation. Thus, the total communication time is independent of the number of nodes. Ring-based allreduce distributes the communication cost across all N_n nodes to avoid a node becoming a performance bottleneck.

Model training is not very demanding for high-precision calculations. Compared to double-precision, using single-precision or even half-precision can increase arithmetic throughput without decreasing accuracy. Low precision training also helps to reduce communication overhead and memory storage requirement since the same number of values could be stored using fewer bits. We use mixed-precision training strategy in ADMMLIB. When optimizing x_i, y_i and z, ADMMLIB uses single-precision or double-precision parameters, depending on the user's choice. When caching and transferring parameters, ADMMLIB uses single precision to reduce memory usage and communication cost.

4   Experiment

In this section, we evaluate the performance and scaling efficiency of ADMMLIB. For comparison, we also use the multi-threading technique to implement the AD-ADMM in master-slave architecture, we call this system MAD-ADMM. We use a cluster of 5 computing nodes interconnected with a Gigabit Ethernet. Each node has two Intel E5-2690 CPU (2.9 GHz/8core) processors and 64 GB memory. In the experiment, we solve sparse logistic regression problem. We consider a large dataset: URL[1]. The URL has more than 2 million samples and 3 million features.

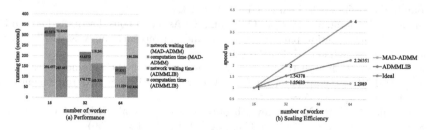

Fig. 2. Performance and scaling efficiency comparisons between ADMMLIB and MAD-ADMM.

We set up three experiments with 16, 32 and 64 workers, respectively. First, we test the performance of ADMMLIB. We compare two systems by running them to reach 20 iterations, and we recorded the computation time and network waiting time of the two systems. Figure 2(a) shows the performance comparison. It can be seen from Fig. 2(a) that as scaling increases the level of parallelism and

---

[1] https://www.csie.ntu.edu.tw/~cjlin/libsvmtools/datasets/binary.html#url

(consequently) reduces the computation time of the two systems, the network waiting time of ADMMLIB changes little, however the network waiting time in the MAD-ADMM increases linearly. Therefore, ADMMLIB outperforms MAD-ADMM. ADMMLIB can reduce network waiting time by 62.9% when testing with 32 workers and reduce network waiting time by 83% when testing with 64 workers. Figure 2(b) shows the scaling efficiency (taking the performance of 16 workers as the baseline). ADMMLIB has higher scaling efficiency thanks to the efficient ring-based allreduce algorithm. For 32 workers, we improved the scaling out efficiency from 62.8% to 77.1%. For 64 workers, we improved the scaling out efficiency from 30.2% to 57.5%.

5   Conclusion

Aiming at building a scalable and high-performance distributed model training system based on the AD-ADMM, this paper uses hierarchical communication architecture, ring-based allreduce algorithm and mixed precision training to reduce communication overhead and memory usage. Experiments show our system has higher performance and scalability than the original AD-ADMM implementation. But scalability of our system still does not reach ideal efficiency. In future work, we will try to optimize the sub-question solving efficiency to solve this problem.

References

1. Boyd, S., Parikh, N., Chu, E., Peleato, B., Eckstein, J., et al.: Distributed optimization and statistical learning via the alternating direction method of multipliers. Found. Trends® Mach. Learn. **3**(1), 1–122 (2011)
2. Chang, T.H., Hong, M., Liao, W.C., Wang, X.: Asynchronous distributed admm for large-scale optimization-part I: algorithm and convergence analysis. IEEE Trans. Signal Process. **64**(12), 3118–3130 (2016)
3. Li, M., et al.: Scaling distributed machine learning with the parameter server. In: 11th {USENIX} Symposium on Operating Systems Design and Implementation ({OSDI} 14), pp. 583–598 (2014)
4. Patarasuk, P., Yuan, X.: Bandwidth optimal all-reduce algorithms for clusters of workstations. J. Parallel Distrib. Comput. **69**(2), 117–124 (2009)
5. Sergeev, A., Del Balso, M.: Horovod: fast and easy distributed deep learning in tensorflow. arXiv preprint arXiv:1802.05799 (2018)
6. Wang, S., Lei, Y.: Fast communication structure for asynchronous distributed ADMM under unbalance process arrival pattern. In: Kůrková, V., Manolopoulos, Y., Hammer, B., Iliadis, L., Maglogiannis, I. (eds.) ICANN 2018. LNCS, vol. 11139, pp. 362–371. Springer, Cham (2018). https://doi.org/10.1007/978-3-030-01418-6_36
7. Xing, E.P., et al.: Petuum: a new platform for distributed machine learning on big data. IEEE Trans. Big Data **1**(2), 49–67 (2015)
8. Zhang, R., Kwok, J.: Asynchronous distributed ADMM for consensus optimization. In: International Conference on Machine Learning, pp. 1701–1709 (2014)

Energy-Aware Resource Scheduling with Fault-Tolerance in Edge Computing

Yanfen Xue[1,2], Guisheng Fan[1(✉)], Huiqun Yu[1], and Huaiying Sun[1]

[1] Department of Computer Science and Engineering,
East China University of Science and Technology, Shanghai, China
{gsfan,yhq}@ecust.edu.cn

[2] Shanghai Key Laboratory of Computer Software Evaluating and Testing,
Shanghai, China

Abstract. Edge computing extends computation and storage resources to the edge of the network, which largely improve the performance problem of cloud computing incurred by the bandwidth limitation. And it still needs to address the challenges of energy and reliability. In this paper, we propose an energy-aware fault-tolerant resource scheduling algorithm to improve system reliability while minimizing the energy consumption. We allocate resources by reliability and energy-aware resource scheduling method for tasks firstly. Then, CPU temperature prediction and time between failures (TBF) prediction are used to trigger proactive fault tolerance mechanism (VM migration). The experimental results show that the reliability is greatly improved and energy consumption generated by VM migration is not very large compared to other methods.

Keywords: Edge computing · Fault tolerance · Energy consumption · Resource scheduling

1 Introduction

Recently, edge computing is seen as an effective solution to the problem of more larger data, which has the advantages of shorter response time and service quality [1]. However, the problems of reliability are still urgent to be solved. The existing fault-tolerant methods can be divided into two categories: reactive and proactive methods. It is well known that reactive schemes will produce low average utilization of resources when the application behavior is highly dynamic. Instead of a reactive scheme, the proactive scheme that adopts a scheme of fault prediction [2–5] can effectively improve the utilization of resources. However, they only consider a single factor when predicting failures, which greatly affects the accuracy of the prediction results.

In this paper, we jointly consider the CPU temperature and time between failures (TBF) of the host to achieve fault prediction and propose an energy-aware fault-tolerant resource scheduling algorithm to improve the reliability

© IFIP International Federation for Information Processing 2019
Published by Springer Nature Switzerland AG 2019
X. Tang et al. (Eds.): NPC 2019, LNCS 11783, pp. 327–332, 2019.
https://doi.org/10.1007/978-3-030-30709-7_28

while reducing the energy consumption. Specifically, we use the reliability and energy-aware resource scheduling [2] to allocate resources for tasks firstly. During the tasks execution, the fault tolerance mechanism (VM migration) will be triggered once the temperature reaches the upper threshold or the predicted failure time.

The rest of this paper is organized as follows. The system model is presented in Sect. 2 and follow is the resource scheduling algorithm. The simulation experiments are conducted in Sect. 4. Section 5 summarizes the paper.

2   Fault-Tolerance Resource Scheduling Model

As shown in Fig. 1, the system is mainly divided into two layers. The Users Layer is the producer and consumer of data. The Edge Cloud Layer is the data processing layer that consists of physical resources. Users submit their application to Edge Cloud layer. Then, the physical resources are allocated to tasks by resource management system (RSM). And in order to improve the reliability of system, the system can migrate the running VM from the deteriorating host to other host by RSM.

In this paper, we use the Bag-of-Task (BoT) application which consists of a set of independent tasks. The tasks in each BoT are defined as $T = \{task_i | 1 \leq i \leq n\}$. l_i is the length of the task $task_i$, which directly affects the execution time, T_i^{ex}. Each task $task_i$ is allocated to a virtual machine $vm_j \in VM$. Each virtual machine vm_j run a set of tasks $T_j \in T$. In addition, $N = \{node_k | 1 \leq k \leq x\}$ denotes the set of the physical hosts on the edge cloud.

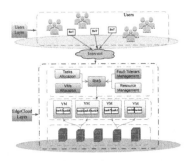

Fig. 1. The System Architecture

2.1   Failure Prediction Model

CPU Temperature Prediction: We use the simulation prediction function model of CPU temperature [3] as one of the methods to predict the host failure time as follow:

$$f(t|A,\omega,t_i,t_{i+1}) = \begin{cases} e^t & 0 \le t \le t_i \\ e^{t_i} & t_i \le t \le t_{i+1} \\ A\sin(\omega t - \omega t_{i+1}) + e^{t_i} & t_{i+1} \le t \le t_{i+2} \end{cases} \quad (1)$$

where i is the positive integer set; t_i is a fixed value calculated by $e^{t_i} = 35$; e^{t_i} is the temperature when CPU is idle, which is always $35\,^\circ C$; t_{i+1} is a random value; t_{i+2} is calculated by $t_{i+2} = \pi/\omega + t_{i+1}$; A is the amplitude(lower than $68\,^\circ C$); ω represents the duration of the CPU execution load.

Time Between Failures Prediction: In addition to the CPU temperature prediction, the method called exponential smoothing [2] is used to predict the TBF. Suppose there is a set of TBFs for the host $node_k$, $TBF_k = \{tbf_t | 1 \le t \le n\}$. Then, the prediction corresponding to tbf_{t+1} can be calculated as :

$$(tbf_k)'_{t+1} = \begin{cases} \alpha \times (tbf_k)_t + ((1-\alpha) \times (tbf_k)'_t), & n > 1 \\ (tbf_k)'_t & otherwise \end{cases} \quad (2)$$

where $(tbf_k)_t$ is the actual value of the TBF, $(tbf_k)'_t$ is the predicted value of the TBF at time t. α is the smoothing constant.

Algorithm 1. Reliability and Energy-aware Resource Scheduling Algorithm

---

Input: Bag of Tasks, B
Output: The result of tasks allocation

1: Sorting R by the ratio of the mean time between failures to the power
2: **for** $j = 1$ to $|V|$ **do**
3:     get the number of CPU cores required of the VM, VM_{cores_j}
4:     **for** $k = 1$ to $|R|$ **do**
5:         **if** $R_k.predictedtoFail()! = true$ and $Cores_k \ge VM_{cores_j}$ **then**
6:             Allocate VM vm_j to the host R_k
7:             $Cores_k = Cores_k - VM_{cores_j}$
8:         EndIF
9:     EndFor
10: EndFor

2.2 Energy Consumption Model

Let vm_j be the VM running on $node_k$ with utilization u_j. Then the energy consumption of the task $task_i$ running on vm_j can be calculated as

$$E_{ij} = (P_k(u_j) \times T_{ij}^{ex}) + E_{extra_{ij}} \quad (3)$$

where $E_{extra_{ij}}$ is the energy generated by VM migration, which can be calculated by the VM migration overhead model in [2], Similar to [6], $P_k(u_j)$ can be calculated by,

$$P_k(u_j) = P_{min_k} + (P_{max_k} - P_{min_k}) \times u_j \quad (4)$$

where P_{min} and P_{max} is the power of node at minimum utilization and maximum utilization, respectively. The utilization u_j of the VM vm_j is the sum of the tasks utilization u_i which is calculated by normalizing the task length l_i with the maximum length l_{max} in B.

3 Energy-Aware Fault-Tolerant Resource Scheduling Algorithm

Given the set of tasks BoT B and the resource configurations of data center. Algorithm 1 is used to configure resources for tasks. Firstly, the Best Fit Bin Packing algorithm [2] is used to allocate the tasks to the VM. Then, the reliability and energy-aware strategy is used to configure physical resources for VMs (lines 1–10). During task execution, once the temperature of the node reaches the upper threshold or the predicted fault time, the VM migration will be triggered. The VM running on deteriorating node selects another node through Algorithm 1 to implement the migration.

4 Performance Evaluation

We do the simulation experiments by extending the simulator 'CloudSim' [3] and download the Grid5000 failure dataset from Fault Tracking Archive (FTA) [2] and select the clusters, G1/site1/c1, as the edge cloud data center. Parameter configuration model in [2] is used to match the configuration for each node and generation the BoTs workload which consist of tasks between 2000 and 3000. In order to evaluate the performance of the proposed algorithm (Tem/Tbf), we compare our method with other fault-tolerant strategies. Specifically, we denote 'NoFT' as the method with no fault tolerance mechanism. 'Restr', 'Pre-Tem', 'Pre-Tbf', 'Tem/Tbf' as the method with resubmission, CPU temperature, TBF, CPU temperature and TBF prediction as the fault tolerant strategy, respectively.

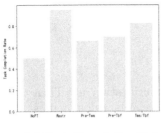
(a) Task Completion Rate

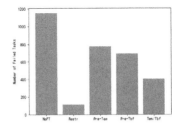
(b) Number of Failed Tasks

Fig. 2. The task completion rate under different fault-tolerant strategies

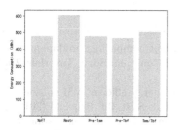

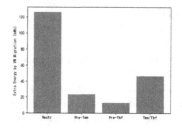

(a) Total Energy Consumption   (b) Extra Energy by using Fault-Tolerant strategies

Fig. 3. The energy consumption under different fault-tolerant strategies

4.1 Experimental Results

Figure 2 shows the task completion rate and the energy consumption is given in Fig. 3. We can see that the task completion rate and energy consumption is the highest when using Restr method. And among using fault prediction as the fault-tolerant strategy, the extra energy by using Tem/Tbf prediction is only 30 Kwh higher than the other two cases. If using task completion rate to measure the reliability of the system, it is the most reliable by using Restr method, but the excessive energy which will greatly influence interests of operators. And when using Tem/Tbf method, the reliability is much higher than the other two proactive strategies and the increased energy is not large. Therefore, the method we proposed(Tem/Tbf) is more effective.

5 Conclusions

In this paper, we study how to improve the reliability of the edge cloud system while reducing energy consumption as much as possible. We use the reliability and energy-aware resource scheduling algorithm to allocate physical resources for tasks firstly. Then, CPU temperature prediction and time between failures prediction are used to achieve fault tolerance. Comparison with other fault-tolerant strategies, the method we proposed is more effective.

Acknowledgements. This work was partially supported by the NSF of China under Grant nos. 61702334 and 61772200, Shanghai Pujiang Talent Program under Grant no. 17PJ1401900, Shanghai Municipal Natural Science Foundation under Grant nos. 17ZR1406900 and 17ZR1429700, Action Plan for Innovation on Science and Technology Projects of Shanghai under Grant no. 16511101000, Collaborative Innovation Foundation of Shanghai Institute of Technology under Grant no. XTCX2016-20, and Educational Research Fund of ECUST under Grant no. ZH1726108.

References

1. Mukherjee, M., Shu, L., Wang, D., et al.: Survey of fog computing: fundamental, network applications, and research challenges. IEEE Commun. Surv. Tutorials **20**(3), 1826–1857 (2018)
2. Sharma, Y., Si, W., Sun, D., et al.: Failure-aware energy-efficient VM consolidation in cloud computing systems. Future Gener. Comput. Syst. **94**, 620–633 (2019)
3. Liu, J., Wang, S., Zhou, A., et al.: Using proactive fault-tolerance approach to enhance cloud service reliability. IEEE Trans. Cloud Comput. **6**(4), 1191–1202 (2018)
4. Charity, T.J., Hua, G.C.: Resource reliability using fault tolerance in cloud computing. In: 2016 2nd International Conference on Next Generation Computing Technologies (NGCT), Dehradun, pp. 65–71 (2016)
5. Liu, J., Wang, S., Zhou, A., et al.: PFT-CCKP: a proactive fault tolerance mechanism for data center network. In: 2015 IEEE 23rd International Symposium on Quality of Service (IWQoS), Portland, pp. 79–80 (2015)
6. Beloglazov, A., Abawajy, J., Buyya, R.: Energy-aware resource allocation heuristics for efficient management of data centers for Cloud computing. Future Gener. Comput. Syst. **28**(5), 755–768 (2012)

DIN: A Bio-Inspired Distributed Intelligence Networking

Yufeng Li[1], Yankang Du[2($\boxtimes$)], Chenhong Cao[1], and Han Qiu[2]

[1] School of Computer Engineering and Science, Shanghai University,
Shanghai 200023, China
[2] National Digital Switching System Engineering and Technology R&D Center,
Zhengzhou 450002, China
duyankang@163.com

Abstract. Software-Defined Networking (SDN) is a promising method to simplify network management and facilitate network evolution. However, SDN is a logically centralized technology with global network-wide view. It faces the problem of scalability and reliability. In this paper, we propose a novel method termed as Distributed Intelligence Networking (DIN). DIN optimizes network management based on distributed coordination of multiple forwarding nodes like the coordination in bird flocking motion, it is a fully physically and logically distributed structure based on neighbor network-wide view. This architecture naturally has the advantage of scalability and reliability.

Keywords: Software-Defined Networking ·
Distributed Intelligence Networking · Neighbor network-wide view

1 Introduction

Software Defined Networking (SDN) offers the chance to introduce Artificial Intelligence (AI) to reduce operational costs and to improve performance and user experience [1]. Its main features are the centralized global network-wide view, programmability, and separation of the data plane and control plane. Now SDN has become a promising method and gets great attention.

As the network size increases, the centralized controller faces to many challenges [2, 3]. Firstly, there will be a bottleneck in the real-time communication capability of the controller with the network scale expanding. Secondly, for large scale networks, the controller needs to be able to process millions of flows per second without sacrificing the service quality. Thirdly, this control plane usually encounters the risk of single point failure. This will disconnect the controller and the forwarding elements.

Many research works have been done to overcome those issues [4–6]. DevoFlow [4] and Software-Defined Counters (SDC) [5] reduced the overhead of the control plane by delegating some work to the forwarding elements. Maestro [6] makes efforts on designing and deploying high performance controllers to increase the performance of the control plane.

X. Tang et al. (Eds.): NPC 2019, LNCS 11783, pp. 333–337, 2019.
https://doi.org/10.1007/978-3-030-30709-7_29

In this paper, we attempt to present a novel architecture termed as the Distributed Intelligence Networking (DIN). DIN optimizes network intelligence based on distributed coordination of neighbor forwarding elements, and acts like the flocking motion such as bird flocking in nature [7]. This can achieve global network coordination and optimization with distributed controller and distributed neighbor network-wide view.

2   Flocking Motion Introduction

Flocking motion exists in the nature in the form of flocks of birds, schools of fish, and so on. The study of consensus problem in flocking motion offers an alternative way to design the intelligent, coordinated and complex systems.

As shown in Fig. 1, DIN has the same physically distributed architecture as SDN flat architecture. However, its logical view is quite different from the SDN flat architecture as described in Onix [8] and HyperFlow [9]. For our proposed DIN, it does not need to maintain the global network-wide view. Each controller in DIN only needs to perceive the state of the neighbor controller. The DIN controller makes decisions based on the coordination control protocol, and adjusts its resources to achieve global coordinated behavior to realize intelligent improvement of network management, performance optimization and service quality.

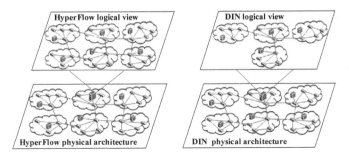

Fig. 1. Physical architecture and logical view in HyperFlow and DIN.

3   Architecture of DIN

There are two conditions to apply flocking motion to the Internet intelligence network. First, network individuals should have basic intelligent attributes. The network individuals can perceive the neighbor information, make decisions according to the control protocol, and make independent adjustments according to the decisions. Second, the network reaches convergence by coordinating among individuals based on the control protocol. This would show global intelligent behavior.

Figure 2 illustrates the architecture of the DIN forwarding node. The node architecture is divided into data-plane, control-plane and application-plane. The data-plane of DIN nodes, like the data-plane of SDN forwarding node, performs basic store and

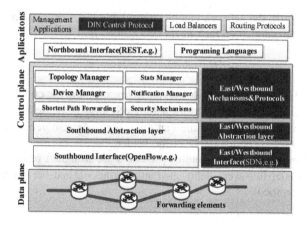

Fig. 2. DIN forwarding node architecture based on SDN

forwarding function. Besides this, it also has the data-plane adjustment function such as rate-limiting, traffic scheduling and so on.

The nodes of DIN can be implemented based on the existing SDN node structure. The DIN network works as follows: DIN node perceives the Coordination Situation (CoS) which means the current node situation, node resource and service request. By doing so, a neighbor network-wide view can be formed. Then it sends the CoS to the DIN control protocol. The DIN control protocol makes service decisions and issues adjustment commands to the data plane through NOS. The data plane adjusts resources for network services as ordered.

4 An Example for the DIN Protocol

A general Internet end-to-end service path is shown in Fig. 3, in which IP flows pass through forwarding nodes hop by hop and get a certain service.

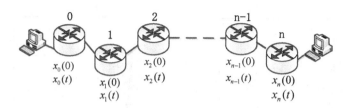

Fig. 3. A general end-to-end service path in network

A leader-followers multi-agent system is a particularly interesting topic in distributed multi-agent coordination theory, where the leader is an agent whose motion is followed by the other agent.

If we take the node labeled 0 in Fig. 3 as the leader, the other nodes indexed by $1, \ldots, n$ are the followers.

According to [10], the network can be controllable if it works under a coordination protocol with the dynamics of each agent

$$\dot{x}_i(t) = u_i(t), \quad i = 1, \ldots, n, \tag{1}$$

In protocol (1), the neighbor node service delay $x_j(t)$ and the communication delay T_{ji} serve as the node i perception information CoS, $u_i(t)$ is the decision rule, thus the node i conforming to DIN architecture can drive the node resource to adjustment and reach consensus on the service delay of the leader.

In this section, two example control protocols are designed based on DIN architecture, one is leader-followers delay guarantee protocol, and the other is leaderless delay guarantee protocol.

5   Simulation Results

As presented in Fig. 3, considering a leader-followers DIN network with control protocol (1), the topology is structured by

$$0 \rightarrow 1 \leftrightarrow 2 \leftrightarrow 3 \leftrightarrow 4 \leftrightarrow 5 \leftrightarrow 6$$

Node 0 works as leader with $x_0(0) = 25$ which is equal to R. It is assumed that the couplings gain is identical and equal to 5. All communication delays are set to 0.2. The initial states $x_i(0)(i = 1, 2, \ldots, 6)$ of the system are different, and each is chosen randomly.

Simulation results of (1) are plotted in Fig. 4. It is shown that under the two conditions without and with communication delay among nodes, all nodes along the path tend to reach a group consensus and remain stable at the initial service delay of the leader. That is to say, with the leader node providing the service level $x_0(0)$ required by the user, each following DIN node will automatically adjust under the protocol (1), and tend to converges to $x_0(0)$ of the leader.

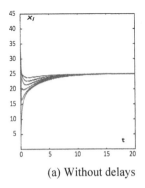

(a) Without delays

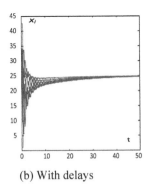

(b) With delays

Fig. 4. Simulation results for leader-followers delay guarantee protocol without and with communication delays.

We can also see that the convergence speed that Fig. 4(b) is obvious lower than the one of Fig. 4(a). This is mainly caused by communication delay among nodes.

Certainly, when the stable state of the group is broken at node i with x_i deviating the stable value $x_0(0)$, the protocol (1) will start a new round of coordination and tend to reach the group consensus again.

6 Conclusion

Inspired by the flocking motion in nature, this paper makes a preliminary exploration of the distributed coordination and control mechanism of Internet resources. We propose an intelligent network resource coordination and fitting scheme called DIN.

The characteristic and advantage of DIN is that it can keep distributed structure in physics and logic while introducing intelligent genes to the network. At present, for large-scale application, SDN moves from single controller to multi-controller, but the scalability and reliability problems caused by the global network-wide view and centralized control of SDN have not been solved well. The DIN proposed in this paper can be implemented completely based on the SDN node structure and specification, and can be deployed jointly with SDN, which can provide a new solution for improving the scalability and reliability of SDN.

Acknowledgments. This research was supported by the Research and Development Program in Key Areas of Guangdong Province under Grants 2018B010113001, and the National Natural Science Foundation of China under Grants 61502528.

References

1. Kim, H., Feamster, N.: Improving network management with software defined networking. IEEE Commun. Mag. **51**(2), 114–119 (2013)
2. Xie, J., Guo, D., Hu, Z., et al.: Control plane of software defined networks: a survey. Comput. Commun. Rev. **67**(1), 1–10 (2015)
3. Yeganeh, S., Tootoonchian, A., Ganjali, Y.: On scalability of software-defined networking. IEEE Commun. Mag. **51**(2), 136–141 (2013)
4. Curtis, A.R., Mogul, J.C., Tourrilhes, J., et al.: DevoFlow: scaling flow management for high performance networks. Comput. Commun. Rev. **41**(4), 254–265 (2011)
5. Mogul, J.C., Congdon, P.: Hey, you darned counters! Get off my asic!. In: Proceedings of the First Workshop on Hot Topics in Software Defined Networks, New York, USA, pp. 25–30 (2012)
6. Cai, Z., Cox, A.L., Ng, T.S.E.: Maestro: A System for Scalable OpenFlow Control. Rice University, Technical report (2011)
7. Vicsek, T., Zafeiris, A.: Collective motion. Phys. Rep. **517**(3–4), 71–140 (2012)
8. Koponen, T., Casado, M., et al.: Onix: a distributed control platform for large-scale production networks. In: Proceedings of the 9th USENIX Conference on Operating Systems Design and Implementation, Berkeley, USA, pp. 1–6 (2010)
9. Tootoonchian, A., Ganjali, Y.: HyperFlow: a distributed control plane for OpenFlow. In: Proceedings of the 2010 Internet Network Management Conference on Research on Enterprise Networking, USA (2010)
10. Rahmani, A., Ji, M., Mesbahi, M., et al.: Controllability of multi-agent systems from a graph-theoretic perspective. SIAM J. Control. Optim. **48**(1), 162–186 (2009)

A DAG Refactor Based Automatic Execution Optimization Mechanism for Spark

Hang Zhao[1], Yu Rao[1], Donghua Li[1], Jie Tang[1(✉)],
and Shaoshan Liu[2]

[1] South China University of Technology University,
Guangzhou 510641, People's Republic of China
cstangjie@scut.edu.cn
[2] PerceptIn, Fremont, USA

Abstract. In today's big data era, traditional disk-based MapReduce big data framework encountered bottlenecks due to its lower memory utilization and inefficient orchestration of complex tasks. With the advantage of fully use memory resources, Spark provides a lot of data manipulate operators and use DAG to express the dependences. Spark split entire job to multi-stage according to DAG and schedule them in a distributed execution environment, which better adapted to the new characteristic of big data processing. However, Spark didn't consider the resource requirement of different operators and schedule them indiscriminately, which could cause load imbalances on different nodes in the cluster and cause some node become bottlenecks due to its extraordinary resource consumption. In the past, solve this problem need developers to have a lot of experience of Spark and write code sophisticated. In this paper, we proposed a DAG refactor based automatic execution optimization mechanism for Spark. The experimental results show that the DAG refactor mechanism can greatly improve Spark performance by up to 8.8X without misinterpretation of original program semantics.

keywords: Big data · Spark · Semantic DAG · DAG refactor

1 Introduction

With the development of information technology, massive data has been generated every day [1]. Traditional big data processing framework, such as Hadoop, use disk to store intermediate data, always encounter disk I/O bottleneck. Spark use RDD (Resilient Distributed Dataset) to store intermediate data and use Linages to archive fault-tolerate [2], which could archive significant performance improvement compared to Hadoop. Therefore, a lot of applications have implemented in Spark, such as Deep Learning [3], smart city [4, 5], and automatically vehicle.

Spark provides a lot of data manipulate operators and use DAG (Directed Acyclic Graph) to express the dependences, then Spark split entire job to multi-stage according to DAG and schedule them in a distributed execution environment. However, extensive experimentations show that different operators have different running characteristic, while Spark didn't consider the resource requirement of different operators and

© IFIP International Federation for Information Processing 2019
Published by Springer Nature Switzerland AG 2019
X. Tang et al. (Eds.): NPC 2019, LNCS 11783, pp. 338–344, 2019.
https://doi.org/10.1007/978-3-030-30709-7_30

schedule them indiscriminately. In this paper, we proposed a DAG refactor based automatic execution optimization mechanism for Spark. This mechanism could reconstruct the DAG of original program automatically into another structure with higher execution efficiency. With the automatic DAG refactor, the overall system resource utilization can be effectively improved and task execution time can be greatly reduced.

2 DAG in Spark

As shown in Fig. 1, in Spark, *DAGScheduler* divides the Job into several stages according to the wide or narrow depends of RDD. The *DAGScheduler* packages each stage into a *TaskSet* and hands it over to *TaskScheduler*, which will dispatch task to *Executor*s. During scheduling, the *SchedulerBackend* is responsible for providing available resources.

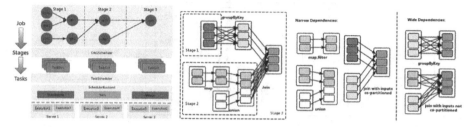

Fig. 1. The architecture of Spark.   **Fig. 2.** Wide vs narrow dependencies

Spark provides a rich set of data manipulation operators to build complex processing logic, them can be divided into two categories: (1) Transformation operator, mainly used to describe the conversion relationship between RDD, such as map, filter, and union in the Fig. 2. (2) Action operator, which will trigger Spark to submit job, such as groupByKey and join in Fig. 2.

3 Observation on Spark Operator

Spark offers great flexibility to application developers by its rich operators set. However, there still lacks theoretical and experimental research on Spark operators. In this paper, we explore different characteristics of operators in Spark through a large number of experiments and get the observation below: (1) Spark operators can be classified into computation intensive operators and Shuffle intensive operators according to the characteristics of operators. (2) Performance of application varies greatly when different operators contributed to the same semantic. (3) Performance of application varies greatly when execution sequence of operators changes. (4) Data volume decides the execution performance and usage of each operator.

4   Automatic DAG Refactor Mechanism

In scheduling, Spark only considers the narrow-dependency or wide-dependency of operators in stages division. It is prone to overlook different resource requirements and runtime feature of operators. Thus resulted schedule decisions can not fully mine the in-memory computing potential. In this paper, we propose a Spark automatic optimization framework based on DAG refactor to take care of such sophisticated work and make execution more efficient automatically.

4.1   System Design

The automatic DAG refactor mechanism proposed is shown as Fig. 3. The mechanism can reconstruct DAG by modification of RDD dependency and the user-defined execution function. It mainly includes a general DAG refactor module and an extensible DAG refactor rule library.

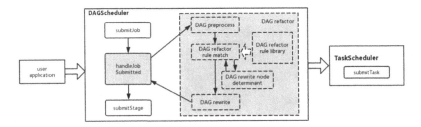

Fig. 3. System architecture diagram

4.2   DAG Refactor Rules Library

In this paper, we extracted the characteristics of different operators in Spark running process through a large number of experiments. Then, conclude the replacement rules of Spark operator, and form a Spark operator replacement rule library through the analysis of characteristics of Spark operator and semantic analysis of DAG. All replacement rules are shown in Table 1.

Table 1. Replacement rules

Rules	Scope of application
map -> mapPartitions	User function overhead too large
foreach -> foreachPartitions	User function overhead too large
groupByKey + map -> reduceByKey	Shuffle data too large
groupByKey + mapPartitions -> reduceByKey	Shuffle data too large
reduce -> treeReduce/treeAggregate	Driver side performance bottleneck
aggregate -> treeAggregate	Driver side performance bottleneck
reduce-side join -> map-side join	Has a broadcastable table
map + filter -> filter + map	Data reduction after filter
filter -> filter + coalesce	Data skew occurs after Filter
union + distinct -> distinct + union + distinct	Very much duplicate data

5   Implementation in Spark

5.1   DAG Refactor

The implementation of proposed DAG refactor mechanism mainly by modifying the function *handleJobSubmitted* in the *DAGScheduler* to handle job submission, and the job submitted by the user can be extracted. Then call the *DAGRefactor* component to refactor user job, form a refactored job, and finally replace the original job with refactored job, and then continue to execute by the Spark.

5.2   DAGRefactor Design and Implementation

The DAGRefactor class diagram is shown in Fig. 4. Origin_job and refactored_job store the original job and refactored job after refactor respectively; adjacency_table and inverse_adjacency_table are intermediate variables of running process, which are used to store the adjacency table and inverse adjacency table of DAG; rule_list lists definable refactor rules, it is convenient to add more refactor schemes later.

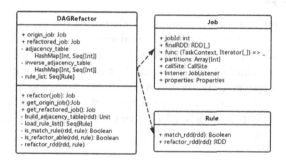

Fig. 4. DAGRefactor class diagram

Depending on their functionality, DAGRefactor provides modules such as DAG Analyse, Rules Match, Refactored Check, DAG Rewrite, and DAG Refactor Rules Library, and provides separate function interfaces for different modules.

6   Experiments and Evaluation

6.1   Evaluation Environment

By running Spark in real environment, the improvement of DAG refactor strategy proposed in this paper can be evaluated and analyzed. Spark is based on its 2.3.0 version and we set 4 Spark worker instances, each have 4 CPU cores and 6 GB memory.

6.2 SQL

Spark-SQL is a typical scenario of *foreach* operator. First, we made experiment by using *foreach* operation to inserting 1,000,000 rows and 100,00 rows respectively. Next, we empty the database and proposed framework make refactor by using *foreachPartitions* operator. Table 2 shows experimental results, compared with *foreach* implementation, *foreachPartitions* gives 8.8X speedup at best meanwhile consumes less bandwidth, and no data loss.

Table 2. Comparison of *foreach* and *foreachPartitions*

	Executor * core	Time (s)	CPU	Bandwidth	Data loss
foreach (1000K)	4 * 4	457.494	10%	0–50 Mbps	79%
foreachPartitions (1000K)	4 * 4	52.089	5%	15 Mbps	0%
foreach (100K)	2 * 1	260.193	5%	6 Mbps	0%
foreachPartitions (100K)	2 * 1	54.054	5%	2 Mbps	0%

6.3 Data Aggregation

Data aggregate summarizes all records of RDDs in two phases. As shown in Fig. 5, 30 s later, the limited computing resources of driver results in slow execution and diver becomes a performance bottleneck, resulting in total execution time of up to 55 s.

Without changing other variables, proposed framework refactor *aggregate* operator into *treeAggregate* operator. As shown in Fig. 6, *treeAggregate* operator adopted a tree-like aggregation strategy in the second phase, thus keeping CPU utilization at a high level consistently. It took only 27 s to complete all tasks and reduced execution time by 51%.

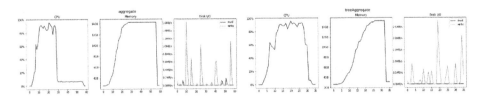

Fig. 5. *aggregate* resource consumption   **Fig. 6.** *treeAggregate* resource consumption

6.4 Merge and Deduplication

We tested the refactor improvement of merging operations and deduplication operations of two RDDs, i.e. *A.union(B).distinct()* is rewritten to *A.distinct().union(B.distinct ()).distinct()*. Figures 7 and 8 shows the resource consumption respectively. It can be seen that by refactor, total running time of program is reduced from 51 s to 47 s, and performance is improved by 7.8%.

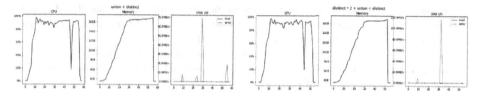

Fig. 7. *union + distinct* resource consumption **Fig. 8.** *distinct + union + distinct*   resource consumption

7 Relation Works

Spark provides rich operators and uses them to organize computational logic, but research on Spark operators is still relatively rare. [6] studied the input-output ratio of different operators, to estimate the size of intermediate data in computing process. [7] talked about incremental calculation, studied the difference between different operators when data increments. According to the difficulty of operator multiplexing, its divided operators into two types: indirect multiplexing which can be directly multiplexed and deduced by predicate. At the same time, implemented FQ-Tree-based reusable fragment matching and DAG refactor. For scheduling shuffle class operators, [8] analyzed memory scheduling algorithm in Spark Shuffle phase. Considering the un-balanced memory requirements of different task, fair memory allocation scheduling algorithm can not meet the demand well, proposed an adaptive scheduling algorithm that could dynamically adjust the memory allocation of tasks based on overflow historically.

8 Conclusions

In this paper, the different characteristics of different operators in Spark are studied by experiment. With this observation, we design and implement a DAG refactor based automatic execution optimization mechanism for Spark. With a large number of experimental analysis of operators in Spark, we summarize several rules for DAG refactor, which can directly optimize the calculation of related operators. Experiments show that the proposed DAG refactor based automatic execution optimization mechanism can improve Spark performance up to 8.8X by DAG refactor without destroying original program semantics.

References

1. Pempek, T.A., Yermolayeva, Y.A., Calvert, S.L.: College students' social networking experiences on Facebook. J. Appl. Dev. Psychol. **30**(3), 227–238 (2009)
2. Zaharia, M., Chowdhury, M., Das, T., et al.: Resilient distributed datasets: a fault-tolerant abstraction for in-memory cluster computing. In: Usenix Conference on Networked Systems Design and Implementation, p. 2. USENIX Association (2012)
3. Hamilton, M., Raghunathan, S., Matiach, I., et al.: MMLSpark: Unifying Machine Learning Ecosystems at Massive Scales. arXiv preprint arXiv:1810.08744 (2018)

4. Agafonov, A., Yumaganov, A.: Short-term traffic flow forecasting using a distributed spatial-temporal k nearest neighbors model. In: 2018 IEEE International Conference on Computational Science and Engineering (CSE), pp. 91–98. IEEE (2018)

5. Nasiri, H., Nasehi, S., Goudarzi, M.: A survey of distributed stream processing systems for smart city data analytics. In: Proceedings of the International Conference on Smart Cities and Internet of Things, p. 12. ACM (2018)

6. Bae, J., Jang, H., Jin, W., et al.: Jointly optimizing task granularity and concurrency for in-memory mapreduce frameworks. In: 2017 IEEE International Conference on Big Data (Big Data), pp. 130–140. IEEE (2017)

7. KanJing: The research of key techniques of incremental computing for DAG-based framework. Beijing University of Technology (2017)

8. Chen, Y.: Analysis and optimization of memory scheduling algorithm of spark shuffle. Zhejiang University (2016)

BTS: Balanced Task Scheduling Strategy Based on Multi-resource Prediction and Allocation in Cloud Environment

Yongzhong Sun[1], Kejiang Ye[1(✉)], Wenbo Wang[2], and Cheng-Zhong Xu[3]

[1] Shenzhen Institutes of Advanced Technology, Chinese Academy of Sciences,
Shenzhen 518055, China
{yz.sun1,kj.ye}@siat.ac.cn

[2] Khoury College of Computer Sciences, Northeastern University,
Seattle, WA 98109, USA
wang.wenbo@husky.neu.edu

[3] Faculty of Science and Technology, University of Macau, Macau, China
czxu@um.edu.mo

Abstract. Cloud computing is a new computing paradigm equipped with large-scale servers to satisfy diverse application demands. Managing and scheduling various application tasks on cloud servers is very challenging. In this paper, we propose a Balanced Task Scheduling (BTS) strategy by combining multi-objective particle swarm optimization and time series prediction model to achieve a better load balance among cloud servers. We not only consider the current server load which is used by most existing scheduling methods, but also take the future load change prediction into account. Experiments on the public Alibaba cluster trace with 1310 servers show that the proposed strategy can achieve a more balanced resource utilization.

Keywords: Load balancing · Workload prediction · Task scheduling

1 Introduction

Despite the adoption of various resource management systems that use typical scheduling algorithms based on instantaneous resource availability during the scheduling, the ability to reliably distribute application tasks among cloud servers remains deficient. According to the analysis of Alibaba cluster data [3], cloud servers have a significant spatial imbalance and time imbalance. Due to the limits of existing task scheduling methods, this paper proposes a balanced task scheduling strategy based on multi-resource prediction and allocation to achieve a better load balance among cloud servers.

The main contributions of this paper are: (i) According to the load feedback sampled periodically, we forecast the future load of servers through a time series prediction model - Prophet [7]. Then we use a multi-objective particle swarm

© IFIP International Federation for Information Processing 2019
Published by Springer Nature Switzerland AG 2019
X. Tang et al. (Eds.): NPC 2019, LNCS 11783, pp. 345–349, 2019.
https://doi.org/10.1007/978-3-030-30709-7_31

optimization algorithm - OMOPSO [8] to determine the mapping relationship between the tasks and the servers from the predicted load, actual load, and load threshold. (ii) We use the Alibaba cluster trace with 1310 servers as the test dataset to evaluate the prediction accuracy and also perform the load balance analysis to verify the effectiveness of the task scheduling strategy. Experimental results show that the proposed strategy can achieve a more balanced CPU and memory utilization.

2   Problem Description

Definition 1. Server and its resource utilization vector. The data center has n servers $S_i, i \in [1, n]$. Vector $\overrightarrow{S_i^{cur}} = (S_{i,CPU}^{cur}, S_{i,Mem}^{cur})$ represents the current resource utilization of different servers in the data center, $S_{i,CPU}^{cur}$ is the current CPU utilization of server S_i, $S_{i,Mem}^{cur}$ is the current memory utilization of server S_i. Vector $\overrightarrow{S_i^{nxt}} = (S_{i,CPU}^{nxt}, S_{i,Mem}^{nxt})$ represents the predicted resource utilization of different servers in the data center at the next time.

Definition 2. Batch task and its resource occupancy rate. The number of batch tasks that need to be deployed to the server at a given time is m, $B_j, j \in [1, m]$ represents a batch task, $B_{j,CPU}$ is the CPU requirement of B_j, $B_{j,Mem}$ is the memory requirement of B_j.

Definition 3. Batch tasks to servers deployment matrix. The deployment relationship between the batch tasks and servers can be expressed as a matrix $E = (e_{ij})_{n \times m}$. When batch task B_j is deployed to server S_i, $e_{ij} = 1$, otherwise $e_{ij} = 0$.

Definition 4. Server and its current utilization estimate. For server S_i, its current CPU utilization estimate is the sum of $S_{i,CPU}^{cur}$ and the CPU resource requested for all batch tasks deployed on it: $EST_{i,CPU}^{cur} = S_{i,CPU}^{cur} + \sum_{j=1}^{m} e_{ij}B_{j,CPU}$. In the same way, its current memory utilization estimate is $EST_{i,Mem}^{cur} = S_{i,Mem}^{cur} + \sum_{j=1}^{m} e_{ij}B_{j,Mem}$.

Definition 5. Server and its next-period utilization estimate. Assume that the batch tasks currently deployed are not finished in the next period. For server S_i, its next-period CPU utilization estimate $EST_{i,CPU}^{nxt}$ is the sum of $S_{i,CPU}^{nxt}$ and the CPU resource requested for all the batch tasks currently deployed on it: $EST_{i,CPU}^{nxt} = S_{i,CPU}^{nxt} + \sum_{j=1}^{m} e_{ij}B_{j,CPU}$. Its next-period memory utilization estimate $EST_{i,Mem}^{nxt} = S_{i,Mem}^{nxt} + \sum_{j=1}^{m} e_{ij}B_{j,Mem}$.

Problem Model. By introducing the above definitions, the server load balancing problem can be modeled as a multi-objective optimization problem, whose objective functions:

$$min(K_{Res}^{cur}) = min\left(\sqrt{\frac{1}{n}\sum_{i=1}^{n}\left(EST_{i,Res}^{cur} - \frac{1}{n}\sum_{i=1}^{n}EST_{i,Res}^{cur}\right)^2}\right), \quad (1)$$

$$Res \in \{CPU, Mem\}$$

K_{Res}^{cur} is the standard deviation of the current resource utilization estimate for servers of the data center.

The constraint functions are as follows:

$$\sum_{i=1}^{n} e_{ij} = 1, j = 1, 2, ..., m \tag{2}$$

indicating that each batch task can only be deployed on one server.

$$EST_{i,Res}^{cur} = S_{i,Res}^{cur} + \sum_{j=1}^{m} e_{ij}B_{j,Res} < T_{i,Res} \tag{3}$$

$$EST_{i,Res}^{nxt} = S_{i,Res}^{nxt} + \sum_{j=1}^{m} e_{ij}B_{j,Res} < T_{i,Res} \tag{4}$$

represent that when the batch tasks are deployed on the servers, the current and next-period resource utilization cannot exceed the server resource threshold. The resource threshold of server S_i is $T_{i,Res}$.

3   Experimental Evaluation

The cluster data released by Alibaba in 2017 is used as the experimental data. It contains 12-h trace information of 1,310 machines, including machine resource usage and batch task workload.

We use the logistic regression model of Prophet for prediction. The model parameters are as follows: capacity is 100%, changepoint_range is 100%, changepoint_prior_scale is 0.2, and n_changepoint is automatically set by the model. The sliding window mechanism was applied to predict the workload and the length of the window is set to 8.

We first verify the prediction accuracy of the proposed method. Figure 1 shows the actual load and predicted load of a server (id = 600) in the sampling period. The figure shows that the prediction can fit the fluctuation of the machine load very well.

Then, we evaluate the effectiveness of balanced scheduling strategy. We select 4 load sampling time periods from Alibaba cluster data, using the first 5,000 batch tasks in all servers for rescheduling in each time period.

We find the solution to problem (1) by the OMOPSO algorithm under constraints (2)(3)(4). By tracking 4 load sampling timestamps, we get the actual resource utilization $S_{i,CPU}^{cur}$ and $S_{i,Mem}^{cur}$ of the machines, and we get the predicted value $S_{i,CPU}^{nxt}$ and $S_{i,Mem}^{nxt}$ of future resource utilization through the Prophet model. The resource utilization threshold $T_{i,CPU}$ and $T_{i,Mem}$ of server S_i are set to 70% and 90% respectively. The parameters for particle swarm optimization are set as follows: $w = rand(0.1, 0.5)$, $c_1, c_2 = rand(1.5, 2.0)$, $r_1, r_2 = (0.0, 1.0)$, $polupationSize = 50$ and $maxEvalution = 1000$.

The load balancing effect is tested by calculating the standard deviation of the load of cloud servers, and the results are shown in Table 1, where K_{CPU}^{orig} and

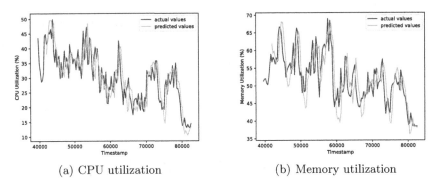

(a) CPU utilization   (b) Memory utilization

Fig. 1. Actual and predicted load comparison of machine id 600

K_{Mem}^{orig} represent the standard deviation of the CPU load and memory load of the machines when the original scheduling strategy is adopted. In the case of using the proposed scheduling strategy, the load balance of each experimental group is improved compared with the original scheduling strategy.

Table 1. Load balancing effect of two scheduling strategies

Group	K_{CPU}^{cur}	K_{CPU}^{orig}	ΔK_{CPU}	K_{Mem}^{cur}	K_{Mem}^{orig}	ΔK_{Mem}
A	6.899	7.255	**−0.356**	10.376	11.179	**−0.803**
B	6.409	6.816	**−0.407**	11.715	12.351	**−0.636**
C	7.140	7.969	**−0.829**	10.726	11.195	**−0.469**
D	7.167	7.520	**−0.353**	12.265	12.884	**−0.619**

4   Related Work

The intelligent algorithms such as simulated annealing algorithm [9], genetic algorithm [6] and particle swarm optimization [4] are powerful in solving the task scheduling problem under multi-resource constraints. LD et al. [1] propose a dynamic load balancing algorithm HBB-LB based on bees' foraging behavior, aiming to achieve load balancing across VMs to maximize throughput. The priority of the task in the waiting sequence in the node is considered to minimize the waiting time of the task in the queue. Li et al. [2] propose a cloud task scheduling policy based on Load Balancing Ant Colony Optimization (LBACO) algorithm. The algorithm selects the best resource to perform a task based on the resource state and the size of a given task in the cloud environment. It balances the overall system and minimizes the completion time for a given set of tasks. Ramezani et al. [5] propose a Task-based System Load Balancing method using Particle Swarm Optimization (TBSLB-PSO) that achieves system load balancing by only transferring extra tasks from an overloaded VM instead of migrating the entire overloaded VM. It significantly reduces the time taken for the load balancing process.

5   Conclusion

In order to solve the load balancing problem, this paper proposes a task scheduling strategy based on the combination of multi-objective particle swarm optimization and time series prediction model. The goal of this strategy is to improve load balancing among the cloud servers, and the impact of the current and future load of the servers on task scheduling is also considered. The experiments based on Alibaba cluster trace with 1310 servers show that this scheduling strategy can effectively achieve the goal of reasonable task allocation with a more balanced resource utilization.

Acknowledgment. This work is supported by the National Key R&D Program of China (No. 2018YFB1004804), National Natural Science Foundation of China (No. 61702492), Shenzhen Discipline Construction Project for Urban Computing and Data Intelligence, and Shenzhen Basic Research Program (No. JCYJ20170818153016513).

References

1. Ld, D.B., Krishna, P.V.: Honey bee behavior inspired load balancing of tasks in cloud computing environments. Appl. Soft Comput. **13**(5), 2292–2303 (2013)
2. Li, K., Xu, G., Zhao, G., Dong, Y., Wang, D.: Cloud task scheduling based on load balancing ant colony optimization. In: 2011 Sixth Annual China Grid Conference, pp. 3–9. IEEE (2011)
3. Lu, C., Ye, K., Xu, G., Xu, C.Z., Bai, T.: Imbalance in the cloud: an analysis on Alibaba cluster trace. In: 2017 IEEE International Conference on Big Data (Big Data), pp. 2884–2892. IEEE (2017)
4. Ramezani, F., Lu, J., Hussain, F.: Task scheduling optimization in cloud computing applying multi-objective particle swarm optimization. In: Basu, S., Pautasso, C., Zhang, L., Fu, X. (eds.) ICSOC 2013. LNCS, vol. 8274, pp. 237–251. Springer, Heidelberg (2013). https://doi.org/10.1007/978-3-642-45005-1_17
5. Ramezani, F., Lu, J., Hussain, F.K.: Task-based system load balancing in cloud computing using particle swarm optimization. Int. J. Parallel Program. **42**(5), 739–754 (2014)
6. Sharma, N.K., Reddy, G.R.M.: Novel energy efficient virtual machine allocation at data center using genetic algorithm. In: 2015 3rd International Conference on Signal Processing, Communication and Networking (ICSCN), pp. 1–6. IEEE (2015)
7. Sierra, M.R., Coello Coello, C.A.: Improving PSO-based multi-objective optimization using crowding, mutation and $\in$-dominance. In: Coello Coello, C.A., Hernández Aguirre, A., Zitzler, E. (eds.) EMO 2005. LNCS, vol. 3410, pp. 505–519. Springer, Heidelberg (2005). https://doi.org/10.1007/978-3-540-31880-4_35
8. Taylor, S.J., Letham, B.: Forecasting at scale. Am. Stat. **72**(1), 37–45 (2018)
9. Yuan, H., Bi, J., Tan, W., Li, B.H.: Temporal task scheduling with constrained service delay for profit maximization in hybrid clouds. IEEE Trans. Autom. Sci. Eng. **14**(1), 337–348 (2017)

DAFL: Deep Adaptive Feature Learning for Network Anomaly Detection

Shujian Ji[1,2], Tongzheng Sun[1], Kejiang Ye[1(✉)], Wenbo Wang[3], and Cheng-Zhong Xu[4]

[1] Shenzhen Institutes of Advanced Technology, Chinese Academy of Sciences, Shenzhen 518055, China
{sj.ji,tz.sun,kj.ye}@siat.ac.cn
[2] University of Chinese Academy of Sciences, Beijing 100049, China
[3] Khoury College of Computer Sciences, Northeastern University, Seattle, WA 98109, USA
wang.wenbo@husky.neu.edu
[4] Faculty of Science and Technology, University of Macau, Macau, China
czxu@um.edu.mo

Abstract. With the rapid development of the Internet and the growing complexity of the network topology, network anomaly has become more diverse. In this paper, we propose an algorithm named Deep Adaptive Feature Learning (DAFL) for traffic anomaly detection based on deep learning model. By setting proper feature parameters θ on the neural network structure, DAFL can effectively generate low-dimensional new abstract features. Experimental results show the DAFL algorithm has good adaptability and robustness, which can effectively improve the detection accuracy and significantly reduce the detection time.

Keywords: Network anomaly detection · Deep learning · Feature learning

1 Introduction

Network attack is a serious problem in the Internet environment. With the rapid development of the Internet and the growing complexity of the network topology, network anomaly has become more diverse. Network anomaly detection is an effective way to deal with different network attacks [1].

Machine learning is a common method for anomaly detection in the network environment, such as Naive Bayes, Support Vector Machine and other shallow learning technologies [2,3]. Although these technologies have improved the detection accuracy to a certain extent, they also face some limitations. For example, expert knowledge is required for data processing, and a large amount of time is needed for data training. Recently, deep learning based methods [4–6] are proposed for anomaly detection due to the better feature learning ability. However,

© IFIP International Federation for Information Processing 2019
Published by Springer Nature Switzerland AG 2019
X. Tang et al. (Eds.): NPC 2019, LNCS 11783, pp. 350–354, 2019.
https://doi.org/10.1007/978-3-030-30709-7_32

they improved the detection accuracy, without taking into account the training time and execution time in high-speed networks.

In this paper, we propose an algorithm named Deep Adaptive Feature Learning (DAFL) which can utilize the feature learning ability of deep learning and the advantages of transfer learning. The contributions of this paper are summarized as follows: (i) the algorithm can determine the structure of the neural network according to the dimension of data. (ii) By combining deep learning with shallow machine learning, DAFL improves the classification performance of anomaly detection and greatly reduces the training time.

2　DAFL Algorithm

We design the DAFL algorithm to determine the number of layers of the network hidden layer and the number of neurons in each layer according to the dimensions of the input data, and construct a pre-trained learning model that can adapt to the dimension of data features, as shown in Algorithm 1.

Algorithm 1. Deep Adaptive Feature Learning

Require: training sample v, feature parameter θ, learning rate η, list N
Ensure: pre-train model $xW + b$
1: D = data dimension of v, layer number: $\mathrm{l} = \lceil D/5 \rceil$
2: initalize $n_1 = D$, calculate neurons number of each layer: $n_l = \lceil \theta * D \rceil$
3: **for** $i = 2$ to $l - 1$ **do**
4:　　neurons: $n_i = \lceil D/i^2 \rceil + \lceil \theta * D \rceil$, save n_i to N
5: **end for**
6: **for** $i = 1$ to l **do**
7:　　use N to build network with l-th layers and n_i neurons.
8:　　output layer: $S(x) = \frac{1}{1+e^{-x}}$
9: **end for**
10: **for all** v_i **do**
11:　　calculate the actual output of the neuron $v_i^{'}$
12:　　$\delta_k = v_i^{'}(1 - v_i^{'})(v_i - v_i^{'})$
13:　　hidden layer h error gradient: $\delta_h = v_h^{'}(1 - v_h^{'})W_{hk}\delta_k$
14:　　update the weights: $W_{ij} = W_{ij} + \Delta W_{ij}$, $\Delta W_{ij} = \eta O_i \delta_j$, update the bias: $b_j = b_j + \eta \delta_j$
15: **end for**

In order to balance the training speed and accuracy of the deep learning model, we design a feature parameter θ (from 0.1 to 1) as the control value in the DAFL algorithm to make the high hidden layer generate abstract features of different dimensions. As shown in Fig. 1, combining the deep network structure based on DAFL with different conventional shallow machine learning classifiers can be used as the detection model.

(a) Pre-train DBN model by using (b) Use the remain structure to gen-
DAFL   erate new feature

Fig. 1. The design of DAFL

3   Experiment

We conduct experiments with NSL-KDD [7] dataset to evaluate our proposed
algorithm. By comparing the performance of the original data and the data
processed by the DAFL algorithm on the classifier, we can verify the validity of
DAFL.

Table 1. Models performance in NSL-KDD dataset

Model	Accuracy	Precision	Recall	$F_1 - score$	Time(s)
Support Vector Machine (SVM)	97.26%	98.03%	96.06%	97.19%	91.73 s
DAFL SVM ($\theta = 0.8$)	**99.17%**	**99.49%**	**98.74%**	**99.15%**	**8.72 s**
K-Nearest Neighbors (KNN)	99.02%	99.34%	98.88%	99.15%	107.19 s
DAFL KNN ($\theta = 0.8$)	**99.21%**	**99.41%**	**98.89%**	**99.19%**	**13.62 s**
Logistic Regression (LR)	95.05%	95.26%	94.05%	94.98%	1.98 s
DAFL LR ($\theta = 0.8$)	**99.15%**	**99.36%**	**98.82%**	**99.13%**	**0.40 s**
Decision Tree (DT)	98.94%	98.98%	98.75%	98.93%	1.19 s
DAFL DT ($\theta = 0.8$)	**99.67%**	**99.72%**	**99.58%**	**99.65%**	**0.47 s**
Naive Bayes (NB)	88.82%	86.72%	89.24%	88.66%	0.10 s
DAFL NB ($\theta = 0.8$)	**98.77%**	**95.25%**	**99.75%**	**97.70%**	**0.06 s**

Experiments show that the classifier achieves the best result when the feature parameter is set to 0.8. Table 1 shows the changes of classifier performance
metrics when DAFL is applied to the classifier on the NSL-KDD dataset. It is
worth noting that the accuracy in the NB classifier increased from 88.92% to
98.77%, and the recall increased from 89.24% to 99.75%. In terms of detection
time, the classifier that has been processed by the DAFL algorithm has a significant reduction in detection time. The most obvious change is that the time of
the SVM classifier is reduced from 91.73 s to 8.72 s, and the detection time of the
KNN classifier is reduced from 107.19 s to 13.62 s. Figure 2 shows the accuracy
and time saving percentage on NSL-KDD.

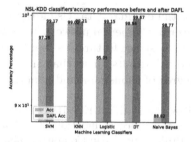
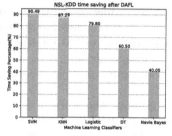

(a) Classifiers' accuracy perfor- (b) NSL-KDD time saving after
mance comparison on NSL-KDD   DAFL

Fig. 2. Accuracy and time saving on NSL-KDD

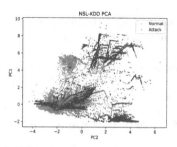
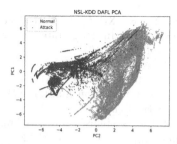

(a) PCA visualization for NSL-KDD (b) PCA visualization for NSL-
                                       KDD after DAFL

Fig. 3. PCA visualization for NSL-KDD before and after DAFL

We perform data scatter visualization by PCA method for normal traffic and
abnormal traffic in the dataset in Fig. 3. It is obvious that the DAFL algorithm
can separate the normal traffic and abnormal traffic.

4   Related Work

There are a lot of work on network anomaly detection. Ibrahimi *et al.* used classification algorithms such as linear discriminant analysis (LDA) and principal
component analysis (PCA) to classify abnormal network traffic [8]. Alrawashdeh
et al. used Restricted Boltzmann Machine (RBM) to perform unsupervised feature reduction [9]. Potluri *et al.* proposed an accelerated DNN structure for identifying network data anomalies [10]. Kang *et al.* proposed an intrusion detection
system based on deep neural network [11]. Our research group also proposed
different algorithms and tool for network anomaly detection [1, 4–6].

5   Conclusion

In this paper, we propose a DAFL algorithm for network anomaly detection that
can determine the number of hidden layers and the number of neurons in each

hidden layer according to the dimension of the original data. Using the idea of transfer learning, we remove the output layer of the neural network and use the residual structure to generate new data with abstract features as input of other machine learning classifiers. The experimental results show that the method achieves good results, and has a certain degree of robustness and adaptability.

Acknowledgment. This work is supported by the National Key R&D Program of China (No. 2018YFB1004804), National Natural Science Foundation of China (No. 61702492), Shenzhen Discipline Construction Project for Urban Computing and Data Intelligence, and Shenzhen Basic Research Program (No. JCYJ20170818153016513).

References

1. Lin, P., Ye, K., Xu, C.-Z.: NetDetector: an anomaly detection platform for networked systems. In: IEEE International Conference on Real-time Computing and Robotics. IEEE (2019)
2. Shon, T., Kim, Y., Lee, C., Moon, J.: A machine learning framework for network anomaly detection using SVM and GA. In: Proceedings from the Sixth Annual IEEE SMC Information Assurance Workshop, pp. 176–183. IEEE (2005)
3. Amor, N.B., Benferhat, S., Elouedi, Z.: Naive Bayes vs decision trees in intrusion detection systems. In: Proceedings of the 2004 ACM Symposium on Applied Computing, pp. 420–424. ACM (2004)
4. Lin, P., Ye, K., Xu, C.-Z.: Dynamic network anomaly detection system by using deep learning techniques. In: Da Silva, D., Wang, Q., Zhang, L.J. (eds.) CLOUD 2019. LNCS, vol. 11513, pp. 161–176. Springer, Cham (2019). https://doi.org/10.1007/978-3-030-23502-4_12
5. Zhu, M., Ye, K., Wang, Y., Xu, C.-Z.: A deep learning approach for network anomaly detection based on AMF-LSTM. In: Zhang, F., Zhai, J., Snir, M., Jin, H., Kasahara, H., Valero, M. (eds.) NPC 2018. LNCS, vol. 11276, pp. 137–141. Springer, Cham (2018). https://doi.org/10.1007/978-3-030-05677-3_13
6. Zhu, M., Ye, K., Xu, C.-Z.: Network anomaly detection and identification based on deep learning methods. In: Luo, M., Zhang, L.-J. (eds.) CLOUD 2018. LNCS, vol. 10967, pp. 219–234. Springer, Cham (2018). https://doi.org/10.1007/978-3-319-94295-7_15
7. NSL-KDD (1999). https://iscxdownloads.cs.unb.ca/iscxdownloads/NSL-KDD
8. Ibrahimi, K., Ouaddane, M.: Management of intrusion detection systems based-KDD99: analysis with LDA and PCA. In: 2017 International Conference on Wireless Networks and Mobile Communications (WINCOM), pp. 1–6. IEEE (2017)
9. Alrawashdeh, K., Purdy, C.: Toward an online anomaly intrusion detection system based on deep learning. In: 2016 15th IEEE International Conference on Machine Learning and Applications (ICMLA), pp. 195–200. IEEE (2016)
10. Potluri, S., Diedrich, C.: Accelerated deep neural networks for enhanced intrusion detection system. In: 2016 IEEE 21st International Conference on Emerging Technologies and Factory Automation (ETFA), pp. 1–8. IEEE (2016)
11. Kang, M.-J., Kang, J.-W.: Intrusion detection system using deep neural network for in-vehicle network security. PloS one **11**(6), e0155781 (2016)

SIRM: Shift Insensitive Racetrack Main Memory

Hongbin Zhang[1], Bo Wei[2], Youyou Lu[1], and Jiwu Shu[1(✉)]

[1] Tsinghua University, Beijing, China
{zhanghb,luyy09,shujw}@mail.tsinghua.edu.cn
[2] Hangzhou Dianzi University, Hangzhou, China
weibo@hdu.edu.cn

Abstract. Racetrack memory (RM) is a potential DRAM alternative due to its high density and low energy cost and comparative access latency with SRAM. On this occasion, we propose a shift insensitive racetrack main memory architecture SIRM. SIRM provides uniform access latency to upper system, which make it easy to be managed. Experiments demonstrate that RM can outperform DRAM for main memory design with higher density and energy efficiency.

Keywords: Racetrack memory · Shift insensitive · Main memory

1 Introduction

Recently, racetrack memory (RM), which is also known as domain wall memory (DWM), has attracted significant attention of researchers. Previous research has demonstrated that this approach can achieve ultra-high density by integrating multiple domains in a tape-like nanowire [1–3]. In addition, it provides SRAM-comparable access latency and high write endurance [4]. In racetrack memory, each cell has a similar architecture and access pattern to an STT-RAM cell. Each racetrack contains one or more access ports, and the data aligned with each port can be read/write by these accessing ports. In order to access other bits that are not aligned with a port, a shift operation must be performed to move these bits to the nearest access port. Owing to the comparable access latency of a SRAM, RM is a promising candidate for on-chip memory or caching [5]. Furthermore, data placement mechanisms for optimizing its access latency and energy cost have been researched intensively [6,7]. All of these studies have focused on how to significantly reduce the shift intensity, either at the system or compiler level, in order to leverage the density, shift latency, and energy cost.

There are also approaches to compose racetrack as the main memory. A shift-sense address mapping policy (SSAM) has been proposed for reducing shift operations in racetrack-based main memory systems [8]. SSAM significantly reduces shift intensity by employing a specified address mapping policy. However, SSAM also introduces complexity to the memory management and system

© IFIP International Federation for Information Processing 2019
Published by Springer Nature Switzerland AG 2019
X. Tang et al. (Eds.): NPC 2019, LNCS 11783, pp. 355–360, 2019.
https://doi.org/10.1007/978-3-030-30709-7_33

design because each request may have a different number of shift steps or length and the read/write latency varies.

In this work, we propose an improved solution for RM serving as main memory, shift insensitive racetrack main memory (SIRM) which successfully hide shift operations, and provide a uniform read/write interface to upper system. With its inherent advantages, SIRM can provide higher density, superior performance, and lower energy main memory compared with DRAM.

2   Motivation

In order to achieve higher bandwidth, modern commodity DRAM generally work according to DDR standard [11] in most computer architecture. The most important performance of DDR is data rate, or data burst latency. As previous work point out [1], the shift latency of RM array is related to the tape length, number of access ports and overlap layout. We propose a specific RM array with appropriate design and keep the shift latency equal or smaller than the DRAM burst cycle, then the shift latency will be covered by the DRAM burst time interval. So SIRM has a good scalability to fit for different DDR standard, which will be discussed in Sect. 4. We design the main memory architecture with multiple RM arrays. Together with SIAM which provides a pipeline mechanism to read data from adjacent racetrack, the shift operation will be hidden under the memory level and invisible to upper system. We design and implement SIRM according to this idea and testify its effectiveness.

3   The Shift Insensitive Racetrack Main Memory Design

3.1   Basic Array of RM

Prior work [1] proposes an organization with overlapped RM cells, called Macro Unit (MU), as a basic building block of RM array, as shown in Fig. 1(a). In this section, we mainly discuss the shift latency caused by MU structure and their application in SIRM.

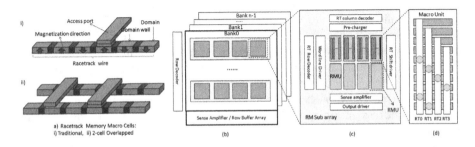

Fig. 1. The RM based main memory Architecture. (a) RM Macro Cells; (b) Overview of a bank; (c) A RM subarray; (d) A Macro Unit;

The basic array composed by MUs with different parameters has different shift latency. Theoretically, the longer the racetrack is, the longer shift latency is. The more access ports the racetrack has, she shorter shift latency is. And the more racetracks MU has, the longer the shift latency is because it needs more energy to sense the data out. According to [1], the area optimized solution for RM data array is MU-64-32-4. According to [8], the MU-64-32-4 and MU-64-16-4 have similar performance. This paper use MU-64-32-4 as basic MU to simulate main memory. MU-64-32-4 has 4 racetracks, 32 access ports and each racetrack has 64 storage domains and 8 access ports. In order to compose an appropriate RM data array which has comparable shift latency with DRAM burst cycle, we simulate several RM data array with different capacity and test their performance with NVsim [10]. According to the result, we choose 8 MB as the basic size of array because data shows that the 8 MB array cost 1.25 ns to read data and shift a step, which is equal to the burst cycle of DDR4-1600 MHz modules, which has a I/O bus of 1600 MHz. Theoretically, the RM array can provide the same read latency with DDR4-1600 and shift operations will be covered.

In this paper, according to Microns data sheet [9], we extend this model to simulate a 128 Mb RM memory chip with 16 banks, and 256B row buffer in 45 nm technology, which is the most advanced one we can get. The RM device level parameters in this paper are similar to the previous work [1].

3.2   Main Memory Architecture

A single rank of main memory contains multiple memory chips, which typically has 4, 8 or 16 data output pins [12]. As shown in Fig. 1, a RM based chip can be organized as three levels: bank, mat, and subarray. Bank is the top level unit, mat is the building block of bank, and subarray is the elementary structure. Subarray is composed by 8 RMU that is described in Fig. 2. One RMU is composed by 16 MU and a MU has a structure of 64-32-4 as described above. Then, one RMU has 4K bits and an subarray has 32K bytes. One bank has 256 subarray and 8M bits. One chip has 16 banks and 128M bits. Multiple RMUs in one subarray share the same corresponding periphery circuitry in order to shrink the energy cost.

3.3   Shift Insensitive Address Mapping

We propose SIAM policy to cover the shift operation through pipeline operation. We take the part of subarray to simplify the discussion. As is shown in Fig. 2, an array has eight RMUs in lateral and each RMU has 64 * 8 bytes in vertical. Data are numbered in cacheline which has 64bytes. For example, cacheline1 is numbered one and cacheline8 is numbered eight. As conventional, data is addressed sequentially along the RMU, as Fig. 2(1) shown. In SSAM, data are addressed across the RMUs and cacheline is spread across RMUs as Fig. 2(2) shown. In SIAM, data also is addressed in cacheline, but each of them is distributed along the diagonal across RMUs as Fig. 2(3) shown. Each cacheline is divided into eight parts and can be read through eight phase in pipeline. As described above, we design the racetrack array which has the exact shift latency equal or smaller than the burst time span in DDR4-1600. Thus, one cacheline can be read out through eight phase, costing the same latency just as DRAM read the data through eight bursts. Then, the shift operations are covered to the upper system.

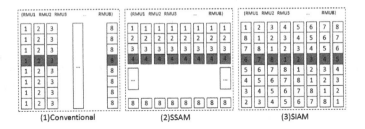

Fig. 2. Shift Insensitive Address Mapping.

4   Experimental Results

4.1   Experimental Setup

We evaluate the SIRM with a full system cycle accurate simulator gem5 [13]. For workload, we select 13 workloads from Parsec3 benchmarks.

4.2   Comparison of Experimental Results

(1) **Performance Evaluation:** We compare the normalized performance between DRAM, SSAM and SIRM in Fig. 3. The results are normalized to baseline of DRAM main memory. SIRM and SSAM has obvious shorter access time than DRAM because of the RM's access characteristic. The SIRM has similar or better access latency than SSAM in most benchmarks except *facesim* and *x264*. Mainly because these two program is data centric and the CPU read several words from memory each time, just as we discussed in the fourth section.

(2) **Energy Evaluation:** We compare normalized energy overhead between DRAM, SSAM and SIRM in Fig. 3. All results are normalized to baseline of DRAM main memory. It is obvious that SIRM reduced much energy than DRAM and SSAM in most benchmarks. For *facesim* and *x264*, the energy cost is similar with SSAM, mainly because SIRM uses more shift operation than other benchmarks.

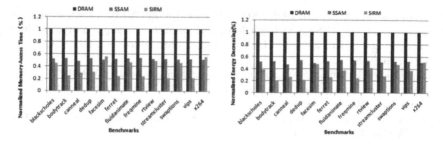

Fig. 3. Normalized memory access time and energy decrease.

5   Conclusion

Racetrack memory is attractive because of its high density and comparable read/write latency with SRAM, and non-volatility. It has the potential to be the replacement of DRAM. In this work, we presents a SIRM architecture based on specific RM array design, SIAM policy and special timing restriction in pipeline. In SIRM, the shift operations are covered and shift latency is insensitive to system level, which make RM memory easy to be managed by operating system. Experimental results show that in most applications SIRM can outperform DRAM or SSAM based racetrack main memory in performance and energy efficiency.

References

1. Zhang, C., et al.: Quantitative modeling of racetrack memory, a tradeoff among area, performance, and power. In: Proceedings of the 20th Asia and South Pacific Design Automation Conference, Chiba, Japan, January 2015, pp. 100–105 (2015)
2. Zhang, Y., et al.: Perspectives of racetrack memory for large-capacity on-chip memory: from device to system. IEEE Trans. Circ. Syst. **63**(5), 629–638 (2016)
3. Sun, G., et al.: From device to system: cross-layer design exploration of racetrack memory. In: Proceedings of the 18th Design, Automation and Test in Europe (DATE), Grenoble, France, 9–13 March 2015, pp. 1018–1023 (2015)
4. Parkin, S.S., Hayashi, M., Thomas, L.: Magnetic domain-wall racetrack memory. Science **320**(5873), 190–194 (2008)
5. Venkatesan R, et al.: TapeCache: a high density, energy efficient cache based on domain wall memory. In: Proceedings of the 2012 ACM/IEEE International Symposium on Low Power Electronics and Design, pp. 185–190. ACM (2012)
6. Mao, H., et al.: Exploring data placement in racetrack memory based scratch-pad memory. In: Proceedings of the 4th IEEE Non-Volatile Memory System and Applications Symposium, Hong Kong, China, August 2015, pp. 1–5 (2015)
7. Chen, X., et al.: Optimizing data placement for reducing shift operations on Domain Wall Memories. In: Design Automation Conference, pp. 1–6. ACM (2015)
8. Hu, Q., et al.: Exploring main memory design based on racetrack memory technology. In: Proceedings of the 26th ACM Great Lakes Symposium on VLSI (GLSVLSI), Boston, MA, USA, 18–20 May 2016, pp. 397–402 (2016)

9. Micron. 8Gb: x4, x8, x16 DDR4 SDRAM Description (2016). www.micron.com
10. Dong, X., et al.: NVSim: a circuit-level performance, energy, and area model for emerging nonvolatile memory. IEEE Trans. Comput. Aided Des. Integr. Circ. Syst. **31**(7), 994–1007 (2012)
11. https://en.wikipedia.org/wiki/DDR4_SDRA#Mcite_note-JESD79-3F-3
12. Jacob, B., et al.: Memory Systems: Cache, DRAM, Disk. Morgan Kaufmann, San Francisco (2010)
13. Binkert, N., Beckmann, B., Black, G., et al.: The gem5 simulator. SIGARCH Comput. Archit. **39**, 1–7 (2011)

PDRM: A Probability Distribution Based Resource Management for Batch Workloads in Heterogeneous Cluster

Jun Zhou[1,2,3], Dan Feng[1,2,3(✉)], and Fang Wang[1,2,3]

[1] School of Computer Science and Technology, Huazhong University of Science and Technology, Wuhan 430074, China
{JunZhou,dfeng,wangfang}@hust.edu.cn
[2] Wuhan National Laboratory for Optoelectronics, Wuhan 430074, China
[3] Key Laboratory of Information Storage System,
Engineering Research Center of data storage systems and Technology,
Ministry of Education of China, Wuhan, China

Abstract. Resource consumption prediction and dynamic resource provision based on historical consumption are common methods to improve cluster resource utilization, however they have to face the challenge of fluctuation in resource consumption for accurate prediction. We propose PDRM, an efficient resource management scheme based on resource consumption probability distribution for batch workloads to deal with this dilemma. Based on the common sense that the same type of tasks have similar resource consumption on the same node, we get the resource consumption probability distribution of each type of task to describe the fluctuations in its resource consumption. Based on the resource consumption distribution function, we can allocate resources precisely for tasks. Experimental results demonstrate that PDRM achieves good performance for various application in the heterogeneous cluster. PDRM can effectively improve resource utilization and reduce job completion time.

Keywords: Resource management · Big data · Gaussian distribution · Heterogeneous

1 Introduction

Low resource utilization is a common issue in cloud platforms. Reiss et al. [5] shows that a Google cluster achieves CPU utilization of 25–35% and memory utilization of 40%. Quasar [3] indicates that the CPU utilization is consistently below 20%, and the memory utilization is (40–50%) on a production cluster at Twitter. Fluctuation of resource consumption and complex heterogeneous environment bring much more challenges to resource allocation in cloud cluster. It is difficult to match resource allocation precisely with resource consumption, and resources are usually over allocated to guarantee task execution.

© IFIP International Federation for Information Processing 2019
Published by Springer Nature Switzerland AG 2019
X. Tang et al. (Eds.): NPC 2019, LNCS 11783, pp. 361–365, 2019.
https://doi.org/10.1007/978-3-030-30709-7_34

In this paper, we propose PDRM, a resource management scheme based on task resource consumption probability distribution. The main idea of PDRM is to adopt the probability distribution of resource consumption to quantify the fluctuation of resource consumption for accurate resource allocation, so as to improve resource utilization and reduce task running time. We evaluate PDRM, and results show that it improves job execution efficiency and resource utilization.

2   Related Work

Several recent researches have tried to address the issue of improving efficiency in allocating resources to applications with varying degree of success. (1) **Dynamic Resource Provisioning.** Mohan et al. [4] have proposed dynamic resource management solutions for applications. These research works are used for resource management of long term services, but not suitable for batch workloads. However the batch workload, which consists of a large number of short tasks usually completed in minutes or seconds, is too important to be ignored. (2) **Resource Provisioning with an Appropriate Configuration.** CherryPick [1] builds a performance model with Bayesian Optimization to distinguish the optimal or a near-optimal configuration from the rest. MrMoulder [2] adopts optimization technique to tuning Hadoop configuration parameter settings. They mostly focus on improving of application performance and pay less attention to the resource utilization of the cluster. (3) **Harvesting Idle Resource from Colocated Jobs.** Zhang et al. [6] schedules related batch tasks on servers to colocate with latency-critical jobs. The main idea is harvesting idle resource from other jobs, but it can't harvest idle resources from the job itself. (4) **Characterizing and Classifying Workloads.** Quasar [3] classifies any new incoming application, assign the application proper resources in a datacenter. Classification techniques cannot fully reflect the differences in resource consumption of various tasks.

3   Motivation

Existing research assumes that the resource consumption of the task is the same as the historical consumption. However, the resource consumption of repeated tasks will fluctuate instead of being absolutely the same. The main reason for fluctuations in resource consumption is that the complexity of the algorithm on different input data content is different. It may cause deviations in resource consumption prediction. Therefore, we have conducted in-depth research on the similarity of resource consumption. We repeat running a variety of different batch workloads with different data sets. We extract resource consumption for all types of tasks when the application is running. It shows that the resource consumption of the same tasks on the same cluster node are similar, and the resource consumption fluctuates within a certain range. By counting the number of tasks in different resource consumption intervals, we can obtain the task resource consumption probability distribution. The probability distribution of resource consumption is in accordance with the Gaussian distribution, and the Gaussian distribution is fitted well.

4   PDRM Design

By extracting resource consumption of big data applications, we can get the task consumption probability distribution. Based on the distribution function of resource consumption, we propose an accurate resource allocation scheme, called PDRM.

We use $Task_0$ to denote a type of task, the resource allocation vector of $Task_0$ is expressed as $[ra_{10}, ra_{20}, ..., ra_{m0}]^T$ and the resource consumption vector of $Task_0$ is expressed as $[rc_{10}, rc_{20}, ..., rc_{m0}]^T$, where m is the total number of the types of resource, $ra_{i0}(i \in N, 1 \leq i \leq m)$ is the amount of the class i resource allocated to $Task_0$ and $rc_{i0}(i \in N, 1 \leq i \leq m)$ is the class i resource consumed by $Task_0$. We perform a Gaussian fitting on the class i resource consumption of $Task_0$, and the Gaussian distribution satisfied by rc_{i0} is expressed as $N_{io}(\mu_{io}, \sigma_{io}^2)$, where μ_{io} is the mean of rc_{i0} and σ_{io}^2 is the variance of the probability distribution of rc_{i0}. The cumulative distribution function of the probability distribution of rc_{i0} is

$$F_{i0}(x) = \int_{-\infty}^{x} \frac{1}{\sqrt{2\pi\sigma_{i0}^2}} e^{-\frac{(x-\mu_{i0})^2}{2\sigma_{i0}^2}} dx. \tag{1}$$

The probability that rc_{i0} is less than ra_{i0} can be denoted as $P_{i0} = F_{i0}(ra_{i0})$. The success ratio of $Task_0$ can be expressed as $min\{P_{10}, P_{20}, ..., P_{m0}\}$. Conversely, only when the resource allocated for the class i resource ra_{i0} is not less than $F_{i0}^{-1}(P_{\text{success}})$, the success rate of $Task_0$ could reach P_{success}.

The resource allocation of class i resource for $Task_j$ is ra_{ij}. P_{success} is the probability that $Task_j$ can be successfully completed. The failed task will be restarted with the default resource allocation which is much larger than the actual consumption of the task to ensure successful execution. The Expectation resource allocation of class i resource for $Task_j$ expressed as $E(ra_{ij}) = ra_{ij} + (1 - P_{\text{success}})ra_{ij\_default}$, where $ra_{ij\_default}$ is the default resource allocation. The average resource utilization of class i resource on a node is $\overline{ut_i} = \frac{\sum_{j=1}^{n} \mu_{ij}}{\sum_{j=1}^{n} E(ra_{ij})}$.

When the derivative of $\overline{ut_i}$ is 0, $\overline{ut_i}$ takes the maximum value. By solving the formula $(\overline{ut_i})' = 0$, we can get the solution of optimal resource utilization and set the values of P_{success}. Then, we get the resource allocation vector for each type of task.

5   Evaluation

We implement PDRM as a component on Hadoop Yarn. In this section, we demonstrate the effectiveness of our approach on a heterogeneous cluster.

We choose four representative applications on Hadoop to show different resources requirement: Terasort, WordCount, TextSearch, and TriangleOfOriented. We select 6 physical nodes to build a heterogeneous environment, named NODE0-NODE5. NODE0 is the master node, NODE1-NODE5 are the slave nodes. NODE0-NODE3 have the same physical configuration (two Intel Xeon E2620 6x cores 2.1 GHz CPUs, 16 GB memory). The number of virtual cores available for container allocation on each node is 8. As a comparative instance of CPU heterogeneity, NODE4 has two Intel Xeon E5620 4x cores 2.4 GHz CPUs. As a comparative instance of Memory heterogeneity, the memory available for container allocation on NODE5 is 3 GB, while the memory available on other nodes are 8 GB.

5.1 Job Completion Time of Heterogeneous Applications

In this experiment, we evaluate the effectiveness of PDRM for reducing job completion time. Figure 1 shows the job completion times with different resource allocation schemes. It can be observed that PDRM reduces the job completion time by 30.4%, 24.3%, 25.1%, 24.7% compared to the default for Terasort, WordCount, TextSearch, TriangleOfOrineted, respectively. PDRM resource allocation schemes can effectively reduce job completion time.

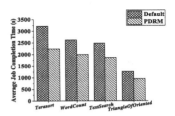

Fig. 1. Job completion time.

5.2 Resource Allocation Ratio and Resource Consumption Ratio of Heterogeneous Cluster

In this experiment, we compare the resource allocation ratio and the resource consumption ratio of different nodes in heterogeneous clusters under the default and PDRM. We normalize the node available resources to 1. We run Terasort on the heterogeneous cluster, the resource allocation ratio and resource consumption ratio of NODE1, NODE4 and NODE5 are shown in Figs. 2 and 3, respectively. It can be seen that the resource allocation ratio of PDRM is less than the that of the default, but the resource consumptions ratio of PDRM are greater that of the default.

The performance of NODE1 and NODE4 are limited by the CPU resources. With the PDRM resource allocation scheme, the CPU resources of NODE1 and NODE4 can achieve higher utilization. The CPU processing power of NODE4 is lower than that of NODE1. The NODE4 CPU can maintain high utilization, but the NODE4 memory resource utilization is less than NODE1. The performance of NODE5 is limited by the memory resources. NODE5 has less memory resources than NODE1. With PDRM resource allocation scheme, the memory resources of NODE5 can achieve higher utilization, far greater than that of NODE1. PDRM can improve resource utilization, and the scarce resources of nodes can be efficiently utilized in heterogeneous clusters.

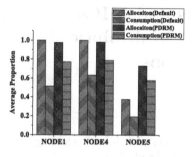

Fig. 2. CPU resource ratio.

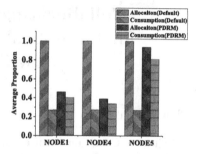

Fig. 3. Memory resource ratio.

6 Conclusion

The resource consumption probability distribution of the task can well describe the fluctuation of resource consumption. We propose PDRM, a resource allocation scheme based on the probability distribution of task resource consumption. Through experimental verification, PDRM can reduce job completion time by over 25%. What's more, PDRM can minimize the gap between resource allocation and resource consumption, and make efficient use of scarce resources in heterogeneous clusters.

References

1. Alipourfard, O., Liu, H.H., Chen, J., Venkataraman, S., Yu, M., Zhang, M.: CherryPick: adaptively unearthing the best cloud configurations for big data analytics. In: 14th USENIX Symposium on Networked Systems Design and Implementation (NSDI 2017), pp. 469–482. USENIX Association, Boston (2017)
2. Cai, L., Qi, Y., Wei, W., Wu, J., Li, J.: mrMoulder: a recommendation-based adaptive parameter tuning approach for big data processing platform. Future Gener. Comput. Syst. **93**(1), 570–582 (2019)
3. Delimitrou, C., Kozyrakis, C.: Quasar: resource-efficient and QoS-aware cluster management. In: Proceedings of the 19th International Conference on Architectural Support for Programming Languages and Operating Systems, pp. 127–144. ACM, New York (2014)
4. Mohan, A., Kaseb, A.S., Lu, Y., Hacker, T.: Adaptive resource management for analyzing video streams from globally distributed network cameras. IEEE Trans. Cloud Comput. 1 (2018)
5. Reiss, C., Tumanov, A., Ganger, G.R., Katz, R.H., Kozuch, M.A.: Heterogeneity and dynamicity of clouds at scale: Google trace analysis. In: Proceedings of the Third ACM Symposium on Cloud Computing, pp. 7:1–7:13. ACM, New York (2012)
6. Zhang, Y., Prekas, G., Fumarola, G.M., Fontoura, M., Goiri, I.n., Bianchini, R.: History-based harvesting of spare cycles and storage in large-scale datacenters. In: Proceedings of the 12th USENIX Conference on Operating Systems Design and Implementation, pp. 755–770. USENIX Association, Berkeley (2016)

Collaborating CPUs and MICs for Large-Scale LBM Multiphase Flow Simulations

Chuanfu Xu$^{(\boxtimes)}$, Xi Wang, Dali Li, Yonggang Che, and Zhenghua Wang

College of Computer Science, National University of Defense Technology,
Changsha 410073, People's Republic of China
xuchuanfu@nudt.edu.cn

Abstract. This paper highlights the use of the OpenMP4.5 accelerator programming model to collaborate CPUs and Intel Many Integrated Cores (MIC) co-processors for large-scale LBM multiphase flow simulationson the Tianhe-2 supercomputer. To enhance the collaborative efficiency among intra-node CPUs and co-processors, we propose a flexible load balance model with heterogeneous domain decomposition for CPU-MIC task allocation, as well as asynchronous offloading to overlap operations of CPUs and multiple MICs. Tests for 3D multi-phase (liquid and gases) problem (about 100 Billion lattices) simulating drop impact with gravity effect using D3Q19 Lattice Boltzmann discretization and Shan-Chen BGK single relaxation time collision model are presented, achieving a weak parallel efficiency of above 80% in going from 128 to 2048 compute nodes.

Keywords: Heterogeneous parallel computing ·
Lattice Boltzmann methods · Many-core processor ·
OpenMP4.5 accelerator programming model

1 Introduction

Lattice Boltzmann Methods (LBM) regard fluids as Newtonian fluids from a microscopic perspective, divide flow field into small lattices (mass points), and simulate fluid evolution dynamics through collision models (lattices collision and streaming) [1]. Currently, LBM has been increasingly used for real-world flow problems with complex geometries and various boundary conditions. Large-scale LBM simulations with increasing resolution and extending temporal range require massive high performance computing resources. It is therefore essential and practical to port LBM codes onto modern supercomputers, often featuring many-core accelerators/coprocessors (GPU, Intel MIC, or specialized ones).

Supported by NSFC under Grant No. 61772542.

These heterogeneous processors can dramatically enhance the overall performance of HPC systems with remarkably low total cost of ownership and power consumption, but the development and optimization of large-scale applications are also becoming exceptionally difficult. Accelerator programming models such as OpenMP4.X [2], OpenACC and Intel Offload aim to provide performant and productive heterogeneous computing through simple compiler directives. Among them, OpenMP4.X is especially attractive since it incorporates accelerator programming with traditional shared memory multithreading into a unified high-level model, and supports major languages (C++, C and Fortran...) and devices (CPU, GPU, MIC, ARM, DSP...).

In this paper, we parallelize an LBM code *openlbmflow* and highlight the use of OpenMP4.5 for large-scale CPU-MIC collaboration on the Tianhe-2 supercomputer [3]. A load balance model with heterogeneous domain decomposition is proposed for CPU-MIC task allocation. We use asynchronous offloading to minimize the cost of halo exchanges and significantly overlap CPU-MIC computation/communication. Our collaborative approach achieves a speedup of up to 5.0X compared to the CPU-only approach. Tests for 3D multi-phase (liquid and gases) problem (about 100 Billion lattices) simulating drop impact with gravity effect using D3Q19 Lattice Boltzmann discretization and Shan-Chen BGK single relaxation time collision model are presented, achieving a weak scaling efficiency of above 80% in going from 128 to 2048 compute nodes.

```
 1: #pragma omp declare target
 2:    //declare variables and functions on MICs
 3: #pragma omp end declare target
 4:    //initialization on CPUs
 5: for (n=0;n<mic_num;n++) //for multiple MICs
 6: {
 7:    //pre-allocate and initialize variables on MICs
 8:    #pragma omp target device(mic_num) data map(alloc...)
 9:    #pragma omp target device(mic_num) data map(to...)
10: }
11: for (iter=1;iter<max_iter;iter++)//time-marching loop
12: {
13:    for (n=0;n<mic_num;n++) //for multiple MICs
14:    {
15:       //gather boundary lattices into the Inbuffer (CPU)
16:       #pragma omp target device(mic_num) map(to...) map(from...) nowait
17:       //H2D the Inbuffer
18:       //scatter the Inbuffer and update halo lattices (MIC)
19:       //MIC caltulation
20:       //gather boundary lattices into the Outbuffer (MIC)
21:       //D2H the Outbuffer
22:    }
23:    //caltulation on CPUs
24:    #pragma omp taskwait //synchronization
25:    for (n=0;n<mic_num;n++) //for multiple MICs
26:    {
27:       //scatter the Outbuffer and update halo lattices (CPU)
28:    }
29:    //boundary conditions
30:    //MPI communication
31: }
```

Fig. 1. Code skeleton for CPU-MIC collaboration with asynchronous offloading and overlapping of CPU-MIC computation/communication using OpenMP directives.

2   CPU-MIC Collaboration and Performance Results

openlbmflow is an LBM code written in C that can simulate both 2D/3D single-phase or multi-phase flow problems with periodic and/or bounce-back boundary conditions. It mainly consists of three phases: initialization, time iteration, and post-processing. During the initialization phase, the geometry of the flow field, flow density and the distribution function are initialized. The time iteration phase includes three important procedures: inter-particle force calculation (as well as velocity and density), collision and streaming. In the post-processing phase, simulation results are collected and saved according to a user-specified iteration interval.

We decompose the original computational domain along the three dimensions evenly into many blocks and distribute them among MPI processes. On each compute node, each block is divided into 4 sub-blocks with one calculated by CPUs and the other three offloaded to the three coprocessors. Figure 1 illustrates the intra-node collaborative programming approach. Before time-marching loops, we use `omp declare target` directive to declare variables or functions which are both available on CPU and MIC (line 1–3). We use `omp target data` directive with `map` clause to pre-allocate device memory and perform initialization of global flow variables and data transfer buffers on each MIC (line 5–10). We design a unified *In/Out-buffer* for PCI-e data transfer among intra-node CPUs and coprocessors. In each iteration, boundary lattices on CPUs are gathered into the *Inbuffer*, and transferred to different MICs using `map` clause with array section syntax (line 15–17). Before MIC calculation, we scatter boundary lattices from the *Inbuffer* and update halo lattices on MICs (line 18). After MIC calculation, boundary lattices on MICs will be gathered into the *Outbuffer* and transferred back to CPUs (line 20–21). We use OpenMP `nowait` to asynchronously dispatch kernels on MIC and overlap CPU-MIC computation/communication. We synchronize CPU-MIC computation using the `taskwait` directive to ensure that both sides have finished their computations before updating halo lattices on CPUs and MPI communications. We use a parameter r to represent the workload ratio on CPU side and r can be configured by profiling *openlbmflow*'s sustainable performance on both sides.

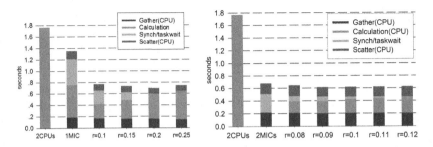

Fig. 2. Performance of CPU+1MIC (left) and 2MICs (right) with problem size 256 × 256 × 256.

We use icc 17.0.1 from Intel composer 2017.1.132 in out tests. Our heterogeneous code was compiled in double precision with option "-qopenmp -O3 -fno-alias -restrict -xAVX". MPICH2-GLEX was used for MPI communications. Figure 2(left) demonstrates the performance of CPU+1MIC with overlapping of both CPU/MIC computation and PCI-e data transfer. We decompose the costs into CPU gather/scater, CPU calculation and CPU-MIC synchronization. Due to overlapping, the synchronization cost decreases with increasing workloads on CPUs, and disappears when $r = 0.2$, indicating a perfect overlapping. Afterwards further increasing r will improve the cost of CPU calculation and degrade the overall performance. The maximum speedup was improved to about 2.5 due to the enhanced overlapping. For CPU+2MICs (Fig. 2(right)), the maximum speedup is about 2.88 ($r = 0.09$), only about 15.2% enhancement compared to the CPU+1MIC simulation. This is mainly due to a relatively small total workload, and the collaborative overhead exceeds more than half of the whole execution time.

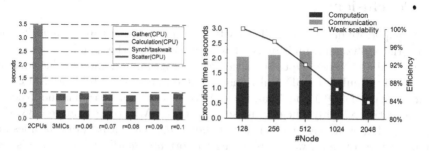

Fig. 3. Performance of CPU+3MICs with problem size of $512 \times 256 \times 256$ (left) and large-scale weak scalability on CPU+MIC nodes (right).

In Fig. 3(left), the maximum speedups are 3.93 ($r = 0.08$) and 4.81 ($r = 0.07$) for the problem set $512 \times 256 \times 256$ with CPU+3MICs. Because the sustainable performance of *openlbmflow* on a MIC outperforms much of that on two CPUs, only less than 10% of the whole workload is allocated to CPUs for collaborative simulations with multiple MICs. Due to the limited device memory capacity (8 GB) on Xeon Phi 31S1P, the maximum problem size for each MIC is about $256 \times 256 \times 256$. As a result, we couldn't achieve ideal load balance in heterogeneous simulations. Figure 3(right) reports the weak scalability results for CPU+MIC collaborative simulations. Although large-scale heterogeneous simulations involve quite complicated interactions, efficiencies stay well above 80%. This is comparable to that of large-scale CPU-only simulations and demonstrates the effectiveness of the overlapping optimization.

3   Related Work

Few researches about parallelizing scientific codes using the new OpenMP4.X accelerator programming model on heterogeneous supercomputers are reported, but many researchers have shown the experiences of porting LBM codes onto

GPUs or MICs using other programming models. Paper [4] ported a GPU-accelerated 2D LBM code onto Xeon Phi, and compared with previous implementations on state-of-the-art GPUs and CPUs. Paper [5] implemented a LBM program using the portable programming model OpenCL, and evaluated its performance on multi-core CPUs, NVIDIA GPUs as well as Intel Xeon Phi. In [6], researchers have also parallelized *openlbmflow* on the Tianhe-2 supercomputer and collaborate CPUs and MICs using Intel Offload programming model. The performance was preliminary evaluated in single precision. To summarize, current reports only involve simple LBM models on small MIC clusters. Paper [7] Collaborated CPU and GPU for large-scale high-order CFD simulations with complex grids on the TianHe-1A supercomputer. This is the first paper, to our best knowledge, reporting CPU-MIC collaborative LBM simulations using complex 3D multi-phase flow models with OpenMP4.5.

4   Conclusions

In this paper, we developed a CPU+MIC collaborative software *openlbmflow* for 3D Lattice Boltzmann multiphase flow simulations on the Tianhe-2 supercomputer based on the new OpenMP accelerator programming model. The software successfully simulated a 3D multi-phase (liquid and gases) problem (100 billion lattices) using D3Q19 and Shan-Chen BGK models on 2048 Tianhe-2 nodes, demonstrating a highly efficient and scalable CPU+MIC collaborative LBM simulation with a weak scaling efficiency of above 80%. For future work, besides fine tuning of the software, we are planning to port *openlbmflow* onto China's self-developed many-core processors/coprocessors based on the power-efficient high performance ARM architecture.

References

1. Succi, S., Benzi, R., et al.: The lattice Boltzmann equation: a new tool for computational fluid-dynamics. Phys. D Nonlinear Phenom. **47**, 219–230 (1991)
2. Martineau, M., Price, J., McIntosh-Smith, S., Gaudin, W.: Pragmatic performance portability with OpenMP 4.x. In: Maruyama, N., de Supinski, B.R., Wahib, M. (eds.) IWOMP 2016. LNCS, vol. 9903, pp. 253–267. Springer, Cham (2016). https://doi.org/10.1007/978-3-319-45550-1_18
3. Xiangke, L., Liquan, X., Canqun, Y.: MilkyWay-2 supercomputer: system and application. Front. Comput. Sci. **8**(3), 345–356 (2014)
4. Crimi, G., Mantovani, F., Pivanti, M., Schifano, S.F., Tripiccione, R.: Early experience on porting and running a Lattice Boltzmann code on the Xeon-Phi coprocessor. Procedia Comput. Sci. **18**, 551–560 (2013)
5. McIntosh-Smith, S., Curran, D.: Evaluation of a performance portable lattice Boltzmann code using OpenCL. In: International Workshop on OpenCL, pp. 1–12 (2014)
6. Dali, L., Chuanfu, X., Yongxian, W., Zhifang, S., et al.: Parallelizing and optimizing large-scale 3D multi-phase flow simulations on the Tianhe-2 supercomputer. Concurr. Comput. Pract. Exp. **28**, 1678–1692 (2015)
7. Chuanfu, X., Xiaogang, D., Lilun, Z., et al.: Collaborating CPU and GPU for large-scale high-order CFD simulations with complex grids on the TianHe-1A supercomputer. J. Comput. Phys. **278**, 275–C297 (2014)

Multiple Algorithms Against Multiple Hardware Architectures: Data-Driven Exploration on Deep Convolution Neural Network

Chongyang Xu[1], Zhongzhi Luan[1], Lan Gao[1], Rui Wang[1(✉)], Han Zhang[2], Lianyi Zhang[2], Yi Liu[1], and Depei Qian[1]

[1] Beihang University, Beijing, China
{xuchongyang1995,07680,lan.gao,wangrui,yi.liu,depeiq}@buaa.edu.cn
[2] Science and Technology on Special System Simulation Laboratory, Beijing Simulation Center, Beijing, China
xia_mei2000@163.com, yzhang117@163.com

Abstract. With the rapid development of deep learning (DL), various convolution neural network (CNN) models have been developed. Moreover, to execute different DL workloads efficiently, many accelerators have been proposed. To guide the design of both CNN models and hardware architectures for a high-performance inference system, we choose five types of CNN models and test them on six processors and measure three metrics. With our experiments, we get two observations and conduct two insights for the design of CNN algorithms and hardware architectures.

Keywords: Convolutional neural network · Hardware architecture · Performance evaluation

1 Introduction

CNN models have large computation and consume much energy, putting significant pressure on CPUs and GPUs. To execute CNN models more efficiently, many specific accelerators are proposed (e.g., Cambricon-1A [11] and TPU [9]).

Due to the complexity of both sides, it is challenging to design high-performance processors for various CNN models and design CNN models with different types of processors. To tackle this, we perform a lot of evaluations, and we get two observations. Based on observations, we get two insights for the design of CNN algorithms and hardware architectures.

Following of this paper includes related work, experiments methodology, experiments result and analysis, conclusion and acknowledgements.

© IFIP International Federation for Information Processing 2019
Published by Springer Nature Switzerland AG 2019
X. Tang et al. (Eds.): NPC 2019, LNCS 11783, pp. 371–375, 2019.
https://doi.org/10.1007/978-3-030-30709-7_36

2   Related Work

Related evaluation work of CNN inference systems is as follows.

AI benchmark [7] measures only latency and one type of processors. Fathom [2] and SyNERGY [13] test two different types of processors and one metric. However, they do not compare the same type of processors with different versions. DjiNN and Tonic [3] measure the latency and throughput. They only use one CPU and one GPU without comparing different versions of same processor. BenchIP [16] and [9] use three metrics and three types of processors. BenchIP focuses on the design of hardware and use prototype chips instead of production level accelerators. [9] focus the performance of hardware architecture only, we give insights for the design of both CNN models and hardware architectures.

3   Experiment Methodology

In this section, we present the principles of workloads choosing, processors choosing, and software environment. We also give details of measurement.

Table 1. Selected models

Model	Conv	FC	Weights (10^6)	Gflops	Input size	Dataset
AlexNet [5]	5	3	60.1	0.62	$227 \times 227 \times 3$	Imagenet [14]
MobileNetv1 [6]	27	1	4.2	0.57	$227 \times 227 \times 3$	Imagenet
ResNet50 [4]	49	1	25.6	3.89	$227 \times 227 \times 3$	Imagenet
Vgg16 [15]	13	3	138.3	15.47	$227 \times 227 \times 3$	Imagenet
Yolov2 [12]	19	0	67.4	17.51	$416 \times 416 \times 3$	COCO [10]

Table 2. Selected processors and software environment

Processor	GHz	TDP (W)	#TFLOPS/s	#core	GB/s	Numeric library
CPU-E5	2.40	240	2.15	28	76.8	Intel MKL 2017 update 4
CPU-I5	3.40	65	0.20	4	34.1	Intel MKL 2017 update 4
GPU-P100	1.33	300	9.30	3584	732	Cuda8, CuDNN7
GPU-970	1.05	145	3.50	1664	224	Cuda8, CuDNN7
Cambricon	-	-	1.92	1	27.8	Libipu
TPU	-	-	180.00	8	600	-

Workloads are chosen from widely used tasks, with different layers, of different depths, of different size and of different topology as shown in Table 1.

Hardware architectures are chosen from scenarios such as user-oriented situation, datacenter usage and mobile devices as shown in Table 2, (1) Intel Xeon E5-2680 v4, (2) Intel Core I5-6500, (3) Nvidia TESLA P100, (4) Nvidia GeForce

GTX 970, (5) Cambricon is a typical neural processor and the actual processor is HiSilicon Kirin 970 SoC in Huawei Mate 10. and (6) TPUv2, a publicly available DL accelerator from Google Cloud.

The same frame framework (tensorflow v1.6 [1]) and pre-trained models (.pb file) are used except Cambricon. Cambricon has its inference API and model format. Tensorflow 1.8 is provided for TPU by Google Cloud.

Three metrics are measured. Latency, the average milliseconds spent for an image. Throughput, the average images processed in a second. Energy efficiency, the amount of computation when a processor consumes 1 joule of energy.

To measure latency, we (1) load 100 images into memory and perform preprocessing, (2) run once to warm up, (3) infer one image each time, record time of 100 times inference and compute average latency. It is similar to throughput but using 1000 images and inferring one batch each time. Max throughput is achieved by tuning batch size. Measuring energy efficiency is similar to measuring the maximum throughput. Power is sampled via sysfs powercap interface at 1 Hz, nvidia-smi at 10 Hz on CPU and GPU respectively. We take energy consumption as energy consumed when the processor is under workload minus when the processor is idle. For Cambricon, we use MC DAQ USB-2408.

4   Experiment Results and Analysis

Figure 1 shows the result. As shown in Fig. 1, for most cases, the more is the computation, the higher is the latency or lower is the throughput. However, there are exceptions; we summarize them into two observations.

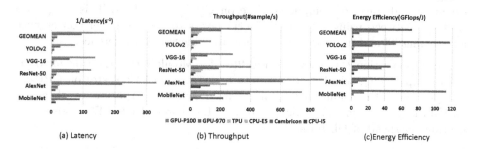

Fig. 1. Measured data. The longer the bar is, the better the performance is.

Observation 1: CNN models that have more computation may not incur higher latency or lower throughput. Models have more computation are expected to take more computing time, thus have higher latency and lower throughput. However, in Fig. 1(a), on CPU-E5, AlexNet with more computation has lower latency than MobileNetv1; on GPU-P100, Vgg16, AlexNet with more computation has lower latency than ResNet50, MobileNetv1 respectively.

Observation 2: Optimizations on CNN models are only applicable to specific processors. As shown in Fig. 1(a), MobileNetv1 has lower latency than AlexNet on CPU-I5 but higher latency on CPU-E5. MobileNetv1 is an optimized model but only performs well on a less powerful CPU.

To explain these two observations, we measure latency breakdown by layer types and functions, the result is shown in Fig. 2.

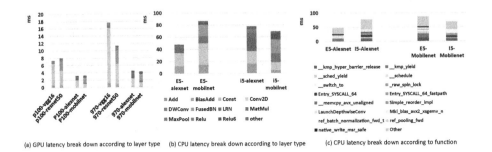

(a) GPU latency break down according to layer type   (b) CPU latency break down according to layer type   (c) CPU latency break down according to function

Fig. 2. Experiments for Observations

For Observation 1, as shown in Fig. 2(a), (b) BatchNorm layers have large execution time with low computation, which cause higher latency of MobileNetv1 than AlexNet on CPU-E5, higher latency of ResNet50 and MobileNetv1 than Vgg16 and AlexNet respectively on GPU-P100. Thus, we give Insight 1.

Insight 1: BatchNorm layers have a low ratio of computation but a disproportionately high ratio of computing time on CPUs and GPUs. This suggests a trade-off between using more BatchNorm layers to achieve faster convergence for training [8] and using less BatchNorm to achieve faster inference.

For Observation 2, as shown in Fig. 2(c), for MobileNetv1, the runtime overhead (kmp_yield(), sched_yield(), switch_to(), raw_spin_lock(), etc) on CPU-E5 occupies more than 40ms of 86.8ms in total, while the runtime overhead on CPU-I5 is about 20ms of 68.8ms in total. More cores of CPU-E5 increase the runtime overhead of DL frameworks.

Insight 2: The runtime overhead of modern DL frameworks increases with the increment of the core number on CPU. This suggests improving the computing capability of individual cores rather than increasing the number of cores to reduce latency.

5   Conclusion

In this work, we choose five CNN models and six processors and measure the latency, throughput, and energy efficiency. We present two observations and conclude two insights. These insights might be useful for both algorithms and hardware architectures designers.

- For algorithm designers, they need to balance the usage of BatchNorm layers for which can accelerate the training process but slow down inference.
- For hardware designers, BatchNorm layers deserve more attention; to reduce latency, it is more critical to improve the computing capability of individual cores than increasing the number of cores.

Acknowledgements. This work is supported by the National Key Research and Development Program of China under grant 2017YFB0203201. This work is also supported by the NSF of China under grant 61732002.

References

1. Abadi, M., et al.: TensorFlow: a system for large-scale machine learning. In: OSDI 2016 (2016)
2. Adolf, R., Rama, S., Reagen, B., Wei, G.Y., Brooks, D.: Fathom: reference workloads for modern deep learning methods. In: IISWC 2016 (2016)
3. Hauswald, J., et al.: DjiNN and tonic: DNN as a service and its implications for future warehouse scale computers. In: ISCA 2015 (2015)
4. He, K., Zhang, X., Ren, S., Sun, J.: Deep residual learning for image recognition. In: Proceedings of the IEEE Conference on Computer Vision and Pattern Recognition (2016)
5. Hinton, G.E., Krizhevsky, A., Sutskever, I.: ImageNet classification with deep convolutional neural networks. In: Advances in Neural Information Processing Systems (2012)
6. Howard, A.G., et al.: MobileNets: efficient convolutional neural networks for mobile vision applications. arXiv preprint arXiv:1704.04861 (2017)
7. Ignatov, A., et al.: AI benchmark: Running deep neural networks on android smartphones. In: European Conference on Computer Vision (2018)
8. Ioffe, S., Szegedy, C.: Batch normalization: accelerating deep network training by reducing internal covariate shift. In: ICML 2015 (2015)
9. Jouppi, N.P., et al.: In-datacenter performance analysis of a tensor processing unit. In: ISCA 2017 (2017)
10. Lin, T.-Y., Maire, M., Belongie, S., Hays, J., Perona, P., Ramanan, D., Dollár, P., Zitnick, C.L.: Microsoft COCO: common objects in context. In: Fleet, D., Pajdla, T., Schiele, B., Tuytelaars, T. (eds.) ECCV 2014. LNCS, vol. 8693, pp. 740–755. Springer, Cham (2014). https://doi.org/10.1007/978-3-319-10602-1_48
11. Liu, S., et al.: Cambricon: an instruction set architecture for neural networks. In: ACM SIGARCH Computer Architecture News (2016)
12. Redmon, J., Farhadi, A.: YOLO9000: better, faster, stronger. In: Proceedings of the IEEE Conference on Computer Vision and Pattern Recognition (2017)
13. Rodrigues, C.F., Riley, G.D., Luján, M.: Fine-grained energy profiling for deep convolutional neural networks on the Jetson TX1. CoRR abs/1803.11151 (2018)
14. Russakovsky, O., et al.: ImageNet large scale visual recognition challenge. Int. J. Comput. Vis. **115**, 211–252 (2015)
15. Simonyan, K., Zisserman, A.: Very deep convolutional networks for large-scale image recognition. arXiv preprint arXiv:1409.1556 (2014)
16. Tao, J.H., et al.: BenchIP: benchmarking intelligence processors. J. Comput. Sci. Technol. **33**, 1–23 (2018)

A Parallel Retinex Image Enhancement Algorithm Based on OpenMP

Shixiong Cheng[1], Bin Liu[1,2,3(✉)], Dongjian He[2,3,4], Jinrong He[5],
Yuancheng Li[6], and Yanning Du[7]

[1] College of Information Engineering, Northwest A&F University,
Yangling, Shaanxi, China
liubin0929@nwsuaf.edu.cn
[2] Key Laboratory of Agricultural Internet of Things, Northwest A&F University,
Ministry of Agriculture and Rural Affairs, Yangling 712100, Shaanxi, China
[3] Shaanxi Key Laboratory of Agricultural Information Perception and Intelligent
Service, Northwest A&F University, Yangling 712100, Shaanxi, China
[4] College of Mechanical and Electronic Engineering,
Northwest A&F University, Yangling, Shaanxi, China
[5] College of Mathematics and Computer Science,
Yan'an University, Yan'an, Shaanxi, China
[6] School of Computer Science and Technology,
Xi'an University of Science and Technology, Xi'an, China
[7] School of Computer Science and Engineering,
Xi'an University of Technology, Xi'an, China

Abstract. Retinex image enhancement algorithm occupies an important position in eliminating image uneven exposure, low contrast, and smog influence. However, with the increasing of image resolution, the real-time performance of the serial Retinex algorithm has not satisfied the requirements of practical applications. This paper proposes an OpenMP-based parallel Retinex algorithm. The parallelism of the Retinex algorithm is first identified by theoretical analyses. Then, the time-consuming sub-algorithms such as Gaussian convolution and exponential transformation, of the serial algorithm are designed and executed in parallel. On Tianhe-2 supercomputer platform, the experimental results show that the speedup of the parallel algorithm is significantly improved, and the test image set achieves an average speedup of 12. It indicates that the parallel algorithm can satisfy the needs of real-time processing in image enhancement field.

Keywords: Image enhancement · Parallel algorithm · Retinex · OpenMP · Agricultural image

1 Introduction

Image enhancement algorithms are basic work in many areas especially in the public security, biomedical field, health service, and marine information field, where significant achievements have been made in [1–3]. At present, researchers have proposed a great diversity of parallel image processing algorithms, such as CUDA-based image enhancement algorithms [4], and image processing algorithms based on multicore

© IFIP International Federation for Information Processing 2019
Published by Springer Nature Switzerland AG 2019
X. Tang et al. (Eds.): NPC 2019, LNCS 11783, pp. 376–381, 2019.
https://doi.org/10.1007/978-3-030-30709-7_37

DSP [5]. However, as one of the most important images processing technologies, the serial single-scale Retinex (SSR) algorithm is still too slow to finish the image enhancement tasks within an acceptable time.

In order to solve above problem, this paper proposes the parallel SSR image enhancement algorithm based on OpenMP. The parallel SSR image enhancement algorithm is implemented on the Tianhe-2 supercomputer using the OpenMP programming model, which is evaluated and achieves an average speedup of 12. The experimental results show that the proposed parallel algorithm can fulfill the needs of real-time processing in image enhancement field.

2 Parallel Design and Implementation

2.1 Parallelism Analyses

SSR algorithm enhances an image through the implementation of sub-algorithms such as Gaussian template, Gaussian convolution, and exponential transformation. As the data processed by these sub-algorithms is independent of each other, SSR algorithm has good parallelism. As illustrated in Fig. 1, the following three aspects are presented to analyze the parallelism of the serial SSR algorithm.

Parallelism 1: the blurred image estimating the incident illumination component is generated by Gaussian convolution operations. During this, each pixel is not associated with others. So, the image can be divided into sub-blocks for parallel computing.

Parallelism 2: the size of the Gaussian template is determined by the input parameters. When the Gaussian weight is normalized, each pixel is divided respectively by the sum of the weights, This process can be calculated in parallel.

Parallelism 3: the operation of exponential transformation can be also executed in parallel because there is no data dependence directly in those operations.

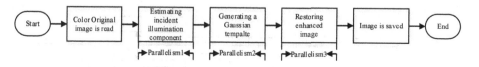

Fig. 1. The parallelism of single-scale Retinex algorithm

2.2 Parallel Design and Implementation of SSR Algorithm

In this section, parallel design of these sub-algorithms are firstly illustrated in Fig. 2 and then parallel implement are presented.

Parallel Design and Implementation of Gaussian Convolution. The subsequent operation of image segmentation into different data blocks is independent in Gaussian convolution serial algorithm. This is consistent with the parallel characteristics of OpenMP because no dependence between the data of non-direct adjacent pixels in the image. And a two-dimensional Gaussian function G(x, y) could be written as the product of two one-dimensional Gaussian functions G(x) and G(y), meaning that G(x, y) can be calculated serially by convolution of G(x)δ(y) and G(y)δ(x). However each of one-dimensional Gaussian functions G(x) could be executed in parallel. So the two-dimensional Gaussian convolution can be generated serially by two one-dimensional Gaussian convolutions performed respectively in parallel in the X and Y directions. And for example, there is an image which size is 7 × 7, and the convolution kernel is 3 × 3. It can be seen that the convolution operation requires 9 multiplications for each element in the image, so the total number of multiplication operations executed in sequential algorithm is 7 × 7 × 9 = 441 times. In contrast, parallel execution requires only 2 × 7 × 9 = 126 operations in the case of sufficient threads. And the execution time will be reduced and the speed will be increased compared with sequential algorithm.

Parallel Design and Implementation of Gaussian Template. The Gaussian template generation is mainly divided into two steps. The first step is that the weight sum is calculated serially, and the second step is that the normalization Gaussian template is generated in parallel. Supposing a 3 by 3 normalization Gaussian template is generated in serial algorithm with one thread and needs to be executed 9 times. However, in the case of parallelized execution with 9 threads, it only needs to be executed once.

Parallel Design and Implementation of Exponential Transformation. The original image and the Gaussian blurred image are set to the logarithmic domain to obtain a logarithmic image. The function of exponential transformation is to extend the image's high gray level and compress the low gray level. The most critical step in the exponential transformation is the linear mapping of each value. Assuming that the image size is 1000 × 1000, it takes a lot of time to go through the linear mapping. If linear mapping is performed in parallel using 24 threads, the image only needs to perform 1737 operations rather than 1,000,000 in the serial algorithm. Therefore, it is very profitable to perform each worthy linear mapping in parallel.

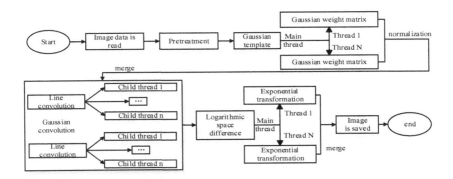

Fig. 2. Parallel design of single-scale Retinex algorithm

3   Experimental Results and Performance Analysis

3.1   Experimental Environment and Test Set

The experiment is performed on Tianhe-2 supercomputer equipped with 16000 nodes, which each note has three coprocessors, two Xeon E5-2692 processors, 24 cores and 64 GB of memory. The experimental environment is shown in Table 1. In this section, 10 different sizes of pictures are used to demonstrate the speedup performance of the parallel algorithm, the minimum size is 1730 × 883, the maximum is 4000 × 3000, and the format is JPG. These pictures are all agricultural images, including apples, pears, kiwis, farmland and mountain forests. They are from the shooting of the Dajiang UAV. The image test set is shown in Table 2.

Table 1. Experimental environment

Name	Description
Computer	Tianhe-2 Supercomputer
Processor	Intel Xeon E5-2692V2
Coprocessor	Intel Xeon phi
OS	Red Hat 4.4.7-4
Compiler	GCC 4.4.7
Cmake	Cmake-3.2.2
OpenCv	OpenCv-2.4.9

Table 2. Image test set

No	Description	Resolution	Size
img1	Farmland 1	1767 × 885	620 KB
img2	Farmland 2	1730 × 883	261 KB
img3	Qingyang Apple	2448 × 3264	2.20 MB
img4	Ruixue Apple	2784 × 1856	1.09 MB
img5	Ruiyang Apple	3088 × 2056	2.07 MB
img6	Kiwi 1	3264 × 2448	2.19 MB
img7	Kiwi 2	3264 × 2448	1.03 MB
img8	Farmland 3	4000 × 3000	4.36 MB
img9	Mountain 1	4000 × 3000	5.38 MB
img10	Mountain 2	4000 × 3000	4.72 MB

3.2   Speedup Comparison

The running time of serial and parallel algorithms respectively in Table 3 which Th represents thread. Within a certain range, the parallel SSR algorithm shortens the image processing time with the number of OpenMP threads increasing, and the average speedup is increased by about 12. After the parallel SSR algorithm are executed in parallel from dual thread to 24 thread, the speedup is obviously improved, and the parallel SSR algorithm can achieve near linear acceleration. The speedup curve is shown in Fig. 3. This experiment was carried out on a single node of Tianhe-2 supercomputer, each node had 24 cores, and the speedup reached a peak at 24 threads, making full use of the performance of multi-core. The speedup start to reduce at 32 threads because the number of threads at this time exceeds the number of CPU cores, but the processing time is still better than the serial algorithm. And the experimental results show that the speedup of the proposed parallel algorithm is significantly improved, and can satisfy the needs of real-time processing in image enhancement field.

Table 3. Comparison of running time (s)

Name	Serial	2Th	4Th	8Th	16Th	24Th	32Th
img1	7	4.71	2.442	1.279	0.726	0.542	0.627
img2	7	4.59	2.501	1.341	0.784	0.531	0.681
img3	39	24.799	12.746	6.733	3.897	2.86	3.445
img4	24	15.977	7.944	4.202	2.325	1.719	1.816
img5	30	19.089	9.985	5.128	2.888	2.072	2.39
img6	39	24.151	12.443	6.995	3.667	2.624	2.809
img7	39	24.437	12.281	6.609	3.619	2.626	2.837
img8	58	38.306	21.468	10.172	5.65	4.188	5.065
img9	59	36.762	18.923	9.598	5.549	4.779	4.743
img10	59	38.968	18.781	10.13	5.525	4.089	4.243

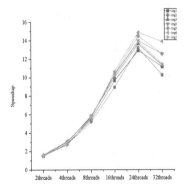

Fig. 3. Speedup comparison

4   Conclusion

This paper proposes a parallel SSR algorithm based on OpenMP. Compared to the serial Retinex algorithm, the proposed parallel algorithm can achieve an average speedup of 12, which represents a significant decrease in execution time. Experimental results show that the proposed parallel algorithm can acquire a significant increase in speedup and can better meet the requirements of real-time processing of the image enhancement algorithm in the image processing field.

Acknowledgment. This research is supported by the National Natural Science Foundation of China under Grant No. 61602388, by the Natural Science Basic Research Plan in Shaanxi Province of China under Grant No. 2017JM6059, by the Fundamental Research Funds for the Central Universities 2452019064, by the China Postdoctoral Science Foundation under Grant No. 2017M613216, by the Postdoctoral Science Foundation of Shaanxi Province of China under Grant No. 2016BSHEDZZ121, by the Fundamental Research Funds for the Central Universities under Grants No. 2452016081, by the Key Program of the National Natural Science Foundation of China under Grant No. 61834005, by China Postdoctoral Science Foundation No. 2018M6 33585, by Natural Science Basic Research Plan in Shaanxi Province of China No. 2018JQ6060, by the Natural Science Basic Research Plan in Shaanxi Province of China under Grant No. 2015JM6355, and by Doctoral Starting up Foundation of Yan'an University YDBK2019-06.

References

1. Salazar-Colores, S., Cabal-Yepez, E., Ramos-Arreguin, J.M., Botella, G., Ledesma-Carrillo, L.M., Ledesma, S.: A fast image dehazing algorithm using morphological reconstruction. IEEE Trans. Image Process. **28**(5), 2357–2366 (2019)
2. Muslim, H.S.M., Khan, S.A., Hussain, S., Jamal, A., Qasim, H.S.A.: A knowledge-based image enhancement and denoising approach. Comput. Math. Organ. Theory **25**, 108–121 (2018)
3. Bhowmik, M., Ghoshal, D., Bhowmik, S.: An Improved method for the enhancement of under ocean image. In: 2015 International Conference on Communications and Signal Processing, Melmaruvathur, India, pp. 1739–1742. IEEE (2015)

4. Li, H., Xie, W.H., Wang, X.G., Liu, S.S., Gai, Y.Y., Yang, L.: GPU implementation of multi-scale retinex image enhancement algorithm. In: 2016 IEEE/ACS 13th International Conference of Computer Systems and Applications, Agadir, Morocco, pp. 1–5. IEEE (2016)
5. Wang, G., Liu, X.: A parallel image processing platform based on multi-core DSP. In: 2017 IEEE/ACIS 16th International Conference on Computer and Information Science, Wuhan, China, pp. 775–779. IEEE (2017)

Author Index

An, Hong 143

Bao, Jingjing 169

Cao, Chenhong 333
Che, Yonggang 366
Chen, Hanhua 132
Chen, Hongkun 156
Chen, Quan 235
Chen, Shuo 3
Chen, Xiaobing 277
Chen, Zhiguang 289
Cheng, Peng 289
Cheng, Shixiong 376
Cheng, Yang 209
Chi, Mengxian 143

Dai, Hua 169
Du, Xiaoyong 93
Du, Yankang 333
Du, Yanning 376
Du, Yunfei 289

Fan, Guisheng 327
Feng, Dan 3, 361
Feng, Ningxuan 93
Feng, Qi 225

Gao, Kaihui 209
Gao, Lan 371
Geng, Jinkun 209
Gu, Lin 182
Guo, Minyi 235
Guo, Shasha 69

He, Chao 56
He, Dongjian 376
He, Jinrong 376
Hu, Junyan 16
Hu, Wenxin 82
Hu, Xiaoyang 132
Huang, Hong 132
Huang, Kaixin 251

Huang, Linpeng 251
Huo, Zhisheng 305

Ji, Shujian 350
Jiang, Wanchun 43
Jin, Hai 132, 182
Jin, Xu 143
Jin, Zongze 120

Kang, Ziyang 69
Kong, Linghe 156

Lei, Yongmei 322
Leng, Jingwen 235
Li, Dali 366
Li, Donghua 338
Li, Feng 143
Li, Jinpeng 132
Li, Kenli 16
Li, Keqin 16
Li, Shaodi 316
Li, Shiming 69
Li, Wei 107, 277
Li, Yuancheng 376
Li, Yufeng 333
Li, Yunchun 107
Lin, Haowei 311
Lin, Jiazao 93
Lin, Tingyu 182
Liu, Bin 376
Liu, Chubo 16
Liu, Shang 3
Liu, Shaoli 277
Liu, Shaoshan 338
Liu, Yang 289
Liu, Yi 371
Lu, Jiaqi 82
Lu, Youyou 355
Lu, Yutong 289
Luan, Zhongzhi 371

Ma, Xiaojing 132
Mu, Weimin 120

Peng, Lijuan 43
Peng, Shaohui 277
Peng, Wei 225
Peng, Yamei 3

Qian, Depei 371
Qin, Guangjun 305
Qiu, Han 333

Rao, Jia 182
Rao, Yu 338
Ruan, Chang 43

Sanic, Mustafa 235
Shen, Guowei 209
Shen, Li 197
Shi, Feng 31
Shi, Guoqiang 182
Shi, Zhan 3
Shu, Bing 316
Shu, Jiwu 355
Shuo, Tian 69
Song, Yang 107
Stones, Rebecca J. 264
Sun, Huaiying 327
Sun, Tongzheng 350
Sun, Yongzhong 345

Tang, Feilong 156
Tang, Jie 338
Tang, Qizhi 182
Tian, Chang 56
Toschi, Alessandro 235

Wang, Chunlin 235
Wang, Fang 3, 361
Wang, Gang 264
Wang, Jianxin 43
Wang, Junwei 120
Wang, Lei 69
Wang, Rui 371
Wang, Shuai 209
Wang, Shuquan 69
Wang, Weiping 120
Wang, Wenbo 345, 350
Wang, Xi 366
Wang, XiaoJun 31
Wang, Xinheng 311
Wang, Zhenghua 366

Wei, Bing 305
Wei, Bo 355
Wu, Jia 43
Wu, Junmin 316
Wu, Shuhan 107
Wu, Song 182
Wu, Xinzhou 197

Xiao, Limin 305
Xie, Jinyang 322
Xie, Xia 132
Xie, Zhidong 56
Xu, Cheng-Zhong 345, 350
Xu, Chongyang 371
Xu, Chuanfu 366
Xu, Weixia 69
Xu, Wenchao 156
Xu, Xiaolong 311
Xue, Yanfen 327

Yan, Baicheng 305
Yan, Ge 251
Yang, Geng 169
Yang, Hailong 107
Yang, Maohu 169
Yang, Yanqin 156
Ye, Kejiang 345, 350
Yi, Liping 264
Yi, Xun 169
Yin, Yan 316
Yu, Huiqun 327
Yuan, Ninghui 197

Zhai, Jidong 93
Zhang, Feng 93
Zhang, Han 371
Zhang, Hong 31
Zhang, Hongbin 355
Zhang, Lianyi 371
Zhang, Runhua 209
Zhang, Shuzheng 69
Zhang, Xingjun 156
Zhang, Yi 316
Zhang, Yingxi 182
Zhao, Hang 338
Zheng, Jun 82
Zhi, Tian 277
Zhou, Bingyu 305

Zhou, Jun   361
Zhou, Shengyuan   277
Zhou, Yawei   316

Zhu, Weilin   120
Zhu, Zhihao   197
Zhuang, Yimin   277

Printed in the United States
By Bookmasters